Two Manuals Combined into One Book

Marine Operations on Urbanized Terrain

And

Marine Scouting and Patrolling

Department of Defense

Pentagon
Publishing

Military Operations on Urbanized Terrain

Table of Contents

Chapter 1

Modern Urban Warfare

> **"... the likelihood is high that in the future, the National Command Authorities will again commit Marines to missions in urban areas."**
>
> *A Concept for Future Military Operations on Urbanized Terrain*

1001. Importance of Urban Areas. Throughout history, military planners have viewed cities as centers of gravity. As such, in war, cities are something to be either protected or taken away, depending upon one's perspective (MCDP 1, *Warfighting*). Cities house the population centers, transportation hubs, seats of government, sources of wealth, centers for industry, information networks, and key nodes of communication within a nation. Recent forecasts based on population statistics and the worldwide migration trend from agrarian to industrialized societies predict that 85 percent of the world's population will reside in urbanized areas by the year 2025. As the world trend toward urbanization increases, the military significance of cities is likely to increase proportionally.

Urbanized areas, themselves, may be significant sources of future conflict. Cities historically are where radical ideas ferment, dissenters find allies, mixtures of people cause ethnic friction, and discontented groups receive media attention. Adversaries may focus on the capture of radio and television stations in an attempt to influence public opinion and attain their political goals. Our political leaders may take advantage to neutralize or stabilize some extremely volatile political situations, or to provide assistance to allies in need of support, by deploying U.S. forces into urban environments,

1002. The Marine Corps Role in Urban Warfare. As the Nation's force in readiness, forward deployed with expeditionary forces, Marines must be prepared to fight on urbanized terrain. In the past two decades, MAGTFs ranging in size from MEFs (Saudi Arabia, Desert Shield/Desert Storm; Somalia, Restore Hope) through Marine expeditionary units (MEUs) (Beirut, Lebanon; Grenada, Urgent Fury; Somalia, Eastern Exit and Restore Hope) have participated in MOUT. The task-organization and combined-arms aspect of the MAGTF makes it well suited for combat on urbanized terrain.

The results of geographical studies show that 60 percent of politically significant urban areas outside allied or former Warsaw Pact territory are located along or within 25 miles of a coastline; 75 percent are within 150 miles; 87 percent are within 300 miles; 95 percent are within 600 miles; and all are within 800 miles. U.S. embassies and diplomatic facilities are primarily located in cities where the host country's political and economic leadership is concentrated. The Marine Corps will continue to play a prominent role in future evacuations of U.S. citizens, as well as the conduct of peace, counterinsurgency, and contingency operations centered on urbanized areas.

Today's Marine air-ground task forces (MAGTFs) are deployed as part of naval expeditionary forces (NEFs) that maintain a *global forward presence for rapid crisis response*. These integrated combined-arms forces are part of the Nation's proven contingency and naval power projection force. Therefore, Marines may find themselves rapidly deployed and employed in actions across the spectrum of military operations. Many of these trouble spots will likely be located in or around large urban centers.

In the years since World War II, the United States has employed military force more than 200 times. Of these, four out of five involved naval forces, and the majority of the naval efforts included Marines embarked in amphibious ships. The reasons are straightforward: availability and adaptability. Availability derives from the loiter time of forward deployed forces embarked on amphibious shipping. Adaptability comes from the Marine Corps' MAGTF organization, doctrine, training, and equipment, which prepare us for expeditionary missions from the sea in support of a variety of missions, including forcible entry. Enhancing our adaptability are the maritime prepositioning forces (MPFs). MPFs provide a rapid buildup of combat and logistics equipment that is joined with Marines on a distant shore, creating a substantial combat force. Despite our availability and adaptability, the prospect of urban warfare combined with an amphibious assault is a complex task which requires special preparation.

At the outset of a developing situation, forward-deployed expeditionary forces can move quickly within range of a crisis that threatens the political stability of a country. Urban intervention operations must often be planned and executed in a matter of hours or days (rather than weeks or months) to take advantage of the internal turmoil surrounding a developing crisis. Navy and Marine forces should anticipate deployment to urbanized areas on a day-to-day basis.

1003. Distinguishing Features of Urbanized Terrain. Urbanized terrain is a complex and challenging environment. It possesses all of the characteristics of the natural landscape, coupled with manmade construction, resulting in an incredibly complicated and fluid environment that influences the conduct of military operations in unique ways.

Military operations on urbanized terrain (MOUT) is defined as *all military actions planned and conducted on a topographical complex and its adjacent terrain where manmade construction is the dominant feature. It includes combat in cities, which is that portion of MOUT involving house-to-house and street-by-street fighting in towns and cities* (Marine Corps Reference Publication (MCRP) 5-12A, *Operational Terms and Graphics*). MOUT effects the tactical options available to a commander. A built-up area is *a concentration of structures, facilities, and populations, such as villages, cities, and towns,* that form the economic and cultural focus for the surrounding area. (MCRP 5-12A)

a. Cities. Cities are centers of finance, politics, transportation, communication, industry, and culture. They generally have large population concentrations ranging from tens of thousands to millions of people. Because of their psychological, political, or logistical value, control of cities have often been the scenes of pitched battles.

(1) Operations in built-up areas are normally conducted to capitalize on the operational or tactical significance of a particular city. In developing nations, control of only a few cities is often the key to the control of national resources. The side that controls a major city usually has a psychological advantage, which can be enough to significantly affect the outcome of a countrywide conflict.

(2) The abundance of guerrilla and terrorist operations conducted in built-up areas (e.g., Santo Domingo, Caracas, Belfast, Managua, and Beirut) demonstrates the importance many insurgent groups place on urban warfare.

(3) In the past 40 years, many cities have expanded dramatically, losing their well-defined boundaries as they extended into the countryside. New transportation systems (highways, canals, and railroads) have been built to connect population centers. Industries have grown along those connectors, creating "strip areas." Rural areas, although retaining much of their farmlike character, are connected to the towns by a network of secondary roads.

b. Multiple Avenues of Approach. Urbanized terrain is a unique battlespace that provides both attacker and defender with numerous and varied avenues of approach and fields of fire. The urban battlespace is divided into four basic levels: *building*, *street*, *subterranean,* and *air*. Operations can be conducted from above ground, on ground level, inside buildings, or below the ground. Most operations will include fighting on all levels simultaneously.

(1) Building Level. Buildings provide cover and concealment; limit or increase fields of observation and fire; and canalize, restrict, or block movement of forces, especially mechanized forces. They provide optimum perches for snipers and antiair weapons. Buildings also provide antitank weapons optimum positioning to allow engagement from above, exploiting an inherent weakness found in most armored vehicles.

(2) Street Level. While streets provide the means for rapid advance or withdawal, forces moving along streets are often canalized by buildings and have little space for off-road maneuver. Because they are more difficult to bypass, obstacles on streets in urbanized areas are usually more effective than those on roads in open terrain.

(3) Subterranean Level. Subterranean systems are easily overlooked but can be important to the outcome of operations. These areas may be substantial and include subways, sewers, cellars, and utility systems (Figure 1-1 on page 1-4). The city of Los Angeles alone has more than 200 miles of storm sewers located under the city streets. Both attacker and defender can use subterranean avenues to maneuver to the rear or the flanks of an enemy. These avenues also facilitate the conduct of ambushes, counterattacks, and infiltrations. (See Appendix E)

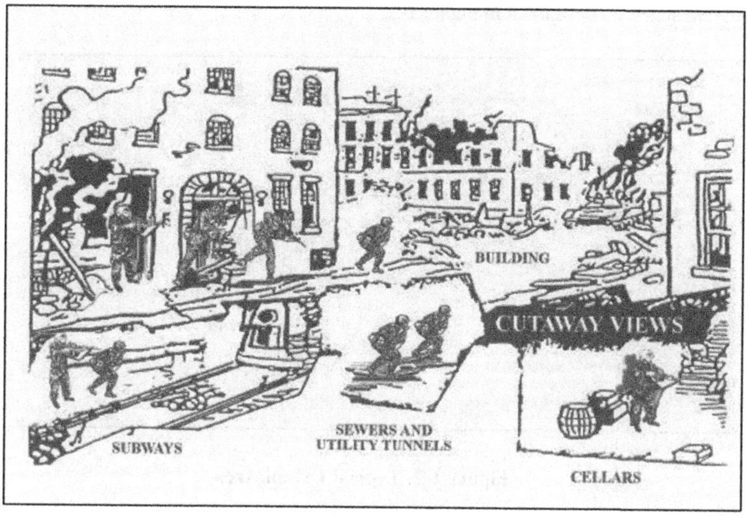

Figure 1-1. Subterranean Systems

(4) Air Level. The air provides another avenue of approach in urbanized areas. Aviation assets can be used for high speed insertion or extraction of troops, supplies, and equipment. While aviation assets are not affected by obstacles on the streets, they are affected by light towers, signs, power lines, and other aerial obstructions. They are also vulnerable to the man-portable surface-to-air missile threat, crew served weapons, and small arms fire.

c. Categories of Built-Up Areas. Built-up areas are generally classified as:

- Villages (populations of 3,000 or less)
- Strip areas (industrialized zones built along roads connecting towns or cities)
- Towns or small cities (populations of up to 100,000 and not part of a major urban complex)
- Large cities with associated urban sprawl (populations in the millions, covering hundreds of square kilometers).

d. Characteristics of Urbanized Areas. A typical urban area consists of combinations of the *city core*, *commercial ribbon*, *core periphery*, *residential sprawl*, *outlying industrial areas*, and *outlying high-rise areas*. Each of the urban area's regions has distinctive characteristics that may weigh heavily in planning for MOUT. Most urbanized areas resemble the generalized model shown in Figure 1-2.

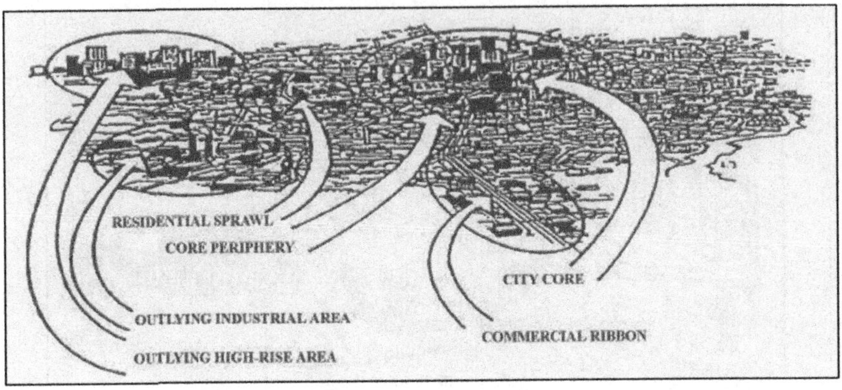

Figure 1-2. Typical Urban Area

(1) City Core. In most cities, the city core has undergone more recent development than the core periphery. As a result, the two regions are often quite different. Typical city cores are made up of high-rise buildings which vary in height. Modern urban planning for built-up areas allows for more open spaces between buildings than in old city cores or in core peripheries (Figure 1-3).

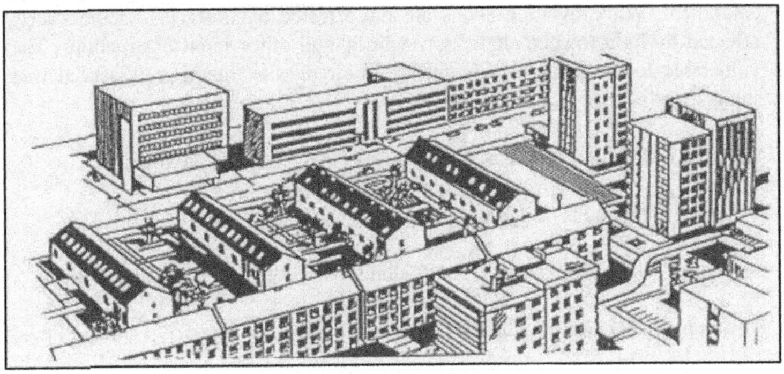

Figure 1-3. City Core

(2) Commercial Ribbon. Commercial ribbons are composed of rows of stores, shops, and restaurants that are built along both sides of major streets through built-up areas.

Typically, such streets are 25 meters wide or more. The buildings in the outer areas are uniformly two to three stories tall—about one story taller than the dwellings on the streets behind them (Figure 1-4).

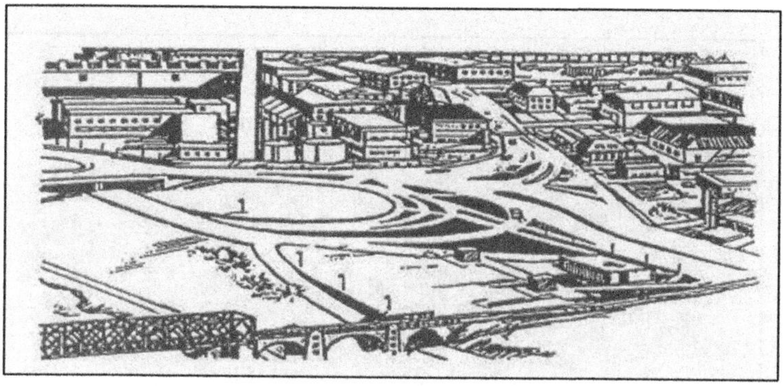

Figure 1-4. Commercial Ribbons

(3) Core Periphery. The core periphery generally consists of streets 12 - 20 meters wide with continuous fronts of brick or concrete buildings. The building heights are fairly uniform—2 or 3 stories in small towns, 5 to 10 or more stories in large cities (Figure 1-5).

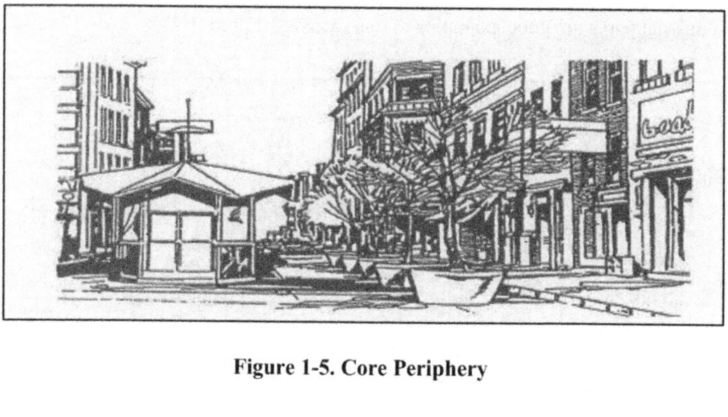

Figure 1-5. Core Periphery

(4) Residential Sprawl. Residential sprawl areas consist mainly of low houses or apartments that are one to three stories tall. The area is primarily composed of detached dwellings that are usually arranged in irregular patterns along streets, with many smaller open areas between structures (Figure 1-6).

Figure 1-6. Residential Sprawl

(5) Outlying Industrial Areas. These areas generally consist of clusters of industrial buildings varying from one to five stories in height. Buildings generally vary dramatically in size and composition to match the needs of the particular businesses they house. Industrial parks are good examples of this category (Figure 1-7).

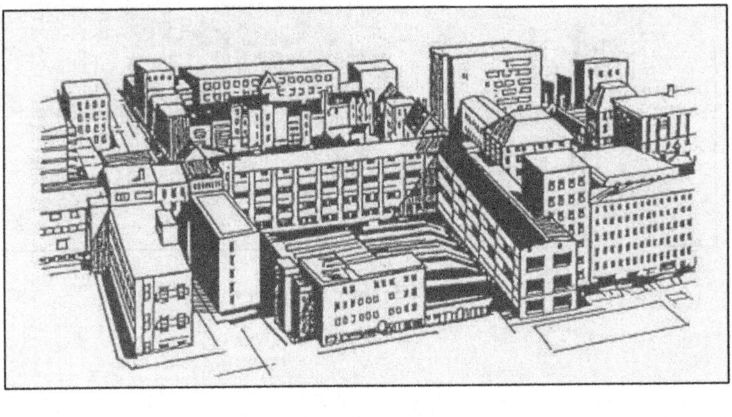

Figure 1-7. Outlying Industrial Areas

(6) Outlying High-Rise Areas. These areas are similar in composition to city core areas, but may be composed of clusters of more modern multistory high-rise buildings in outlying parts of the city. Building height and size may vary dramatically (Figure 1-8). Generally, there is more open space between buildings located in the outlying high-rise areas than is found within the city core area.

Figure 1-8. Outlying High-Rise Area

1004. Modern Battles for Urbanized Terrain. Urban warfare is as old as war itself. Since man began building villages, he has fought battles in and around them. Geography, politics, and economics dictate that cities will continue to be an objective of armies in warfare. From the armies that invaded and liberated Europe twice during the 20th century, to the forces that fought in Korea and Vietnam, to our most recent urban battles in the Middle East and Southwest Asia, *the basic principles of combat in built-up areas have essentially remained unchanged in this century.* While the principles remain the same, the introduction of helicopters, fixed-wing aircraft, armor, and precision-guided munitions (PGMs) has altered some of the techniques associated with urban combat.

Military Operations on Urbanized Terrain

Twenty two modern urban battles are discussed to illustrate the trends, dominant factors, and principles of combat in urbanized areas. (The majority of this material is extracted from the 1987 study, *Modern Experience in City Combat,* produced by Abbott Associates.)

- Aachen
- Arnhem
- Ashrafiyeh
- Ban Me Thuot
- Beirut Port/Hotel (I)
- Beirut 1982 (II)
- Berlin
- Cherbourg
- Hue
- Jerusalem
- Khorramshahr
- Manila
- Ortona
- Quang Tri City I
- Quang Tri City II
- Seoul
- Sidon
- Stalingrad
- Suez City
- Tel Zaatar
- Tyre
- Zahle

A brief description of each battle is provided to illustrate that battle's significance.

a. Stalingrad (1942 - 1943). The tenacious Soviet defense of Stalingrad cost the attacking Germans dearly in every way and set up conditions for a decisive counteroffensive. This classic urban battle involved large forces and resulted in innovative urban combat techniques and the creation of the highly successful storm groups (task-organized assault units). (Length of battle: greater than 30 days) (Casualties: 1,630,000+)

b. Ortona (1943). In this Italian town, determined resistance by a battalion of the elite German 3rd Parachute Regiment against Canadian Army attackers demonstrated the difficulty of overcoming a well-prepared defense. The Canadians were unfamiliar with urban combat and had to develop urban fighting techniques during the battle. After the town was largely destroyed and the defender had extracted a high cost in time and casualties from Canadian forces, the German parachute battalion withdrew. (Length of battle: 6 - 13 days) (Casualties: estimated in the hundreds)

c. Aachen (1944). The battle for Aachen, Germany, in the fall of 1944 developed during the U.S. First Army's offensive to breach the Westwall fortifications and the vaunted Siegfried Line. Aachen, the ancient capital of Charlemagne, had symbolic political and psychological significance to both the Germans and Americans. Furthermore, it was the first city on German soil to face an assault by the Allies. This first major battle on German soil foreshadowed bitter resistance against the American attackers in subsequent battles. The German defenders surrendered only after the city was destroyed. Although the U.S. Army had achieved a clear tactical victory, the German defense of Aachen cost the U.S. First Army valuable time and delayed the planned attack to the Rhine river. (Length of battle: 14 - 30 days) (Casualties: 8,000+)

d. Arnhem (1944). On September 17, 1944, Operation Market-Garden, the largest airborne operation in history, was launched in the Netherlands. The plan was to land three airborne divisions to seize key bridges along a 100-kilometer-long corridor through which allied mechanized forces would pass as the first step in the final offensive into Germany. The British 1st Airborne Division made a surprise landing near the Dutch city of Arnhem in order to seize a bridge over the Rhine river for advancing British forces. An unexpected German armor force counterattacked and eliminated all footholds, virtually destroying the British division before a linkup could occur. (Length of battle: 6 - 13 days) (Casualties: estimated in the thousands)

e. Cherbourg (1944). By June 17, 1944, U.S. forces advancing toward Cherbourg from the Normandy beachhead succeeded in cutting off defending German forces in the Cotentin Peninsula. Four German divisions withdrew to a perimeter surrounding Cherbourg. After much fighting, particularly in strongpoints outside the city, the German garrison surrendered to the Americans. Unfortunately for the Americans, the port facilities were destroyed which denied their early use by Allied forces. (Length of battle: 6 - 13 days) (Casualties: estimated in the thousands)

f. Berlin (1945). The long, bloody Soviet offensive to seize the German capital city effectively concluded the last battle of World War II in Europe. Bitter fighting occurred, but the defense was never well coordinated due in part to poor preparation by the Germans. (Length of battle: 14 - 30 days) (Casualties: estimated in the thousands)

g. Manila (1945). Japanese Army troops evacuated Manila under pressure from advancing American forces, but the local Japanese naval commander independently decided to hold the city at all costs. Despite defending Manila with poorly trained and equipped personnel, the determined resistance resulted in a high number of casualties to attacking U.S. forces as well as the destruction of the city and much of its population. (Length of battle: 14 - 30 days) (Casualties: 22,000+)

h. Seoul (1950). Following the Inchon landing, U.S. and Republic of Korea (ROK) forces recaptured the South Korean capital from the North Koreans. The fighting was unusual in that combat was largely centered on seizure of street barricades rather than buildings. (Length of battle: 6 - 13 days) (Casualties: Marines, 2,383; others, estimated in the thousands)

i. Jerusalem (1967). Israeli forces seized Jerusalem in a well prepared and well executed operation. Despite an uncoordinated Jordanian defense, Israeli casualties in this battle were the highest of those encountered during the Six Day War. Regular Jordanian forces withdrew during the latter stages of the battle, effectively ending organized resistance. (Length of battle: 48 hours to 5 days) (Casualties: Israeli forces, 400+; Jordanian forces, estimated in the hundreds)

j. Hue (1968). On January 31, 1968, the North Vietnamese Army (NVA) and Viet Cong (VC) forces launched their Tet Offensive at targets throughout South Vietnam. As part of this operation, two NVA/VC regiments and two sapper battalions conducted a surprise attack and seized part of the walled city (Citadel) of Hue. The NVA/VC held this part of the walled city for about three weeks against determined U.S. and South Vietnamese forces before finally succumbing. The battle for Hue is considered one of the most intense and savage battles of the Vietnam War. (Length of battle: 14 - 30 days) (Casualties: Marines, 433; others, 5000+)

k. Quang Tri City I and II (1972). An objective of the North Vietnamese 1972 winter-spring offensive was the capture of Quang Tri, the northernmost major city in South Vietnam. The NVA overwhelmed the Army, Republic of Vietnam (ARVN) defenders (I). Later, the city was recaptured (II) by a smaller ARVN force using extensive artillery and air support. The large conventional forces involved on both sides made Quang Tri I and II the major urban battles of the Vietnam War. (Length of battle: Quang Tri I, 6 - 13 days; Quang Tri II, 30 days or greater) (Casualties: battles combined, 30,000+)

l. Suez City (1973). Israeli Defense Forces (IDF) attempted to seize this Egyptian city before the anticipated United Nations (U.N.) cease-fire to end the Yom Kippur War. IDF armored shock tactics led to disaster against a well-prepared Egyptian defense. High casualties forced the IDF to withdraw. (Length of battle: less than 24 hours) (Casualties: Israeli forces, estimated 100 - 500; others, unknown)

m. Ban Me Thuot (1975). This South Vietnamese highlands town was the first strategic city to fall in the final North Vietnamese general offensive in 1975 South Vietnamese forces were surprised and overwhelmed. The fall of Ban Me Thuot resulted in a rout that the North Vietnamese exploited to achieve total victory in Vietnam. (Length of battle: 24 - 48 hours) (Casualties: estimated in the hundreds)

n. Beirut I (1976). When the Lebanese civil war broke out in the spring of 1975, combat in the capital city of Beirut assumed a central role. The battle for Beirut I was a series of small, local operations between largely irregular Christian and Muslim forces fighting over control of the hotel and port districts. Combat was not decisive, but led to changes in the boundary, called the "Green Line." This separated the antagonists and lead to the stagnation of the Lebanese conflict. (Length of battle: greater than 30 days) (Casualties: estimated in the hundreds)

o. Tel Zaatar (1976). Lebanese Christian attackers encircled and methodically besieged this Palestinian camp before overcoming its defenders with a final assault. (Length of battle: greater than 30 days) (Casualties: estimated in the hundreds)

p. Ashrafiyeh (1978). The Syrian forces occupying portions of Lebanon faced a complex political situation in which the power of the Christian militia was seen as a clear threat to stability. In an attempt to weaken the Christian militia by an attack on their center of power, the Syrians laid siege to the Christian militia stronghold of East Beirut (Ashrafiyeh). This urban battle was essentially an artillery bombardment without air attacks. Syria failed to break the will of the defenders and final positions remained unchanged. (Length of battle: greater than 30 days) (Casualties: estimated in the hundreds)

q. Khorramshahr (1980). Iranian regular forces initially evacuated this port city in the face of an Iraqi offensive. Irregular Iranian forces, however, continued to fight. They offered prolonged resistance and inflicted heavy casualties. Iraq eventually won this battle, but at a high cost in time and resources that ultimately served to halt the entire offensive against Iran. The intensity of fighting during the battle for Khorramshahr earned the city the nickname, "City of Blood." (Length of battle: 14 - 30 days) (Casualties: Iraqi, 3,000 - 9,000; Iranian, estimated in the thousands)

r. Zahle (1981). Syria laid siege to the Lebanese regular forces and militia in this key crossroads town. Fighting was inconclusive and ended in a negotiated settlement whereby the Lebanese defenders evacuated the town. (Length of battle: greater than 30 days) (Casualties: estimated in the hundreds)

s. Beirut II (1982). The siege of Beirut culminated the Israeli campaign to evict the Palestine Liberation Organization (PLO) from Lebanon. Fighting under domestic and world political pressures, the IDF besieged the PLO, selectively applying heavy ground and air firepower in conjunction with psychological warfare and limited-objective ground operations. The fighting resulted in a negotiated PLO evacuation from the city. (Length of battle: greater than 30 days) (Casualties: 2,300+)

t. Sidon (1982). Israeli forces easily seized this PLO southern headquarters during the invasion of Lebanon. The IDF was fully prepared for major urban combat using lessons learned from earlier battles, but resistance was unexpectedly light as PLO forces had largely withdrawn from the city. (Length of battle: 48 hours to 5 days) (Casualties: unknown)

u. Tyre (1982). The Israeli attack on the PLO in this Lebanese coastal city was well planned, with excellent intelligence on the target. All branches of the IDF participated in an operation that included naval fire support and amphibious landings. PLO resistance was uncoordinated and easily overcome. (Length of battle: 48 hours to 5 days) (Casualties: Israeli forces, 120+; others unknown)

1005. Modern Urban Battle Analysis and Observations. The 20th-century urban warfare experience has caused us to reevaluate old factors and consider new developments that affect the way we fight in this environment. The following discussion, based on the historical analysis of these select urban battles, illustrates the importance of maneuver warfare and combined arms philosophies in the urban environment. Factors that have had an impact on the manner in which urban warfare has been conducted are:

Intelligence is imperative to success in urban warfare. "Maneuver warfare requires a firm focus on the enemy" (MCDP 2, *Intelligence*). Few subsequent tactical changes can overcome the far-reaching impact of a major intelligence error.

a. Intelligence. The historical review of modern urban battles discloses that the attacker will usually win. Failures to win generally reflect classic military errors not characteristically unique to cities. *However, of the battles studied, battles lost were attributed to errors in* *initial intelligence*. The battles of **Arnhem** and **Suez City** probably would never have occurred had the attacker known the strength and locations of the defender's forces. At **Stalingrad**, the attacking Germans were aware of the defending forces facing them in the Sixth Army's zone, but the buildup of Soviet forces in other areas of the front was not anticipated and resulted in tactical surprise at those points, diluting the German offensive push to seize the city.

Surprise is a combat multiplier and can substantially reduce the cost of urban warfare.

b. Surprise. Surprise is a combat multiplier and can substantially reduce the cost of urban warfare. It can be achieved through deception, stealth, and ambiguity (MCDP 1-3, *Tactics*). Surprise was achieved by the attacker at **Aachen** and **Ban Me Thuot** and by the defender at **Suez City**. Surprise can be an important asset to increase leverage, but, as the failure at **Arnhem** shows, not necessarily a decisive one. When surprise is employed as a means to overcome other disadvantages, it is important to maintain accurate intelligence. In urban areas, tactical surprise by the attacker can be used to preempt effective defensive preparation of a city.

c. Combined Arms. The MAGTF must capitalize on one of the key means for gaining advantage in maneuver warfare — the use of combined arms (MCDP 1-3, *Tactics*). The use of combined arms places the enemy in a dilemma. Any action the enemy takes to avoid one combat arm makes him more vulnerable to another. An analysis of categories of weapons systems found in a MAGTF helps to illustrate it is a warfighting organization well suited for MOUT:

Combat in urban areas is primarily a small unit, infantry intensive operation

(1) Infantry. Combat in urban areas is primarily a small-unit, infantry intensive operation. Restrictions on maneuver, particularly for mechanized units, increases oportunities for infiltration. Urban combat requires small-unit leadership, initiative, and skill. Decentralized actions and difficulties in command , control, and communications are typically encountered. Built-up areas, like close terrain found in other operational environments, are generally considered to be most suited for operations conducted by infantry. Infantry units can be organized, trained and equipped to negotiate urbanized terrain that restricts observation, fields of fire, and mechanized movement.

(2) Armor. The role of armor in urban warfare can be significant. Of the 22 battles studied, armor participated in 21. In three-fourths of these battles, organic tank support was a central element when special assault teams were employed. *Overall, special assault units supported by tanks were more successful than any other task organization.*

The use of tanks to the attacker inside a city has been effective only when they were protected by infantry. Tanks in support of infantry act as an "assault gun" that delivers concentrated, sustained fires to reduce held strongpoints.

The U.S. experience in **Hue** demonstrated the key role that armor can play on a combined-arms team fighting inside the city. The Marines' most effective weapons during the battle were the M48A1 Patton tank and the M-50 Ontos. Both were protected by infantry. The M48A1, with its 90-mm main gun, was used extensively to reduce fortified positions. The Ontos, an armor-protected tracked vehicle mounting six 106-mm recoilless rifles, was highly effective against concrete and steel structures. The munitions of these armored systems provided breaches that the infantry could exploit.

Armor providing direct-fire support inside a built-up area requires protection by infantry.

In contrast, during the **Suez City** battle, Israeli armor forces attacked on "armor thrust avenues" into the city, outpacing their armored personnel carrier (APC) mounted paratroop/infantry support. The Egyptian defenders lacked organic artillery (except limited antiaircraft artillery (AAA) and mortars) and had no air support and virtually no armor support. The Egyptians prepared "kill zones" on the principal avenues down which the IDF armored forces attacked. As the lead IDF armor battalion entered the second of the three road intersection objectives, the Egyptians engaged with Sagger missiles, RPGs, ZSU-23 antiaircraft guns, antitank grenades thrown from balconies, and small arms. All of the tank commanders in the lead battalion were killed or wounded. Disabled vehicles blocked the road. Vehicles veering into the narrow side streets became trapped and were destroyed. *The lack of infantry to protect the armor proved disastrous to the attacking armor battalion.*

(3) Artillery. Artillery has played an important role in most major urban conflicts. At **Aachen,** U.S. forces combined infantry with antitank teams and artillery (in a direct-fire role) down to the squad and fire team level. Also, artillery firing shells with "delay" fuzes in an indirect-fire role were used to penetrate one or more floors before exploding, thus driving the enemy to the ground where infantry and armor could attack. Artillery was also positioned to fire *perpendicular* to the direction of movement of assaulting forces. Thus, fratricide from artillery range errors was alleviated.

Artillery has two distinct roles: outside the built-up area to isolate or prevent isolation with indirect-fire; and within the built-up area to provide direct-fire support.

Artillery proved most useful for interdicting enemy supplies, enemy evacuation, and the movement of reinforcements in the enemy's rear; for physically and psychologically harassing the enemy; and in direct-fire roles within a city.

Artillery employed in the indirect-fire role has been effective in disrupting defenders in half of the studied battles. However, some indirect-fire roles have proved more effective than others in urban combat. Artillery was most effective in the interdiction of supplies, enemy evacuation, movement of reinforcements in the enemy's rear (outside the city), and for indiscriminate physical and psychological harassment of the enemy. Artillery used as an indirect-fire siege weapon, as was done at **Ashrafiyeh** and **Zahle**, proved ineffective. Artillery can also cause problems for the attacker. The rubble resulting from indirect artillery fires can create considerable obstacles for the attacker while providing the defender with obstacles, materials, cover, and concealment.

Artillery employed in the direct-fire role proved useful in the reduction of strongpoints. Self-propelled artillery was used effectively inside **Aachen** and **Stalingrad** and more recently by the IDF at **Beirut II**.

(4) Mortars. The mortar is the most used indirect-fire weapon in urban combat. The mortar's high angle of fire allows the round to reach the street level accurately without being masked by surrounding structures. During the battle for **Hue** city, the most effective fire support provided to Marines was the indirect fire from 60-mm, 81-mm, and 4.2-inch mortars. Approximately 20,000 rounds of high explosive (HE) 60-mm and 81-mm mortar ammunition were expended during the battle. The 4.2-inch mortar was used primarily to deliver riot control agent munitions. It was discovered that 4.2-inch riot control agent munitions could be fired with great effectiveness through the tops of buildings to drive the enemy out. When the enemy emerged from their concealed positions, HE rounds were fired to complete the attack.

(5) Antiaircraft Artillery. Antiaircraft artillery (AAA) was extremely useful in a ground fire role in some urban battles. AAA was used only rarely in World War II, and then generally against assaulting personnel rather than against structures. AAA has been used more frequently in more recent events, but against buildings rather than people. *The high rates of fire of modern AAA make it excellent in terms of shock and destructive potential.* However, ammunition supply can be a problem because the volume of fire. During **Hue**, the Duster, an AAA vehicle mounting twin 40-mm guns, was provided to the Marines by the U.S. Army. These guns were extremely effective in the suppression of enemy positions due to the lethality of the 40-mm round and the quantities in which it could be delivered. During the **Suez City** battle, the Egyptians used the ZSU-23 antiaircraft gun in the direct-fire role against armored columns. This gun proved to be frightening and effective. Its high volume of explosive power created shock among IDF armor personnel akin to the shock that the IDF armor thrust tactics were designed to create. During **Sidon,** the IDF used 20-mm antiaircraft Vulcans in the direct-fire role against ground targets with great success.

(6) Aviation. Historically, aviation assets have played an important role in helping to isolate the objective and to interdict the flow of the defender's supplies and reinforcements. However, aviation has been relatively ineffective when not used in conjunction with ground forces. In past wars, bombing operations have been used in attempts to reduce the defender's will to resist and destroy their physical capabilities. In the majority of urban battles, aerial bombing, by itself, did not erode the defender's will to resist, nor did it significantly degrade the defender's military capabilities.

Aviation plays an important role in interdicting movement of the defender's supplies and reinforcement into the built-up area.

The Marine air-ground task force is employed as a combined arms team. As such, the aviation combat element will not usually be called upon to conduct independent operations in MOUT. It will function as part of a MAGTF and, regardless of operating on or over urbanized terrain, will still execute the six functions of Marine aviation (i.e., offensive air support, assault support, antiair warfare, electronic warfare, air reconnaissance, and control of aircraft and missiles) in support of the MAGTF.

In future urban warfare, aviation will be even more effective due to advances in fixed- and rotary-wing aircraft, unmanned aerial vehicles, precision guided munitions, improved munitions, communications, sensors, and targeting systems. Our battle study already indicates a trend toward more extensive aviation participation in MOUT. As an example, the IDF in **Beirut II** employed bombing by fixed-wing aircraft using cluster bomb units, "smart" bombs, phosphorous, and other munitions. Attack helicopters operated on the outskirts of the built up area with impunity, and medical evacuation (MEDEVAC) proceeded swiftly and efficiently using helicopter support.

d. Combat Forces. Whether attacking or defending, the size of the force relative to the enemy can be critical to success. When provided with adequate forces, the attacker can isolate and encircle the defender and prevent a breakout or linkup. The defender can use them to create a mobile defense or to create strong reserves for counterattacks.

Other factors that impact on the size of forces required are the degree of surprise achieved and the firepower utilized (aviation, armor, artillery, mortars, etc.). They should be weighed against the sophistication of the prepared defense. Ample consideration should be given to the local population, degree of external support, and utilization of existing services (communications, water, etc.).

Regardless of the size or quality of defensive forces, the defender usually extracts large costs from the attacker in time, resources, and casualties.

In the historical examples reviewed, the defender was usually outnumbered by the attacker, the quality of the defender's available forces was inferior, and defeat of the forces defending the city was usually certain. However, regardless of the size or quality of defensive forces, the defender can extract enormous costs to the attacker in time, resources, and casualties. As was seen at **Khorramshahr**, the Iranian defenders, outnumbered 4 to 1, still held the city for 26 days. Although the Iranian defenders eventually lost the city, its defense allowed the remaining Iranian forces time to organize and redeploy. Furthermore, the winter rains that followed the battle turned much of the region into a sea of mud and largely halted further Iraqi efforts. The Iraqi army's offensive thrusts into Iran lost momentum as a result of the defense of Khorramshahr.

e. Special Assault Teams. In these battle studies, "shock units" or "special assault teams" have been used by attackers (and often by defenders) with great success. These organizations were characterized by the integration of combined arms at the battalion level and below. Control was decentralized to lower echelons (down to the squad level in some cases). Assault teams typically contained infantry with various combinations of armor, artillery, and engineers. Hence, all combat arms should plan, train, and develop common tactics, techniques, and procedures for use on urbanized terrain. Before the battle of **Aachen**, U.S. ground forces conducted intensive urban training with antitank teams and artillery pieces down to the squad and fire team levels. In contrast, the defender, in most cases having to fight with whatever forces were locally available, had no opportunity to conduct combined-arms training.

The use of combined arms warfare on urbanized terrain is imperative.

f. Time. In most cases, successful conclusion of an urban battle took two to three times longer than the initial estimates. Consequently, the additional tome resulted in the expenditure of more logistics and the loss of more personnel then initially anticipated. This often had adverse affects on the overall campaign. Well-planned urban defense, even if the defender is isolated or lacking in aviation, armor, or artillery weapons, can be time consuming to the attacker. Time can allow the defender to reorganize, redeploy, or marshal resources in other areas.

Generally, urban warfare is time consuming.

Three battles where time played an unanticipated critical role in the attacker's strategic timetable were **Aachen**, **Khorramshahr**, and **Stalingrad**. In these battles, the defenders delayed the attacker longer than was estimated, resulting in the modification of operational or strategic plans.

g. Isolation. *The attacker won all urban battles where the defender was totally isolated.* Even the partial isolation of the defenders resulted in attackers enjoying a success rate of 80 percent. Conversely, attackers won only 50 percent of the battles in which defenders were not significantly isolated, and those victories came at great cost.

No single factor is more important to the attacker's success than isolation of the urban area.

In most urban battles, some form of isolation occurred as a result of the attacker's actions. While it is unrealistic to envision complete isolation of a city until enemy forces to the rear of the city are pushed far beyond its outermost boundaries, *total isolation does not appear necessary.* The key to the attacker's success is in stemming the *unimpeded* flow of manpower, supplies, and weapons to replace the defender's losses.

h. Cost. The cost of conducting urban warfare is relative to the percentage of total expended resources, the time elapsed, and the results achieved. The cost to the attacker was considered high in the majority of urban battles. A high cost does not necessarily imply that the results were not worth the price. *The attacker and defender must thoroughly evaluate the overall cost prior to committing to an urban battle.*

From the offensive point of view, several factors can be associated with cost. First, and most importantly, isolating the urban area is critical to the attacker's success. Second, overwhelming superiority is needed if all costs are to be minimized. Third, the operation should be carefully planned. Fourth, intelligence is invaluable. Knowing where and how the city has been prepared for defense is also important. Fifth, attacking forces should understand the unique nature of urban combat. Clear tactics, techniques, and procedures for urban combat is required. Every aspect, from taking a building to using destructive technology and coordinating combined arms, must be thoroughly understood. Careful consideration of these five factors can minimize the cost of urban warfare to the attacker.

Attacker cost was generally high in casualties, time, and resources in the majority of urban battles studied.

From the standpoint of the defender, the critical variable is the defensive preparation of the city. Defensive preparations should include measures to prevent isolation of the city by the attackers. The capture of a prepared city can be made to be extremely costly. Preparations can include creating kill zones, clearing fields of fire, constructing canalizing obstacles, establishing reinforcing and fall-back positions, decentralizing command and control, and organizing multiple movement routes above ground, between rooftops, and below ground in subways and sewers. Artillery and aviation support can also be valuable force multipliers in defense.

i. Rules of Engagement (ROE). The nature of the military operation may restrict our use of weapons. The majority of urban battles since 1967 (such as **Beirut II**, **Hue**, **Jerusalem**) have had one or more of the following restrictions imposed on the attacking force:

(1) Minimizing civilian casualties and/or collateral destruction in order to:

- Avoiding alienation of the local population
- Reducing the risk of adverse world or domestic opinion
- Preserving facilities for future use
- Preserving cultural facilities and grounds.

(2) Limiting the use of specific ground or air weapons.

j. Logistics. Timely combat service support, particularly in the areas of ammunition resupply and casualty treatment and evacuation, is a critical element in MOUT.

(1) Logistics Support. Historically, combat in urban environments has seen a dramatic increase in the amounts of Class IV (shoring, sandbags, concertina wire), Class V (ammunition), and Class VIII (medical material). Unique items, such as rope, grappling hooks, and ladders are required for operations on urbanized terrain. Intense close-quarter combat requires a continuous flow of ammunition, particularly small-arms, tank, antitank, mortar, and artillery ammunition, as well as mines, grenades, and demolition explosives. Medical supplies must be readily available to treat the anticipated increase in casualties. Once battle the has been initiated, combat forces will require continuous supply.

Urban warfare requires a flexible, balanced logistics system capable of sustaining the close-quarter fighting evolutions within the built-up area.

One method of providing continuous logistics support is to establish a "push system." Essentially, this system pushes supplies to fighting units without their having to request them. A "push system" should negate the inherent delays of a "pull system," which requires units to request supplies then await their arrival. Optimally however, a balanced approach should be taken to sustain engaged forces (MCDP 4.). Combat service support, in terms of its timeliness and anticipation of the true needs of the combat units, should sustain the tempo of operations. It should avoid the delivery of unwanted or unnecessary supplies. To achieve this balanced approach, logisticians must develop flexible support plans that readily adapt to the ebb and flow of urban combat.

Mobile aid stations and surgical teams should be positioned as far forward as the tactical situation allows and be provided with dedicated evacuation vehicles.

(2) Health Service Support (HSS). Responsive treatment and evacuation plans should be established to handle the expected increase in casualties. Both immediate treatment and evacuation are critical in maintaining the morale and confidence of forces engaged in urban warfare. To meet casualty treatment and evacuation needs, mobile aid stations with surgical teams should be placed as far forward as the tactical situation allows. Dedicated vehicles and aircraft for the evacuation of casualties should be provided.

In addition to casualties resulting from physical injury on the battlefield, there are often considerable psychological casualties in modern battles, particularly urban battles. This can be attributed to the intense, sustained, and close-quarter combat associated with urban warfare. In many battles, attacking troops were not properly trained for urban warfare and felt inadequate in their ability to fight on equal terms with the defender. Many felt moral or ethical dilemmas associated with collateral damage and civilian casualties. Finally, the constant threat of being killed from any quarter coupled with the extreme fatigue of fighting in an urban environment can create psychological casualties. Leaders at all levels should be attuned to the symptoms associated with psychological casualties in order to get effected individuals prompt treatment so that they can be returned to their units.

1006. Implications of Urban Warfare. The commander charged with making decisions needs to understand the operational and strategic implications of a tactical struggle in an urban area (MCDP 1-1, *Strategy*, and MCDP 1-2, *Campaigning*.). Three urban battles (**Stalingrad**, **Hue**, and **Beirut II**) illustrate the importance of seeing beyond the tactical nature of the battle.

Commanders must see beyond the immediate tactical implications of fighting an urban battle.

The battle for **Stalingrad** had major operational and strategic implications. The entanglement of German forces at Stalingrad bought time for the Soviets to mobilize and prevent a major combined effort of two German Army Groups in the south. While Stalingrad resulted operationally in the destruction of the German Sixth Army, the strategic results were even greater. Stalingrad caused a complete change of German strategy in the east. Hitler made major changes in his General Staff, and from this point on in the war, he was a man estranged from his military leadership. The loss of enough men and equipment to field one-fourth of the German Army shook the foundations of the Third Reich.

The Battle for **Hue**, although only one of over one hundred different attacks of the Tet Offensive of 1968, had a negative impact on the will of both the American people and their political leadership. Hue marked a revolution in the coverage of war by modern mass media. It was the first time Americans could and watch an ongoing battle from their living room on the evening news. Hue was a television bonanza for almost a month. When North Vietnamese leadership directed that Hue be held for at least seven days, it was clearly *not* their intent to win a tactical battle, but to strike at the *strategic center of gravity*—in this case, the will of the American people. Although the battle for Hue was a tactical victory for the U.S., the North Vietnamese had achieved their strategic goal of making the American public question the costs associated with the war.

During the battle of **Beirut II**, the IDF's objective was to drive the PLO from Lebanon. The ability of the PLO to leverage the media to gain an advantage was one of the most significant weapons in their arsenal. Despite an Israeli tactical victory, the costs in image, prestige, allies, and, most importantly, its own national will were enormous. The results of this battle eventually caused a change of political leadership at the highest levels of government.

1007. Key Insights. Analysis of modern urban battles provides insight into the major factors affecting the course and outcome of combat. Consideration of these key insights are required in order to effectively and efficiently plan and execute MOUT:

a. MOUT is infantry intensive.

b. A tactical battle may have far-reaching implications. No longer are the strategic, operational, and tactical levels overlap, creating the situation where tactical actions can have operational and possibly strategic repercussions (MCDP 1-1, *Strategy*, and MCDP 1-2, *Campaigning*).

c. Commanders at all levels must understand the impact that media representation will have on the accomplishment of operational and strategic objectives.

d. Maneuver warfare doctrine must be applied to the environment.

e. Intelligence is imperative to success in urban warfare.

f. Surprise is a combat multiplier.

g. Armor, artillery, and aviation are effective at the outer perimeter of built-up areas for causing isolation or preventing reinforcement.

h. Armor operating inside a built-up area must be protected by infantry.

i. Artillery providing direct fire inside a built-up area can be effective in the reduction of strongpoints.

j. As force ratio increases in favor of the attacker, combat duration decreases.

k. Urban warfare is time consuming.

l. Isolation of an urban defender ultimately ensures his defeat.

m. Attack of an urban area is costly to the attacker in terms of resources and casualties.

1008. Necessity for Preparation. Since MAGTFs are usually forward-deployed forces, it is inevitable that MAGTFs will be tasked to deploy to urban areas and conduct military operations. Fighting in this environment will be violent, close, and personal. Fighting in villages, towns, and cities are likely to become more frequent for military forces as the United States responds to global crises. Forces will be expected to exercise adaptability in responding to a wide variety of missions in the world's built-up areas. Marines may soon find themselves operating in the conceptual "three block battlefield" within a city grid: feeding refugees in one block, separating combatants in the adjoining block, and returning hostile fire in a third block.

MOUT is infantry intensive. The remaining chapters and appendices of this publication focus on how the GCE fights and integrates supporting elements in an urban environment.

Chapter 2

Offensive Operations

Urban combat isolates and separates units. Operations are reduced to a series of small-unit actions, placing a premium on small-unit leadership, initiative, and skill.

2001. Introduction. The Marine Corps' manuever warfare philosophy guides the conduct of offensive operations in MOUT. This warfighting philosophy serves to guide commanders through the development, planning, and execution of missions. A detailed discussion of maneuver warfare can be found in MCDP 1, *Warfighting*, MCDP 1-3, *Tactics*, MCWP 3-1, *Ground Combat Operations* (under development), and other Marine Corps doctrinal publications.

This chapter focuses only on offensive operations in an urban environment where collateral damage is of minimal concern. It describes tactics, techniques and procedures to be employed for seizing and clearing buildings and other urbanized areas. Chapter 6 addresses some considerations associated with fighting under constraints and restraints on urbanized terrain.

Section I

Planning

2101. Considerations. Identification of the adversary's centers of gravity and critical vulnerabilities will allow the commander to focus his efforts on those portions of the built-up area essential for mission accomplishment.

a. Reasons for Attacking a Built-Up Area. A commander considers the following before deciding to attack a built-up area:

(1) Tactical Advantage. Cities control key routes of commerce and provide a tactical advantage to the commander who controls them. Control of features such as bridges, railways, and road networks can have a significant impact on future operations. Urbanized areas may be used by the enemy as a base of operations from which they launch their own offensive operations. It may be advantageous to attack those bases and separate the enemy from their support infrastructure.

(2) Political Advantage. The political importance of a built-up area may justify the use of time and resources to liberate it. Capturing a city could destroy the seat of local and national government. At the very least, it could deal the enemy a decisive psychological blow.

(3) Economical Advantage. The destruction or capture of key industrial and commercial cities with the resulting denial of production and distribution of equipment and supplies strikes at the enemy's future ability to wage war. The requirement for a logistics base, especially a port or airfield, may play a pivotal role in the enemy's ability to continue the conflict. Capture of such cities may prove extremely beneficial to the attackers, who can use these resources to their advantage.

(4) Potential Threats to Operations

(a) Enemy Threat Too Great To Bypass. Though the terrain around a built-up area may facilitate its bypass, the enemy within that urbanized area may remain a threat capable of interdicting lines of communications. This may require the enemy force to be contained or destroyed.

(b) Terrain Does Not Allow Bypass. The urbanized area may sit between two natural slopes on the avenue of approach and thus require capture in order to secure the main supply route. Additionally, the urbanized area, itself, may sit on dominating terrain that threatens combat support and CSS elements.

b. Reasons for Not Attacking a Built-Up Area. The commander considers the following reasons for not attacking a built-up area:

(1) Built-Up Area Not Required To Support Future Operations. The attacker may have adequate supply support and resources established at another site from which combat forces can be supported. Therefore, during the estimate process, commanders may assess that the urbanized area is not necessary to support future operations.

(2) Enemy Not a Threat. The commander may decide to bypass upon determination that no substantial threat exists in the built-up area that could affect the unit's ability to accomplish its mission.

(3) Time or Risk Unacceptable. The commander's intent may dictate that speed of movement is essential to the mission. Because MOUT can be time-consuming, the commander may choose to bypass the urbanized area to maintain tempo. Furthermore, the potential for numerous casualties, the expenditure of critical resources, or the restrictions placed on attacking forces may result in unacceptable risks to the commander's primary mission.

(4) Declared an Open City. The area may have been declared an "open city" because it is undefended or of religious or historical significance. By international agreements, open cities are demilitarized and must be neither defended nor attacked. (See HR, 25, Annex to Hague Convention No. IV, *Embodying Regulations Respecting the Laws and Customs of War on Land*, 18 October 1907.) The attacking force must assume civil administrative control and treat the civilians as noncombatants in an occupied country. The defender must immediately evacuate and cannot arm the civilian population. A city can be declared open only before it is attacked. Other reasons for not defending could be the presence of large numbers of noncombatants, hospitals, or wounded personnel or the city's cultural, religious, or historical significance.

2102. Commander's Estimate. Once assigned the mission to attack a built-up area, a thorough mission analysis is conducted. Marines follow the same planning process in MOUT as for other operational environments by evaluating the mission, enemy, terrain and weather, troops and support available, and time available (METT-T). Based on METT-T and application of maneuver warfare concepts, the commander can decide on a plan that defeats the enemy by exploiting identified critical vulnerabilities. The commander may not have sufficient forces to fight everywhere and so must focus on the enemy and concentrate resources to decisive ends. This may involve making conscious decisions to accept risk at some place or time and to seize opportunities at others. The challenges of an urbanized environment dictate an even greater need to fully understand maneuver warfare.

This section addresses METT-T considerations as they pertain to the offense in urban warfare. METT-T is covered in detail in MCRP 3-11.5A, *Marine Troop Leader's Guide.*

a. Mission. Mission analysis is the first step of the estimate process. One of the important mission analysis considerations in urban conflict is the scope of clearance required to accomplish the mission. Commanders and planners should ask themselves the following questions:

- Do I need to clear every building?
- Should I clear only certain blocks?
- Should I only control certain areas?
- What level of protection is required for my lines of communication?

b. Enemy. Maneuver warfare focuses on the enemy. The intelligence preparation of the battlefield (IPB) process should be used to analyze the enemy and terrain in detail. (See Field Manual (FM) 34-130, *Intelligence Preparation of the Battlefield.*) The commander decides whether threat forces are conventional or unconventional.

(1) Conventional Forces. Most potential adversaries have adopted techniques of urban combat from either the United States or the former Soviet Union. Therefore, potential adversaries will build their urban defense to counter the attack of a combined arms force. Potential enemy forces will normally organize their defense in depth. Prepared strong points will usually form the perimeter of a larger defense while reserves locate in a separate position within the perimeter. You can expect ambushes to fill gaps in the perimeter while dummy strongpoints are established to deceive the attacker. Positions for securing the entrances to and exits from underground structures and routes are normally established. Security positions will normally be located forward of first-echelon defensive positions.

(2) Unconventional Forces. Urban areas have become a haven for unconventional forces. The large noncombatant population provides cover and concealment for unconventional force operations. Conventional forces operating in MOUT will normally be placed under restrictive ROE to minimize collateral damage. Unconventional forces will often use our restrictive ROE and the noncombatant population to their advantage when devising an urban defense. (See FM 34-130, FM 7-98, *Operations in a Low-Intensity Conflict*, and Chapter 6 of this manual.)

c. Terrain and Weather

(1) Terrain. Offensive operations should be tailored to the urban environment based on a detailed analysis of urbanized terrain (see Appendix I). Commanders and subordinate leaders use **KOCOA** (**k**ey terrain, **o**bservation and fields of fire, **c**over and concealment, **o**bstacles, and **a**venues of approach) to identify important terrain factors:

(a) K: Key Terrain. Key terrain surrounding an urban area can facilitate entry or deny escape. Within the city, airports or airfields, stadiums, parks, sports fields, school playgrounds, public buildings, road junctions, bridges, or industrial facilities may be key terrain.

Critical public buildings are identified during the terrain-analysis phase of an IPB. Hospitals, clinics, and surgical facilities are important because the laws of war prohibit their attack when not being used for military purposes other than medical support. The locations of civil defense, air raid shelters, and food supplies are critical in dealing with civilian affairs. Additionally, population size, locations, and density; density of the built-up area; firefighting capabilities; the location of hazardous materials; police and security capabilities; civil evacuation plans; and key public buildings should be identified.

(b) O: Observation and Fields of Fire. Urbanized terrain is characterized by restrictive observation and fields of fire. Weapon ranges can be greatly reduced because of buildings and other manmade structures. On the other hand, high ground or tall buildings can provide perches which enhance line of sight (LOS) for observation and communications as well as for individual and crew-served weapons. This includes man-portable surface-to-air missiles.

(c) C: Cover and Concealment. Buildings, sewers, and subways can provide excellent cover and concealment for enemy and friendly forces. They also provide covered and/or concealed maneuver routes within the built-up area. The civilian population can also offer cover and concealment to enemy forces.

(d) O: Obstacles. Natural or manmade obstacles restrict or deny maneuver within the urban area. Bridges, walls/fences, canals, streams, rivers, as well as rubble created by the effects of weapons should be thoroughly analyzed. Construction sites and commercial operations such as lumberyards, brickyards, steelyards, and railroad maintenance yards are primary sources of obstacle and barrier construction materials.

These sites can also supply engineers with materials to strengthen existing obstacles or to set up antitank hedgehogs or crib-type roadblocks.

(e) A: Avenues of Approach. Avenues to the city should support maneuver and be concealed either by terrain, darkness, smoke, or a combination of the three. Avenues that canalize or choke maneuver, due to the density of built-up area or natural terrain, should be identified and avoided. Roads, rivers, streams, and bridges provide high-speed avenues for movement. **Generally, military maps do not provide enough detail for urbanized terrain analysis.** They usually do not show the underground sewer system, subways, underground water system, mass transit routes, and power plants. Local maps of intracity road networks and subway systems and city hall/department of public works blueprints of all city buildings (if available), coupled with aerial photos, should be used.

(2) Weather. As in any military operation, weather affects equipment, terrain, and visibility, but its greatest impact is on the individual Marine. Snow, ice, dust, wind, rain, humidity, and temperature extremes reduce human efficiency. Weather extremes coupled with stress and the physical strain of urban combat can be minimized with effective small-unit leadership. Weather factors include:

(a) Precipitation. Rain or melting snow may flood basements and subway systems. This is especially true when automatic pumping facilities that normally handle rising water are deprived of power. Flooding makes storm and other sewer systems hazardous or impassable. In an NBC environment, chemical agents can be washed into underground systems by precipitation. As a result, these systems may contain chemical agent concentrations that are much higher than surface areas and thus become contaminated "hot spots." Hot spot effects become more pronounced as agents are absorbed by brick or unsealed concrete sewer walls.

(b) Fog. Many major cities experience fog, especially those located in low-lying areas and along canals or rivers. Industrial and transportation areas are the most likely to be affected by fog because of their proximity to waterways. Fog adversely affects vision and optical aids. It may also be used to help conceal friendly movement.

(c) Inversion Layers. Air inversion layers are common over cities, especially cities located in low-lying "bowls" or in river valleys. Inversion layers trap dust, smoke, chemical agents, and other pollutants, reducing visibility and often creating a greenhouse effect, which raises ground and air temperatures.

(d) Temperature. Built-up areas are often warmer than surrounding open areas during both summer and winter. This difference can be as great as 10 to 20 degrees hotter throughout the year.

(e) Wind Effects. Generally, wind chill is not as pronounced in built-up areas. However, the configuration of streets, especially in the city core and outlying high-rise areas, can cause wind tunneling. This increases the effects of the wind along streets that parallel the wind direction, while cross-streets remain relatively well protected. Wind tunneling can have a negative impact on aviation support within the urbanized area by creating dangerous wind effects down streets and between buildings.

(f) Day/Night Differences. Night and periods of reduced visibility favor surprise, which in turn, may facilitate infiltration, detailed reconnaissance, attacks across open areas, seizure of defended strongpoints, and reduction of defended obstacles. Because of the difficulties of night navigation in restrictive terrain (usually without reference points and in close proximity to the enemy) forces may consider using simple maneuver plans with easily recognizable objectives.

Rotary-wing aircraft are best utilized at night when operating on urbanized terrain. The most effective method of protecting helicopters from a successful engagement by hostile ground forces is to remain unseen. During night flights, the use of night vision goggles (NVGs) should be weighed carefully. Artificial lights can render NVGs partially ineffective and can increase the possibility of mid-air collisions by degrading the goggles to the point where nearby aircraft and aerial obstructions are not seen.

(g) Aviation Weather Considerations. Weather conditions can be a critical factor in determining the amount of aviation support the aviation combat element can provide in an urbanized area. Aviation forces should consider the following:

- Presence or absence of fog, industrial haze, low clouds, heavy rain, and other factors that limit visibility for pilots
- Illumination and moon phase/angle during night vision goggle operations
- Ice, sleet, and freezing rain that degrade aerodynamic efficiency
- Updrafts and downdrafts in the urban canyon that may present an unpredictable hazard
- High temperatures and/or high-density altitudes that degrade aircraft engine performance and lift capability
- High winds (large gust spreads) and crosswinds, to include the tunneling effect, that may create localized and unpredictable hazards to aviation
- Weather conditions that create hazards on pick-up zones and LZs, such as blowing dust, sand, or snow.

d. Troops and Support Available. The commander must be aware of all of his combat power (e.g., types of weapons systems, their numbers, and their capabilities) when performing analysis for operations in a built-up area. In the initial planning phase, force size must be evaluated in relation to urban size, enemy forces, and mission assignment. In the attack of a built-up area (population 100,000+), the GCE of a MEF would be a Marine division. However, in the initial stages of an amphibious or MPF operation, a reinforced infantry battalion or regiment may provide the initial forces. No matter what the size GCE, operations will be conducted by task-organized, reinforced battalions and their companies/platoons. Therefore, by using the infantry battalion as the basis for projecting the size of an overall force, the number of regiments or divisions required to secure a built-up area can be determined. Whatever size MAGTF is required for operations in an urbanized area, much of the fighting may be generally conducted by small teams of Marines. Consequently, our success in urban fighting may largely depend upon small unit and individual tactical skills. Specifically, leaders at all levels must analyze the following factors:

- Number and type of available units
- Task organization for urban combat (See Appendix A.)
- Availability of critical weapons systems
- State of training and discipline (training for urban warfare is imperative)
- Strength in terms of men and materiel in relation to enemy and size of built-up area
- Aviation support available
- State of maintenance and supplies
- Available combat support (based on availability and mission requirements)
- Available CSS (critical logistical and maintenance items)
- Host nation support available. (If the city's occupants are friendly to the attacker, then support in intelligence, deception, and diversion may be possible.)

e. Time Available. The following issues should be considered when analyzing the time available for an attack on urbanized terrain:

- A significant amount of time is required for clearing buildings, blocks, or axes of advance.
- Marines tire more quickly when clearing buildings because of stress and additional physical exertion.

- Adequate time should be allocated for thorough reconnaissance, planning, and rehearsals. Ultimately, this can result in higher tempo in the execution of the plan.

Once an estimate of the situation (using METT-T) has been conducted, planning for offensive operations can proceed. It is important to remember that the initial estimate of the situation must be continually updated throughout the planning process and the conduct of operations to account for changes in METT-T. (See MCWP 5-1 *Marine Corps Planning* [under development])

2103. Phases of the Attack. Attacks are categorized as either hasty or deliberate. Both hasty and deliberate attacks should take advantage of as much planning, reconnaissance, and coordination as time and the situation permit. Regardless of the size of the attacking force or of the objective to be secured, the phases of the attack (i.e. reconnoiter, isolate, secure foothold, and control the objective) remain constant.

a. Deliberate Attack. A deliberate attack *is a type of offensive action characterized by preplanned coordinated employment of firepower and maneuver to close with and destroy or capture the enemy* (Jt Pub 1-02). It is a fully coordinated operation that employs the MAGTF's combined arms team against the enemy's defense. It is used when enemy positions are well prepared, when the built-up area is large or severely congested, or when the element of surprise is lost. Given the nature of urbanized terrain, the deliberate attack of a built-up area is similar to the assault of a strongpoint (see MCWP 3-1, *Ground Combat Operations* [under development]). The deliberate attack of a built-up area is conducted in the following phases:

(a) Phase I: Reconnoiter the Objective. Intelligence gathering and reconnaissance/ surveillance are critical to the planning process and success of the operation. All available intelligence resources should be used to gain vital information on the enemy. Whenever possible, a leader makes a personal reconnaissance of the objective area to collect first-hand information regarding the area to be attacked. A thorough reconnaissance yields a wealth of information to use in the continued development of the plan. Avenues of approach, observation posts (OPs), supply routes, and the emplacement positions of direct- and indirect-fire weapons systems are all examples of information that may be amassed during the reconnaissance of the objective area. Composition and structure of buildings and roadbeds, cover and concealment opportunities, and other information not apparent in a map study may have a significant impact on the plan.

(b) Phase II: Isolate the Objective. The objective can be isolated by seizing natural and man-made features that dominate the area. Isolation may also be accomplished by coordinated use of supporting arms to seal off enemy lines of communication. This phase may be conducted simultaneously with Phase III (securing a foothold). Figure 2-1 shows a task-organized, reinforced infantry battalion isolating the objective.

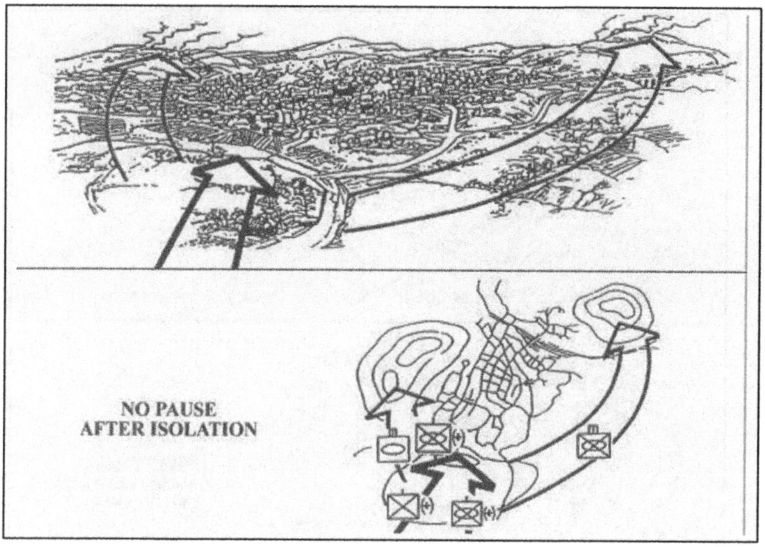

(c) Phase III: Secure a Foothold. Once the objective is isolated, a foothold should be secured as soon as possible in order to maintain tempo. The seizure of a foothold provides the attacking force with a position from which they can continue the assault through the objective area. The assault is supported by direct and indirect suppressive fires and smoke. Figure 2-2 (pg. 2-10) shows the battalion objective being isolated. One company is providing suppressive fires while another company seizes an initial foothold. Fire support assets are providing suppressive fires on the objective.

Figure 2-1. Isolation of the Objective Area

(d) Phase IV: Seizing the Objective. Once a foothold is seized and consolidated, supporting forces move to the built-up area to support the seizing of the objective area. To maintain tempo, the transition between the phases should be seamless. Once the foothold has been established, forward units continue the attack through the objective area. Supporting units assist as required. The momentum of the assault is continued until the objective area is cleared or controlled.

The assault force should establish limited objectives to ensure that the attacking forces do not get strung out along the axis of advance. Gaps may give the enemy the opportunity to infiltrate along the line of advance or make isolated friendly forces vulnerable to attack.

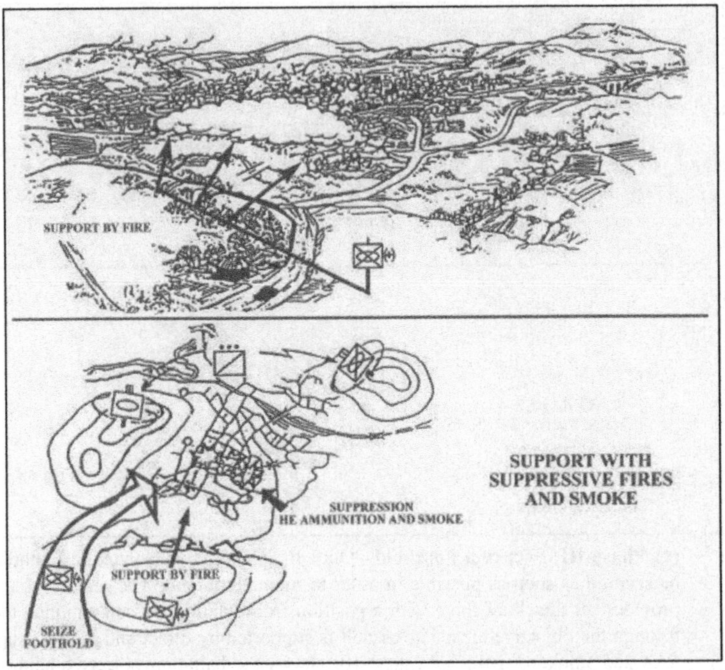

This phase also includes consolidation of the objective area and reorganization for future taskings. By setting limited objectives, the attacking forces have an opportunity to reorganize and defend against counterattacks while maintaining momentum. Once the limited objectives have been secured, the urbanized area may be sectored for detailed clearing. Clearing by sectors may help in the allocation of forces and in the prevention of fratricide. In addition to defensive preparation against counterattack, preparation should be made for follow-on missions (e.g., restoration of civilian facilities, marking and clearing minefields, etc.).

Figure 2-2. Seizing a Foothold

Seizing may consist of a systematic house-by-house, block-by-block advance through the entire zone of action, or it may be a rapid advance through a lightly defended area to seize a key objective. The difference between the two techniques is made by the commander on the basis of mission requirements and METT-T analysis. Seizing an objective in urbanized terrain requires detailed planning, coordination, decentralized command and contol, and small-unit execution. It is a continuation of the assault until all mission requirements are met.

(1) Rapid Advance. Rapid advance (Figure 2-3) may be used to maintain

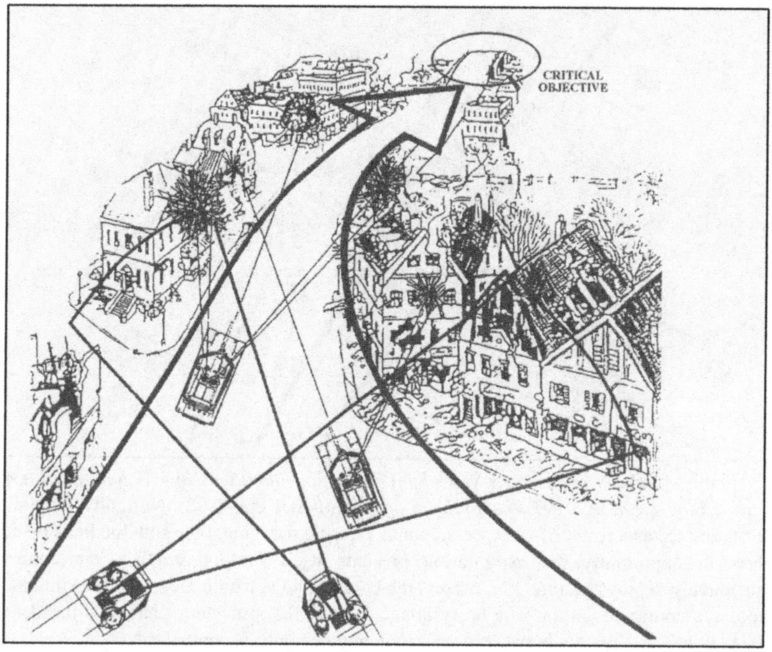

momentum through the zone of action and to seize key objectives. Some enemy positions may be isolated or bypassed in order to maintain tempo. This procedure may be used when:

- A key objective is identified (an enemy critical vulnerability or an objective that supports the friendly scheme of maneuver)
- Time is critical in reaching the objective
- Only selected buildings on the route of advance require clearing.

Figure 2-3. Rapid Advance En Route to a Key Objective

(2) Systematic Clearance. Systematic clearance (figure 2-4 on pg. 2-12) involves the deliberate reduction of all significant enemy positions throughout the objective area. It is usually used when time is not a critical factor.

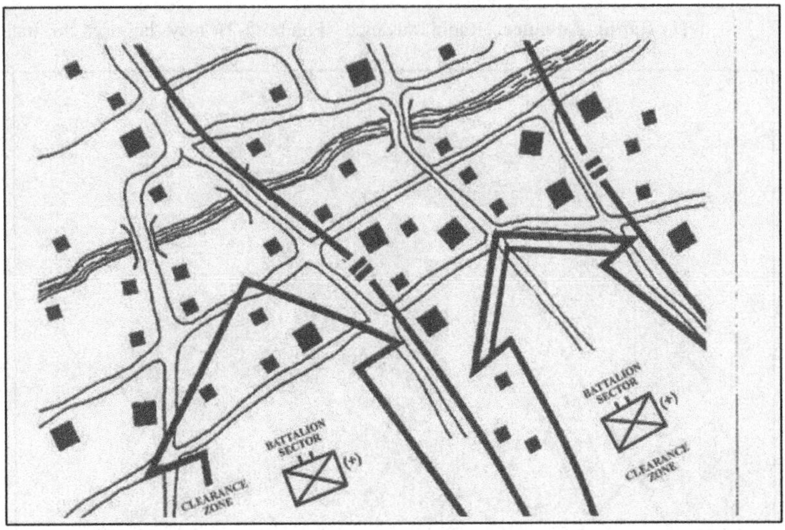

Figure 2-4. Systematic Clearance Within Assigned Sectors

b. Hasty Attack. A hasty attack is, *in land operations, an attack in which preparation time is traded for speed in order to exploit an opportunity* (Jt Pub 1-02). Normally, hasty attacks are conducted as a result of unexpected contact during movement through the built-up area or when an opportunity for exploitation presents itself. Upon contact, the commander immediately deploys, returns fire, reports the contact and situation, develops the situation, and chooses a course of action. The hasty attack follows the same four phases as the deliberate attack; however, the time between preparation and response is compacted.

2104. Warfighting Functions. The warfighting functions (WFs) encompass all activities in the battlespace. Planners consider and integrate the warfighting functions when analyzing how best to accomplish the mission. The WFs include maneuver, intelligence, fires, logistics, command and control, and force protection.

a. Command and Control. The commander positions himself where he can best observe and influence the battle. Positioning is relative to the level of command. For example, a squad leader locates where he can observe and direct his fire teams, while the battalion commander locates where he can observe and direct his companies.

Close-quarter fighting in and around buildings makes command and control exceedingly difficult. Commanders and small-unit leaders rely on guidance and intent provided by higher

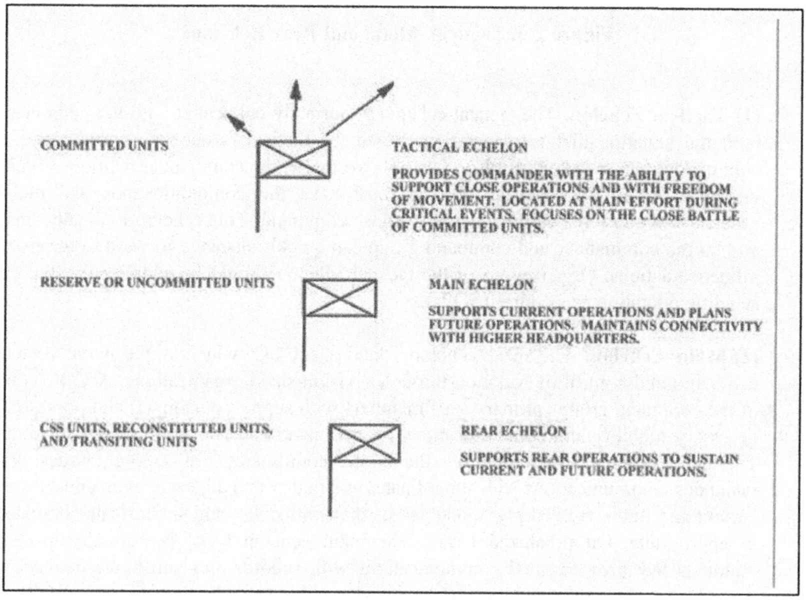

headquarters to facilitate decentralized execution while maintaining control and direction. In close-quarter fighting, commanders may have to rely upon messengers to communicate with their units and higher headquarters.

At battalion and higher levels, command and control is accomplished through three command echelons, which can be collocated or echeloned across the battlefield. The three echelons are tactical, main, and rear (see fig. 2-5). Each echelon may have a combat operations center (COC) capability. The COCs are the facilities (buildings, vehicles, or tents) used by the commander or staff at each echelon to plan, direct, control, and coordinate operations.

Figure 2-5. Tactical, Main, and Rear Echelons

(1) Tactical Echelon. The tactical echelon is normally collocated with the main echelon until the situation dictates separation. When the tactical echelon separates, designated command group members displace forward. Normally, the commander will move with the tactical echelon. The tactical echelon must have the communications and mobility necessary to allow the commander to exercise command. This echelon is usually small in size so the commander and command group can quickly displace forward to observe and influence actions. Organization of the tactical echelon should be addressed in the unit's standing operating procedures (SOPs).

(2) Main Echelon. The main echelon contains a COC, which is the nerve center for command and control of battalion through division-sized organizations. A COC consists of the command group (primary staff members with support personnel) and possesses the necessary mobility and communications for command and control of tactical operations. The main echelon COC monitors the battle, coordinates fire support, issues orders, maintains communications with subordinates and higher and adjacent commands, conducts current and future tactical plans, and assists the commander and subordinate commanders as appropriate. On urbanized terrain, the main echelon COC is normally located in buildings that provide good communications with subordinates and higher and adjacent commands. The command post (CP) is where the commander is physically located. Therefore, the CP can be located at any of the three echelons based on the commander's presence.

(3) Rear Echelon. The rear echelon's COC consists of the logistical and administrative personnel required to plan, coordinate, and execute logistics support. During urban warfare, the rear echelon COC is usually closer to combat forces than normal as it uses the city's captured buildings and facilities to provide cover and concealment for stocking supplies, conducting maintenance repair, and treating the injured.

Attack of a built-up area requires detailed planning due to restrictive terrain, close proximity to opposing forces, fire coordination requirements, and reduced communications capability. The GCE will frame a centralized plan of attack, fire support, and logistics support to conduct urban operations. However, the actual attack of a built-up area is a series of isolated, close-quarter battles carried out by small units. The execution of these attacks should be decentralized.

Communications planning procedures, as outlined in MCWP 3-40.3, *Communications and Information Systems* (under development), are applicable to the urban battle. Planners should be meticulous in their scrutiny of the environment and the effect it will have on communications. Manmade structures can create problems for single-channel radios. These structures inhibit LOS radio communications by absorbing or reflecting transmitted signals. However, the urban environment may have exploitable advantages such as the availability of electrical power, commercial telecommunications networks, and environmental control systems. Electrical power generation stations and other emergency power systems are normally found in protected structures and are probably usable. Due to their value to the enemy, these areas are more likely to be heavily defended. Enclosed areas offer excellent concealment and protection of communications and other command and control support equipment. Extensive commercial communications networks composed of miles of underground protected cable connecting central telephone exchanges are likely to be available, as well as a multitude of public service radio nets (police, fire, civil defense, taxi, etc.) complete with existing antenna systems and retransmission stations. To communicate effectively and continuously, commanders must minimize limitations imposed by the urban environment and maximize the advantage of existing civil communications.

b. Intelligence. Collection of intelligence information on urbanized terrain is difficult. Urbanized terrain provides overhead cover and concealment from observation by aircraft, reconnaissance personnel, and satellites.

(1) Information Requirements. In addition to information about the location of enemy units, nature of the terrain, trafficability of roads, etc., that all operations require, urban warfare requires detailed information on the following:

- Population density, characteristics, and culture
- Location and quality of water supply
- Width and construction of streets
- Layout of road networks
- Location, width, and load capacity of bridges
- Layout of underground networks such as subways, sewers, and utility ducts
- Types and construction of buildings
- Location of key communications and transportation facilities
- Location of canals and waterways.

Information on port facilities should include details on harbor sites, pier networks, unloading capabilities, warehouse/dockside facilities, harbor currents, and sandbars or other midstream obstructions.

(2) Collection. Accurate, detailed, and timely information is vital to the success of any urban operation. Collection operations conducted by national intelligence organizations may provide valuable information on the layout of the city, recent changes to transportation routes, political affiliations of various groups, and locations of military units. This information needs to be current and must be verified and updated as attacking forces close on the city. Reconnaissance activities such as the study of large-scale maps, aerial photographs, and background intelligence reports keep the commander's information current. The latest information on the enemy and the terrain should be continuously fed to the intelligence section using long-range reconnaissance patrols, human intelligence (HUMINT) reports, unmanned aerial vehicle (UAV) missions, and electronic intelligence. Use of reconnaissance assets should be tempered with a risk-versus-gain judgment. Reconnaissance assets should not be risked indiscriminately if the information can be collected by other means.

(3) Maps. Tactical maps should be supplemented by city street maps that contain street names and information on the location of important buildings and transportation terminals. These city street maps should be distributed to all units operating within the city. Although tactical maps are key in controlling fire support missions, simple street maps assist ground troops in maintaining their orientation within the city and in tracking buildings and areas that have been cleared.

c. Maneuver. Maneuver is the employment of forces on the urban battlespace through movement in combination with fire or fire potential to achieve a position of advantage (see MCWP 3-1). Maneuver is an essential element of combat power. Maneuver in itself cannot produce decisive results. Combined with mass, offensive action, economy of force, and surprise, maneuver provides favorable conditions for closing with the enemy. Maneuver contributes significantly to sustaining the initiative, exploiting success, preserving freedom of action, and reducing vulnerability. It is through maneuver that an inferior force can achieve decisive superiority at the necessary time and place. In many cases, maneuver is made possible only through the control of tempo and effective employment of firepower. The commander integrates supporting fires with the scheme of maneuver to create a dilemma for the enemy. Likewise, movement without fires exposes the force to effective enemy counteraction and risks losing the initiative and momentum. Maneuver that does not include violent action against the enemy will not be decisive. At all levels, successful application of this principle in the urban environment requires flexible and innovative plans and actions.

(1) Infantry/Armor Movement. Armored vehicles are vulnerable in built-up areas where streets and alleys provide ready-made fire lanes for defenders. Motorized traffic is greatly restricted, canalized, and vulnerable to ambush and close-range fire. Tanks are at a further disadvantage because their main guns cannot be depressed sufficiently to fire into basements or be elevated to fire into upper floors of buildings at close range (Figure 2-11). Whenever tanks are used in the direct-fire role, infantry must provide security against enemy ground attack. Movement of armor down narrow streets or down narrow paths through debris requires close coordination with infantry. Infantry and armor may use bounding overwatch techniques during movement (See MCWP 3-12, *Marine Tank Employment* [under development]). With this technique, infantry moves ahead of armored vehicles to clear the buildings on each side. Lead infantry units should not get so far ahead of armor that mutual support is lost. Conversely, if armor gets too far ahead of infantry, it may be quickly damaged or destroyed. Coordinated movement also contains a rear security element to protect armor from behind.

During movement down a street tanks should deploy with one tank on each side of the street focusing on the street and lower levels of buildings. *In MOUT, tanks are best deployed in pairs.* Other tanks of the tank platoon or section should move behind the infantry and fire at targets in the upper stories of buildings. When space is inadequate to deploy abreast, tanks may deploy in single file. Movement across open areas should be well coordinated to ensure mutual support. Suppression fire and smoke are used to cover force movements.

Armor can also be used in the reduction of obstacles. For example, grappling hooks and ropes attached to a tank can be used to pull down wire obstacles.

(2) Cover and Concealment. Walls and buildings can provide cover, concealment, and protection from enemy ATGM or heavy machine gun fire. Before armor enters a building, infantry should clear the building and check the ground floors to ensure that there is no basement into which an armored vehicle could fall and become trapped.

Tank and AAV generated smoke may be useful in concealing the location and movement of assaulting forces. Additionally, the thermal sights of the tank are very effective as they can use heat differentials to detect the movement or presence of concealed enemy soldiers, weapons, and vehicles.

(3) Tactical Considerations for Helicopterborne Movement. Helicopterborne operations are planned and conducted using the same combat fundamentals that apply to ground operations. However, there are considerations that apply only to helicopterborne operations. See MCWP 3-24, *Assault Support* (under development) for a detailed discussion on helicopterborne operations. These considerations include:

(a) Helicopterborne forces, once they have landed, lack vehicular mobility and heavy weapons. For this reason, it is important that the forces land on or near the objective. Such a force may be isolated for a time pending linkup with ground forces.

(b) Helicopters may be used for tactical deception. Helicopters may be used to make demonstration landings in several different zones to deceive the enemy as to the true objective of an operation.

(c) Helicopterborne assaults are normally conducted against undefended or lightly defended objectives. When attacking a well-defended objective, an LZ should be selected which provides cover and concealment for the assault force and is close to the objective.

(d) Helicopterborne forces are vulnerable to attack helicopters, fixed-wing aircraft, and surface-to-air missiles. To counter this vulnerability, friendly attack helicopters

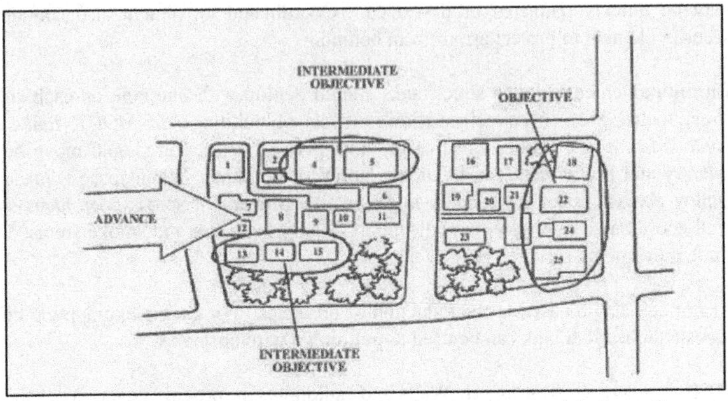

escort the helicopterborne forces during air movement, and indirect fires are used to suppress or neutralize enemy air defenses.

(4) Control Measures. The following control measures assist in the manuever of forces in MOUT:

(a) Objectives. When attacking to seize a foothold, the infantry unit may assign subordinate units the first block of buildings as their first objective. When an objective extends to a street, only the near side of the street is included. The final objective may be buildings or key terrain at the far edge of the built-up area. Key buildings, or groups

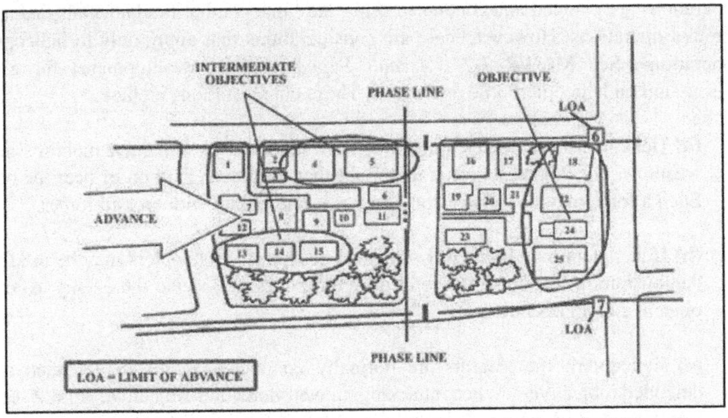

of buildings, may also be assigned as intermediate objectives. Buildings along the route of attack should be identified by numbers for clarification (Figure 2-6 on page 2-18).

During the seizing phase, bypassing buildings may increase the risk of attack from the rear or flank. It may be necessary for the unit to enter, search, and clear each building in its zone of action. A single building may be an objective for a rifle squad or, if the building is large, for a rifle platoon or company. When the commander's concept of operations is based on speed (rapid advance method) or when the force is conducting a hasty attack, a unit may be directed to bypass certain positions within its zone.

Figure 2-6. Objectives and Numbering System

(b) Phase Lines. Phase lines are control measures used to report progress or to control the advance of attacking units. Phase lines should be oriented on readily identifiable terrain features such as principal streets, rivers, and railroad lines. They should be placed on the near side of a street or open area. In an attack, a unit may have the mission to seize positions in its zone of action up to a particular phase line. (Figure 2-7).

Figure 2-7. Phase Lines

(c) Boundaries. Unit boundaries are used to define zones of action and are usually set within blocks so that a street is included in the zone. Both sides of a street should be included within the same unit's boundries.

(d) Checkpoints and Contact Points. Checkpoints aid in reporting locations and controlling movement. Contact points are used to designate specific points where units make physical contact. Checkpoints and contact points are designated by each unit as appropriate for command and control at street corners, buildings, railway crossings, bridges, or any other easily identifiable feature.

(e) Attack Position and Line of Departure (LD). A designated attack position may be occupied by forward units for last-minute preparation and coordination. The attack position is often behind or inside the last large building before crossing the LD. The LD should be located on the near side of an open area running perpendicular to the direction of attack, such as a street or rail line.

(5) Time of Attack. Ideally, the time of attack should be conducted during hours of darkness or limited visibility. Marines can exploit the poor visibility to cross open areas, to gain access to rooftops, to infiltrate enemy areas, and to gain a foothold. When attacking in unrestricted visibility, units should use smoke to conceal movement.

(6) Reserve. Reserves should be mobile and usually follow in trace of forward units. Battalion reserves normally follow one to two blocks to the rear of the lead company. If a company reserve is available, it follows within the same block so that it can immediately influence the attack. A unit with a reserve mission can be called upon to:

- Exploit an enemy weakness or friendly success
- Attack from another direction
- Attrit and/or clear bypassed enemy positions (normally requires designation of a new reserve)
- Secure the rear or a flank
- Maintain contact with adjacent units
- Counterattack

(7) Security. Each unit should ensure security of its flanks and rear (Figure 2-8 on page 2-20). Fighting in a built-up area is unique in that attacks can come from all three dimensions. Units assigned a mission to isolate can also provide security. Small units conduct security patrols and man OPs to supplement higher level reconnaissance and security operations.

Figure 2-8. Enemy Firing From Flank

(8) Engineers. Leading units should have engineers attached for immediate support. Tasks commonly assigned to engineers include:

- Preparing and using explosives to breach walls and obstacles
- Finding and exploding mines in place or helping to remove them

- Cratering roads and performing other countermobility measures
- Clearing rubble and obstacles

Requirements for engineer support in MOUT normally exceed capabilities. To compensate, all Marines should receive basic demolition training prior to conducting urban operations.

d. Fires. The attack of a built-up area may require extensive air and artillery preparation prior to the ground attack. Supporting fires suppress the defender's fire, restricts his movement, and possibly destroys his position. Consideration should be given to the rubbling effect produced by aerial and artillery bombardment. The assault should closely follow air and artillery fire to exploit its immediate effect on the defender. Maneuver units move near the final coordination line while the enemy is engaged by supporting fires. As the attacking force assaults, supporting fires lift and/or shift to block enemy withdrawal or to prevent the enemy from reinforcing their position. (See Chapter 4 for detailed discussion on air and artillery support during urban warfare.) Fire can be categorized into indirect and direct fires.

(1) Indirect Fire

(a) Artillery. Indirect artillery fire is planned to isolate objectives, to prevent reinforcement and resupply, to neutralize known and suspected CPs and OPs, and to suppress enemy defenders. Urbanized terrain requires that most indirect artillery fire use a high-angle trajectory.

(b) Mortars. Mortars are the most responsive indirect-fire weapon for the urban environment. They can be employed to hit targets of opportunity at the close ranges that are typical of combat in built-up areas.

(c) Other. Urban conflict is conducive to innovation. Many direct fire weapons can be employed in the indirect fire mode. Indirect fire weapons may also be effective at clearing the tops of buildings using variable-timed fuses.

(2) Direct Fire.

(a) Armor. Tanks are normally task organized with mechanized units for protection and security operations. The commander can employ tanks to take advantage of their long-range lethality, high-speed mobility, and survivability. They can be used outside the built-up area to cover high-speed armor avenues of approach. Tanks are normally employed in this role during the isolation phase. Tanks also support by fire the assault to seize a foothold and may support attacks to seize objectives within the built-up area. In MOUT, tanks can also be organized into special assault teams. The tank cannon's direct fire is very effective against structures. (Figure 2-9 on page 2-22 and Appendix B)

(b) Howitzers. Artillery rounds delivered by direct fire can be very effective for destroying targets in buildings. When used in this manner, the mobility, flexibility, responsiveness, and survivability of gun crews may be reduced. Artillery used in the direct fire role normally use HE shells.

(c) Infantry. Direct-fire systems, mainly machine guns, antitank guided missiles (ATGMs), shoulder- launched multipurpose assault weapons (SMAWs), and AT4s—are initially employed to support the seizure of a foothold. The positioning of antitank weapons in buildings must allow enough space for the effects of backblasts. Antitank weapons have limited as effectiveness when neutralizing targets behind walls. Designed primarily as armor penetrators, they neutralize a target only if that

Figure 2-9. Tank Direct-Fire Support Within the Built-Up Area

target is located directly behind the point of impact. Consideration must be given to weapons with minimum arming distances. Many of these factors can combine to limit the effects of weapons in the urban environment.

(d) Aviation. The vertical nature of large urban areas may block the pilot's line of sight with the target unless careful consideration is given to inbound headings and altitudes. Plan attack headings to take advantage of large open areas and street axis. Friendly ground forces must be clearly distinguishable from targets. The use of smoke, visible or infrared strobes, or other visual devices must be effectively employed by ground forces and clearly understood by friendly aircraft.

e. Logistics. Logistics support in urban operations should be timely, flexible, and responsive. Logistics planners, both in the supported units and the CSSE, should be fully integrated in the planning process. The final logistics plan must be thoroughly coordinated with the scheme of maneuver. Units may need special equipment such as ropes, grappling hooks, ladders, sandbags, and wire; medical supplies must be readied in anticipation to support increased casualties; and additional supplies must be prestocked. (See Chapter 5 for a detailed discussion on logistics and urban warfare, and MCWP 4-1, *Tactical Logistics*, for additional planning considerations.)

f. Force Protection. Force protection encompasses all those measures that preserve the fighting potential of the command. It is integrated throughout all the warfighting functions. It does not entail security operations alone. It involves the continuous implementation of active and passive measures by all units during all phases of an operation.

Some particular active and passive measures in MOUT include:

-Manmade structures in the urban environment provide cover and concealment for both the attacker and defender.

-Lines of communication can be more easily interrupted due to multiple avenues of approach.

-All units must be prepared to implement force protection in a 3-dimensional environment.

-Proximity of civilian population makes identification of non-combatants more difficult.

-ROE may be more restrictive

-Snipers and man-portable shoulder lauched air defense missiles are usually more difficult to detect.

Section II

Infantry Battalion Operations

2201. Overview. This section addresses some of the operations that a task-organized and/or reinforced infantry battalion could conduct in MOUT. These operations are seldom conducted independently and are usually part of a larger joint operation. In the example given below, the battalion has seized a foothold and is operating in its sector of a town of 100,000 people. Concurrent operations are on-going through other parts of the town. The threat in the battalion sector is a company (-) with a capability to reinforce.

2202. Seizure of a Key Objective. The tactical situation may require the immediate seizure of a building, industrial complex, or transportation site that is deemed decisive terrain. Examples of such objectives may include key bridges, intersections, railheads, or facilities that constitue the center of gravity for the enemy. As an example, a battalion may be assigned the mission of seizing the only bridge that is capable of supporting armored vehicles and resupply by heavy transports along a river bisecting a town. The intent is to isolate enemy forces and secure the bridge for follow-on forces. Intelligence indicates that the bridge is lightly defended at this time. In this case, speed of attack and surprise is critical to capture the objective. Some key intelligence information is required prior to the attack. This would include the number of enemy defending the bridge, their weapons, disposition, their ability to reinforce their defense, location of LZs near the bridge, whether the bridge has been set for demolition, and the location of enemy indirect fire assets that can range the objective.

The commander decides to conduct a penetration using a combination of helicopterborne attack to rapidly seize the bridge followed by a mechanized ground attack to link-up with the helicopterborne forces (Figure 2-10). The helicopterborne force can achieve surprise and speed of movement required while the mechanized force can thrust quickly along high-speed avenues to minimize the time the helicopterborne force must hold the objective against enemy counterattack.

For the ground attack, the battalion commander makes a determination that speed of movement outweighs the risks of rapidly extending his lines of communication. He directs, on the basis of his knowledge of the enemy situation and his mission, that his maneuver forces avoid decisive engagement with the enemy along the route to minimize the time it takes to link up with the helicopterborne force. The commander provides guidance on those enemy forces that should be bypassed and those that should be isolated and reduced.

The commander task organizes his battalion for rapid movement on two axes to allow flexibility in reacting to unexpected enemy contact. Lead and rear security is conducted throughout the operation. As lead units locate enemy positions, they fix them by fire, and quickly reduce or bypass them. They also report the contact so that follow-on forces can then deal with the threat.

As the ground attack force reaches the edge of the built-up area, Marines remain mounted so as not to slow the advance. All vehicle machine guns fire to flanks and into doors, windows, and

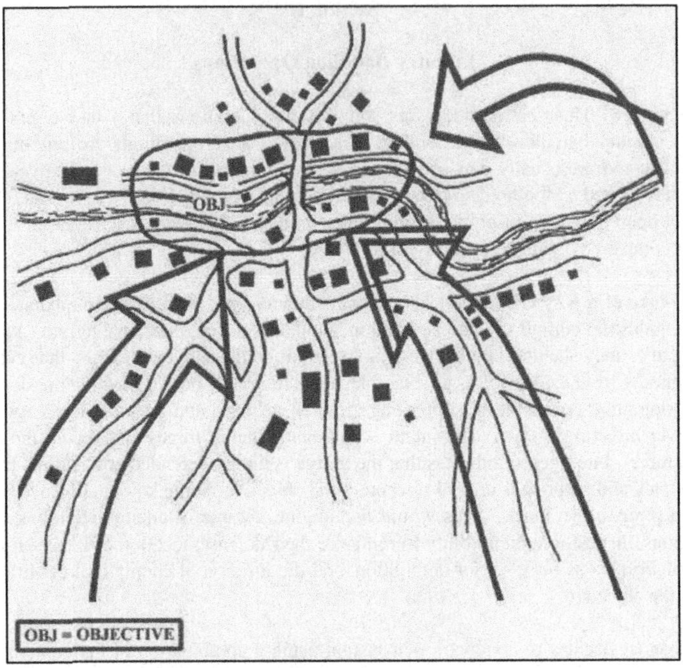

Figure 2-10. Seizure of a Key Objective

alleys to provide suppressing fires (Figure 2-11 on page 2-26). If heavy resistance is met and the column is halted, Marines dismount and the enemy and/or obstacles are cleared as quickly as possible. Once resistance and/or obstacles are cleared, the attack is immediately resumed. If necessary, units are dropped off to assume blocking positions and await follow-on clearing forces.

Once the objective is seized or linkup is accomplished, the battalion establishes a perimeter defense. The companies seize buildings around the objective and expand the size of the perimeter until it is large enough to secure the bridge. Attached engineers can then examine the bridge and clear it of any explosives.

2203. Infiltration. Infiltration is *"the movement through or into an area or territory occupied by either friendly or enemy troops or organizations. The movement is made, either by small groups or by individuals, at extended or irregular intervals. When the term is used in connection with the enemy, it infers that contact is avoided"* (Joint Pub 1-02).

Figure 2-11. Rapid Advance Movement

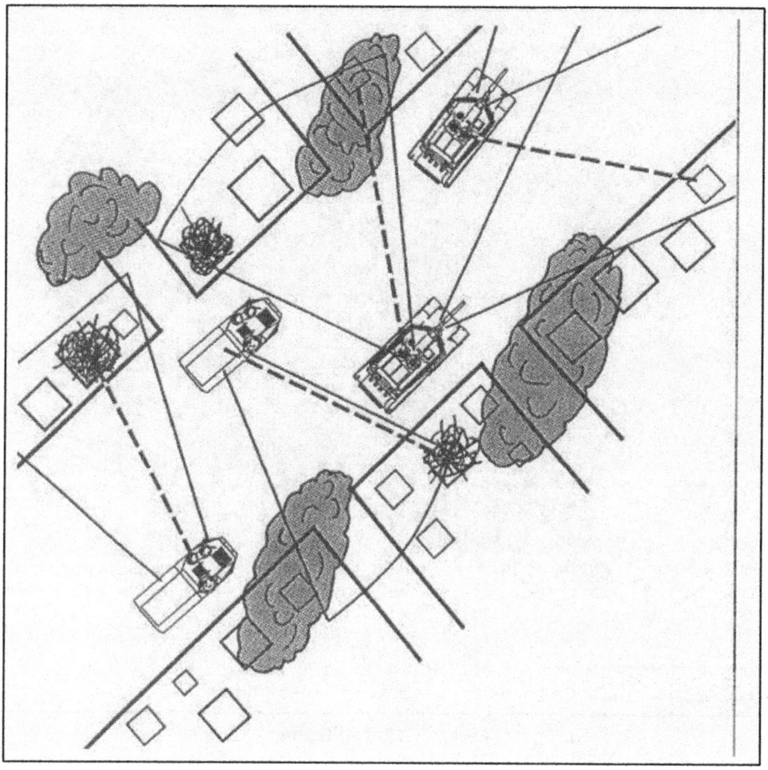

The outskirts of a town may not be strongly defended. The town's defenders may have only a series of antiarmor positions, security elements on the principal approach, or positions blocking the approaches to key features in the town. The enemy strongpoints and reserves may be positioned deeper in the city. In such a situation, an infantry battalion may be able to seize a part of the town by infiltrating companies and platoons between the enemy positions on the outskirts (Figure 2-12).

Moving by stealth on secondary streets, using the cover and concealment of alleys and buildings, the battalion may be able to seize key street junctions or terrain features, isolate enemy positions, and help following units move into the built-up area. Such an infiltration should be performed when visibility is poor.

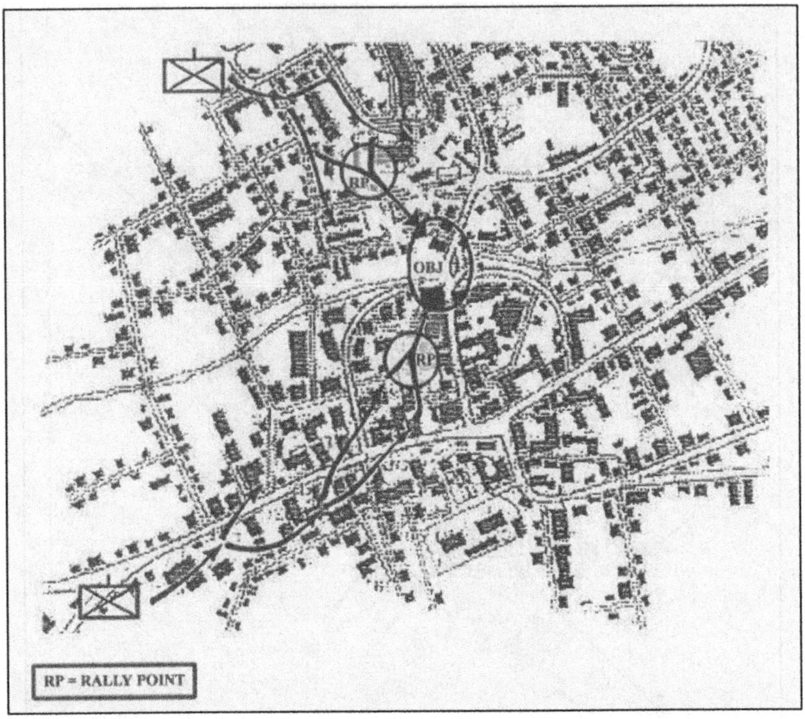

Figure 2-12. Infiltration

The infantry battalion is organized into infiltration companies with engineers attached and a designated reserve. Each company is assigned an infiltration lane that allows for infiltration by company-sized or smaller units. Depending on the construction of the built-up areas and streets, the infiltration lane may be 500 to 1,500 meters wide.

Mortar and artillery fire can be used to divert the enemy's attention and cover the sound of infiltrating forces. Armored vehicles and antitank weapons should be positioned to cover likely avenues of approach for enemy armored vehicles. Designated unit(s) provide isolation and security for infiltrating rifle companies. As the companies move into the built-up area, they secure their own flanks. Security elements may be dropped off along the route to warn of a flank attack. Engineers assist in breaching or bypassing minefields or other obstacles. Enemy positions are avoided but reported. The infiltrating companies proceed until they reach their rally points. At that time, they reorganize and make final preparations for attack of the objective.

2204. Route Security. In MOUT, a battalion may be tasked to provide route security. The process of securing the route may consist of a series of small-unit actions to clear buildings along the route. Fires should be planned to isolate the route as well. It may also require the reduction of obstacles and the emplacement of flank security along the entire route. In an outlying area, the forward units may proceed by bounds from road junction to road junction. Other units could augment flank security by moving down parallel streets and by probing the flanks.

The speed of movement required and the enemy situation determine whether the unit should be mounted or dismounted. For mechanized forces, narrow streets should be avoided if possible to avoid being trapped. Designated armored vehicles overwatch the squad to its front, keeping watch on the opposite side of the street. While armored vehicle movements and positioning provide mutual support, armored vehicles should be protected by dismounted infantry.

When contact is made with the enemy, armored vehicles seek covered positions and provide direct-fire support. Supporting fires fix and isolate enemy positions. Infantry dismount then maneuver to attack. Phase lines are used to coordinate movement along the route. For example, at each phase line, the forward companies might reestablish contact, reorganize, and continue clearing (Figure 2-13).

2205. Battalion Considerations. Some of the concerns for the commander to consider include:

a. Command and Control. CP location can be critical due to line of sight restrictions in the urban environment. Consideration should be given to maintaining a highly mobile CP with links to airborne retrans stations.

b. Intelligence. Organic collection capability resides primarily from units in contact with the enemy. The Scout Sniper Platoon also provides information but they may have obstructed visibility in the urban environment. Commanders may chose to employ them more often in a sniper role. Links to higher command intelligence sources may be key to maintaining current information on the enemy. The same line of sight restrictions encountered with command and contol will also restrict intelligence dissemination. Commanders may have to rely on messengers.

c. Manuever. Maneuver of mechanized forces in an urban environment may be restricted to the size, width, and load capabilities of the streets. Consideration must be given to plan for alternate routes for any attack. Significant manmade obstacles can be encountered throughout the battlespace. Additionally, the reserve is normally kept closer to the assault force. Movement of forces by aviation is less restrictive, but may also be vulnerable to surface to air missiles and small arms threat. Availability of LZs may also be a limiting factor.

d. Fires. A key consideration on the use of fires will be the ROE and the collateral damage that can result by their use. Rubbling of buildings can complicate an already restricted maneuver capability. Weapons effects can be more pronounced and can create a hazard for friendly forces. IPB determines potential firing positions for indirect fire assets that accounts

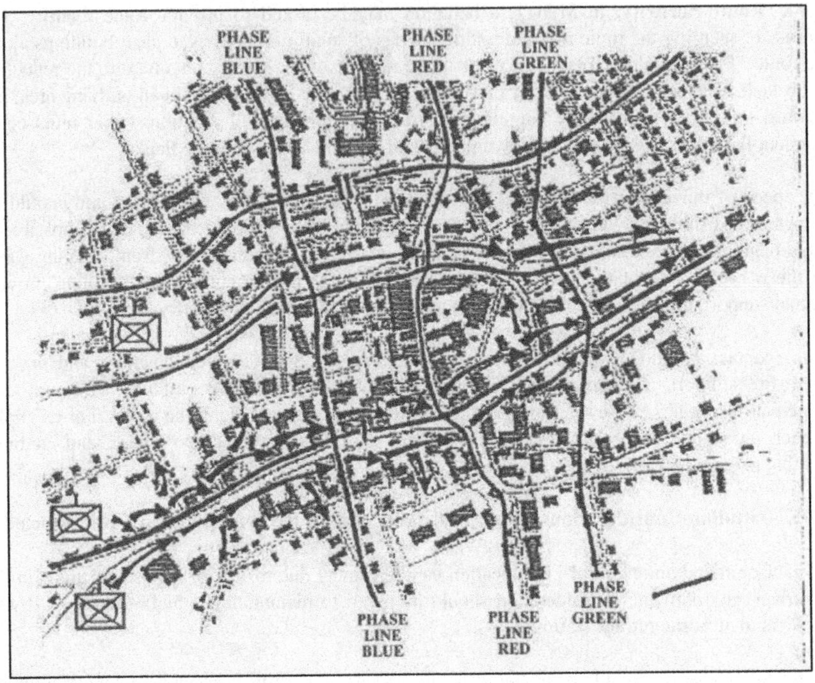

Figure 2-13. Clearing Along a Route

for the overhead masking by buildings and the ability to seat baseplates. Traditional indirect fire assets, such as artillery, may be used in the direct fire role.

e. Logistics. Historically in MOUT, the utilization of class I-V and VIII supplies is much greater. More vehicular assets should be allocated to the small unit level to transport ammunition and support the increased casualty evacuation requirements.

f. Force Protection. Lines of communication require extensive flank security in the urban environment. The commander must determine where his critical security requirements are and concentrate his force protection efforts in those areas while economizing in other areas. Manpower may further be depleted by security requirements for refugees and EPWs.

Section III

Rifle Company Operations

2301. Overview. This section addresses some of the operations the infantry battalion could assign to the rifle company in MOUT. For our focus, the rifle company is a task-organized and/or reinforced unit. Company operations will normally be conducted as part of a larger battalion operation.

2302. Attacking within a Built-up Area. Attacking within a built-up area is a task the rifle company may be assigned (Figure 2-14). This task involves the isolation, attack, and clearing of a specified block of structures. It is usually conducted in conjunction with adjacent friendly attacks or security operations. (See Appendix A.)

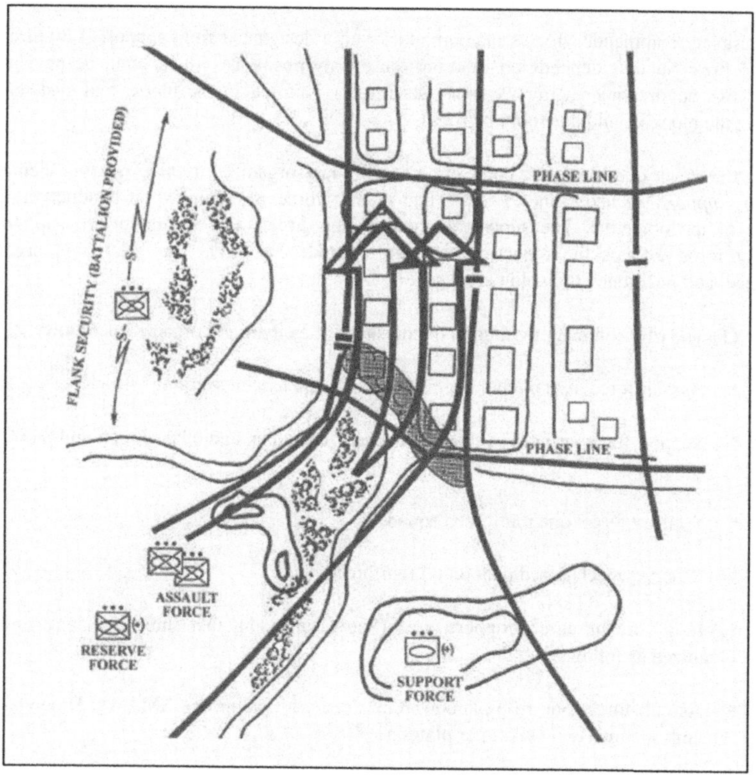

Figure 2-14. Company Attack of a Block

Before an assault, the company commander employs smoke to conceal the movements of assaulting platoons. He provides suppression fires to their flanks with direct-fire weapons. The battalion may provide additional security forces for flank protection.

(1) Concealed by smoke and supported by direct-fire weapons, an assaulting platoon attacks the first isolated building. The platoon must close on the building quickly while the enemy is still stunned by supporting fire. The company commander must closely coordinate the assault with supporting fire so that the fire is shifted at the last possible moment.

(2) The platoons continue to seize each building until the block is cleared. (See Appendix C for a detailed discussion on clearing of buildings.) After seizing the block, the company consolidates and reorganizes to repel a counterattack or to continue the attack.

The company commander directs platoon attacks on a designated front supported by direct and indirect fires. Success depends on isolating the enemy positions (which often become platoon objectives), suppressing enemy weapons, securing a foothold in the block, and systematically clearing the block's buildings room by room.

During the attack of a block, the company is usually task organized usually into four elements: *assault*, *support*, *security,* and *reserve*. The assault force is organized to conduct the actual assault of the objective. The support force provides direct- and/or indirect-fire support. The security force isolates the objective area and provides security. The reserve is organized, equipped, and positioned to exploit success. For example:

(1) The rifle company fighting in the outskirts of a city might organize as follows:

- Assault force: two platoons reinforced with the assault section
- Support force: armored vehicles, mechanized squad, and other direct- and indirect-fire weapons
- Security force: one reinforced squad.
- Reserve: mechanized platoon (-) reinforced.

(2) In a core or core periphery area (See Chapter 1), that same company might be organized as follows:

- Assault force: one rifle platoon, reinforced with engineers, SMAWs, Dragons, and a tank section (two tanks) per platoon
- Support force: One squad with direct- and indirect-fire weapons

- Security force: one rifle platoon (-)

- Reserve: one rifle platoon, mechanized.

c. Armored vehicles, machine guns, and other direct-fire weapons under the company commander's control are assigned to isolate the objective and provide suppressive fires from covered positions. These weapons should not be fired for prolonged periods from one position. The gunners should use a series of positions and displace from one to another to gain better fields of fire and to avoid being targeted by the enemy. Direct-fire support tasks are assigned as follows:

- Machine guns fire along streets and into windows, doors, alleys, etc.; however, innovative use of direct fire weapons can be key in MOUT

- Tanks, ATGMs, and AT4s fire at enemy tanks and other armored vehicles.

- SMAWs can be used to make entrances in buildings and destroy identified bunker/fighting positions.

- Tanks can fire at targets protected by walls, make entrances in buildings, and engage enemy armored vehicles with their main gun and other targets with their machine guns.

- Riflemen engage targets of opportunity.

2303. Attack of an Enemy Outpost. A motorized or mechanized force may encounter enemy outposts, checkpoints, or strongpoints. In a hasty attack on an enemy outpost, the company commander makes a quick assessment based on the mission and situation, and then reacts in accordance with the commander's intent. If not able to bypass an enemy outpost, he may have to attack and reduce it.

The company occupies positions of advantage. The tanks or other armored vehicles support the dismounted infantry in the attack by assuming support-by-fire positions from which they can fire on the outpost, prevent the enemy from escaping, and destroy any reinforcements.

The infantry platoons dismount and move to engage the enemy (Figure 2-15). They should attack from a covered route to hit the outpost at a vulnerable point. As the platoons approach the outpost, smoke may be employed to screen their movement, and supporting fires are shifted. After platoons close on the outpost, they clear the buildings quickly and consolidate. The company is then ready to continue operations.

2304. Seizure of a Traffic Circle or Major Intersection. A company may have to seize a traffic circle or major road or traffic intersection (Figure 2-16 on page 2-34). This operation consists of seizing the buildings adjacent to the traffic circle or intersection while bringing direct-fire weapons into position to fire on likely avenues of approach.

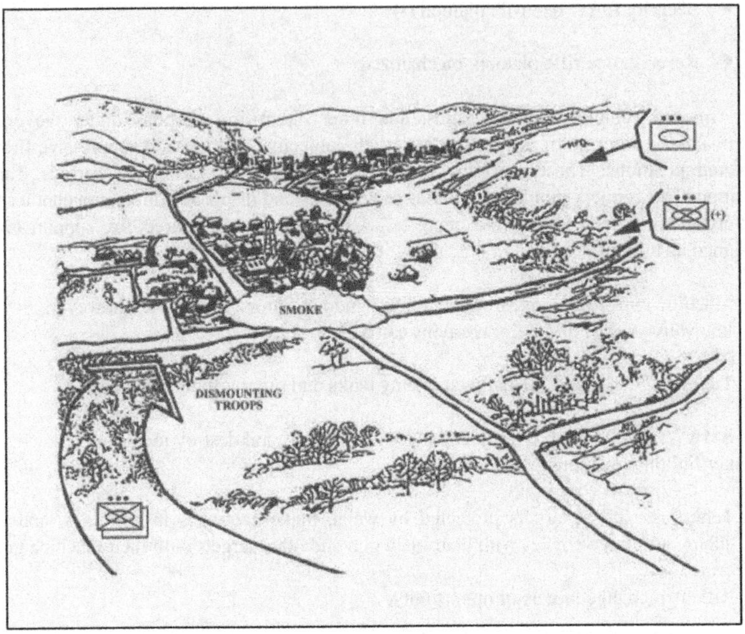

Figure 2-15. Hasty Attack of an Enemy Outpost

The company commander takes the following actions to seize a traffic circle or major intersection:

- Isolate the objective area.
- Seize and clear the buildings overlooking the traffic circle or major intersection under cover of direct-fire weapons.
- Secure the flanks against enemy attack.
- Consolidate and prepare for counterattack.

2305. Seizure of Key Terrain. The seizure of certain key terrain such as a bridge, intersection, facility, or structure may be critical to the success of an operation. Seizing key terrain for friendly use is a possible mission for a rifle company.

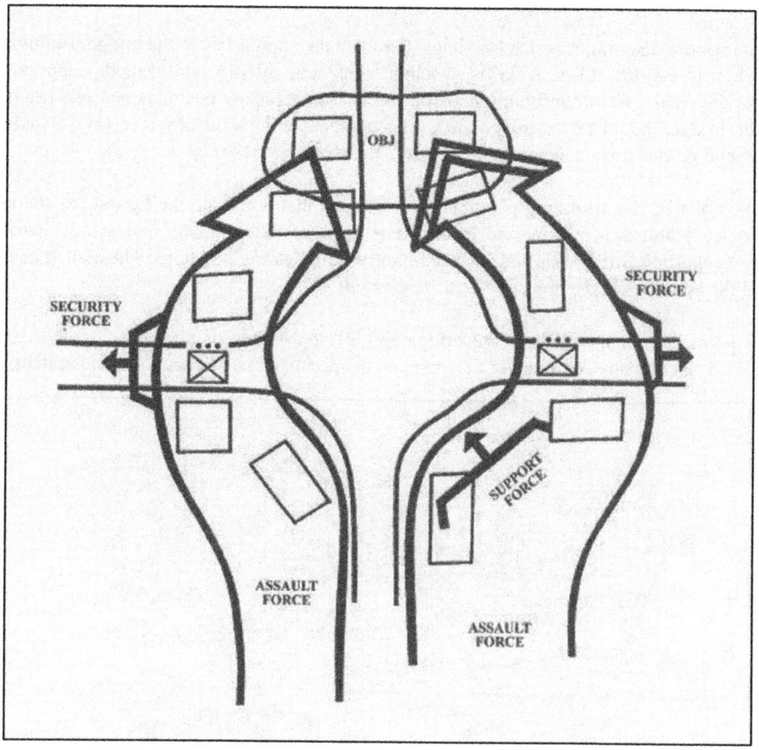

Figure 2-16. Seizure of a Traffic Circle or Major Intersection

To seize a bridge, a company should:

- Clear the buildings on the *near bank* that provide clear fields of fire for supporting weapons
- Suppress enemy weapons on the *far bank*
- Use screening smoke to obscure enemy observation
- Clear the buildings on the *far bank* that dominate the bridge
- Secure a perimeter around the bridge so that the engineers can clear obstacles or remove demolitions from the bridge.

The commander determines which buildings dominate the approaches to the bridge. Buildings that permit him to employ AT4s, ATGMs, machine guns, and riflemen are cleared. Supporting fire prevents the enemy from reinforcing units on the far bank. Enemy positions that can fire directly onto the bridge should be destroyed first. The company's forward observer (FO) should plan artillery and mortar fires to suppress the enemy's positions on the far bank.

The objectives of the assaulting platoons are buildings that dominate the far side of the bridge. One or two platoons are supported by the remainder of the company and assault across the bridge. Supporting fires are shifted and the infantry start clearing buildings. The assault continues until all the buildings in the objective area are cleared.

At this point, the engineers clear the bridge and its approaches of all mines, demolitions, and obstacles. The company commander may expand his perimeter to prepare for counterattack. Once

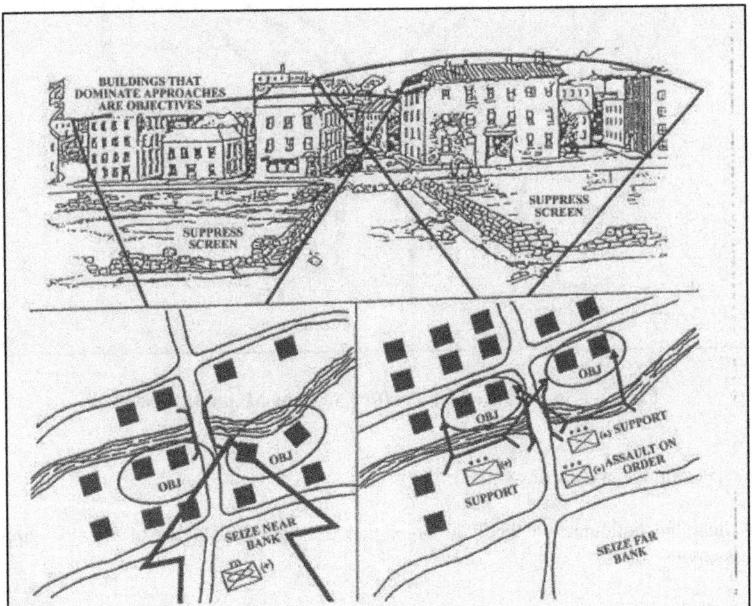

the bridge is cleared, the armored vehicles are brought across to the far bank (Figure 2-17).

Figure 2-17. Seizure of a Bridge

2306. Movement to Contact. A rifle company may conduct a movement to contact along a road. Similarly, the company may have to reconnoiter such a route to prepare for a battalion attack. This type of mission can best be accomplished by a mechanized company with appropriate attachments. (Figure 2-18 on page 2-36).

Figure 2-18. Movement to Contact Along a Highway (Commercial Ribbon)

The speed of movement depends on the mission, terrain, and enemy situation. In open areas where rapid movement is possible, tanks should lead. In close terrain, the infantry may lead while being overwatched by the tanks. Artillery fire should be planned along the route. Engineers may accompany the lead platoon on the main route to help clear obstacles and mines as they may be encountered. Between key terrain, when contact is not likely, the company remains on their vehicles. At key terrain or when enemy contact is likely, the company moves on foot to clear enemy positions or to secure the key terrain. Armored vehicles support the infantry.

Section IV

Rifle Platoon Operations

2401. Overview. This section discusses some of the various tactical missions that might be assigned to a rifle platoon.

2402. Attack of a Building. One offensive mission assigned to a rifle platoon in MOUT is the attack of a building. The attack requires isolating the building and entering the building at the least-defended point or through a breached hole (Figure 2-19 on page 2-38). When possible, clear the building from top to bottom. Close coordination between the assault and support forces is required.

When attacking a building, a platoon should be organized into an assault force, a support force, and a security force. The assault force provides the forces that actually conduct the assault. A support force provides supporting fires and efforts that assist the assault force. A security force covers the platoon's flanks and rear. (See Appendix A.) The platoon may be augmented by armored vehicles and from the company's weapons platoon or the battalions weapons company. If a reserve is required it should be provided by the company.

The assault consists of three steps:

(1) Isolate the building and deny the enemy the ability to maneuver or reinforce.

(2) Enter the building to secure a foothold for further action.

(3) Clear the building systematically.

Clearing is performed by individual squads and teams, which leapfrog as rooms and floors are secured. Platoons tasked to clear buildings may be reinforced with combat engineers.

2403. Movement Down a Street. When moving in urban terrain, a rifle platoon normally follows the same principles of movement as in other environments. However, some movement techniques must be modified to adjust to the unique dangers encountered in an urban environment. (See Appendix A, Section VII.) This section describes the movement down the street of the lead platoon of a company.

The speed of movement depends on the enemy situation. In outlying or lightly defended areas, a mechanized platoon may proceed along the street mounted but send dismounted infantry forward to reconnoiter danger areas. In the center of a built-up area the platoon dismounts to move on foot with two squads leading—one on each side of the road, using all available cover. They move through the buildings, if feasible, to avoid exposure on the streets. The squads coordinate to provide each other with all-around observation and mutual support.

Figure 2-19. Attack of a Building

The platoon should organize and practice to move as two forces: a *maneuver* force and an overwatch force. The maneuver force (normally of one squad on narrow streets or two squads on wide streets) moves forward, scouts danger areas, and closes with the enemy. An *overwatch* force (the remainder of the platoon and its attached or supporting crew-served weapons), moves behind the maneuver force, secures the platoon's flanks and rear, and provides direct-fire support when needed. These two forces, can exchange roles from time to time to avoid overfatigue of the maneuver force (Figure 2-20).

2404. Utilization of the Reserve. In the offense, a rifle platoon may be assigned as a reserve. The reserve can be used to reinforce and attack or exploit success. Reserves are normally held at the company level.

Figure 2-20. Movement Down a Street

In outlying areas, where the terrain is relatively open, a tank platoon reinforced with mechanized assets (tank-heavy task force) may be tasked as the reserve. However, if inside the urban area, a mechanized platoon reinforced with tanks (mech-heavy task force) would most probably be directed to provide the reserve. Regardless of what unit is designated as the reserve, the following items must be considered for their employment in an urban environment:

- The reserve should plan to use a different axis of advance
- Route selection for the reserve may be more constructive since it may have to use building, side streets, sewers, etc.
- Reserves are kept closer to assault forces in MOUT than on open terrain
- Engineers should normally be attached
- Reserves should maintain a high degree of mobility

Chapter 3

Defensive Operations

The essence of defensive tactics is to place the enemy into a position that permits his destruction through the intelligent use of terrain and firepower.

3001. Introduction. In urban combat, the defender possesses key advantages over the attacker. The defender can shape the battlespace to his advantage by maximizing the natural restrictions and obstacles found in the urban environment. ROE for the attacker can add to the defender's advantage by placing restrictions on the application of force by the attacker. Knowledge of the terrain and time available for preparing defensive positions are advantages which may enable the defender to successfully resist a numerically superior force. A spirited and stubborn defense may persuade an attacker to abandon the attack. In some battles, urban defenders can be credited with repelling or decisively delaying an attacker which ultimately influenced the course of the war in their favor.

Marines may be called upon to defend a port city, an embassy located on or near a coastline, a transportation link, or the capital of a foreign government. They may also be tasked to defend from an urban area in an economy-of-force role in order to support operations elsewhere. In combat, offensive and defensive operations are inseparable. Marines need to be capable of conducting both offensive and defensive operations in an urban environment.

The fundamentals of defense do not change in an urban environment. The defenders of a city, however, usually have detailed knowledge of the terrain. This will allow them to establish an extensive defensive network that is designed to force an attacker to expend exorbitant amounts of time, supplies, equipment, and manpower. Commanders need to recognize both the advantages and disadvantages of defensive operations in an urban environment.

3002. Decision to Defend

a. Reasons for Defending Built-Up Areas. Historically, the following are some of the reasons for defending built-up areas:

(1) Denial of Important Strategy/Political Objectives. Capitals and cultural centers can be defended strictly for psychological or national morale purposes even if they do not offer a tactical advantage to the defender. Defending a city can cause an attacker to commit a significant amount of his forces which reduces his capability to attack elsewhere. The defense of a city can delay the overall offensive capability of the attacker.

(2) Retention of Key Economic Centers. In many countries, the entire nation's economic well-being may be tied to a few key cities. These key cities usually contain the country's primary industrial, transportation, and communications base. The capture of these key centers could result in the overthrow of the current government, or deny that

.3

government the ability to adequately support combat operations against enemy regular or insurgent forces.

(3) Control of Avenues of Approach. Most avenues of approach to large cities are straddled by small towns every few kilometers which must be controlled by defending forces. These areas can be used as battle positions or strongpoints to control the avenues of approach. For the attacker to utilize these avenues of approach he must sacrifice speed and expend resources. The defense of these cities or towns may constitute a large part of the overall defense of a city.

(4) Economy of Force. The tactical advantages provided by urbanized terrain allows the defender to engage a numerically superior force. The defender can conserve the bulk of his combat power so that it is available for use in other operations.

(5) Concealment of Forces. Reconnaissance and combat identification is more difficult in an urban environment. CPs, reserves, CSS complexes, and combat forces emplaced in built-up areas are much harder to detect.

3003. Reasons for Not Defending Built-Up Areas. The commander considers the following reasons for deciding not to defend built-up areas.

a. Unnecessary to the Defensive or Offensive Plan. If the built-up area is too far forward or back in a unit's defensive sector, is isolated, or is not astride an enemy's expected avenue of approach, the commander may choose not to defend it.

b. Bypassable. If the nature of nearby terrain allows the enemy to bypass the city, then it will not make a good blocking position. Some built-up areas, mainly smaller ones, are easily bypassed by existing main road and highway systems. A built-up area that can be easily bypassed normally will be, thereby effectively isolating that area's defenders from the remainder of their forces.

c. Inadequate Structures for Defense. Extensive areas of lightly built or flammable structures offer little protection to the defender. Built-up areas near flammable or hazardous industrial areas, such as refineries or chemical plants, also may not lend themselves to successful defense.

d. Adjacent Dominating Terrain. If the built-up area is small and dominated by close, prominent terrain, the commander may choose to defend on that terrain rather than in the built-up area.

e. Better Fields of Fire Elsewhere. The commander may choose to base all or part of his defense on the better fields of fire that exist outside built-up areas. An example would be an armor-heavy force defending in sectors from multiple, small, built-up areas surrounded by open or farm-type areas.

f. Open City. See chapter 2.

Section I

Defensive Planning

3101. Commander's Estimate. As in any other defensive operation, planning begins with the commander's estimate of the situation. The complxities involved with the defense of a built-up area require detailed and centralized planning. This section details the general procedures and principles for planning and organizing the defense of an urban area.

3102. METT-T

a. Mission. Mission analysis is the first step in conducting the estimate of the situation. When conducting mission analysis commanders and their staffs should analyze their assigned tasks and commander's intent to determine the scope of the defense. For example:

- Do I need to control every building or can I control the frontage from key buildings?

- Are certain parts of the urban area more significant? (i.e. main transportation centers, communication nodes government buildings)

b. Enemy. The commander must focus on the enemy and build his defense to concentrate his strength against enemy weakness. The IPB process is used to analyze the terrain and enemy capabilities. (See FM 34-130.) For example:

- If the attacker relies heavily on aviation for mobility, the defender may defend potential LZs and emplace manpad teams to counter the air threat

- If the enemy relies heavily on mechanized movement, the defender may employ additional countermobility assets.

c. Terrain and Weather

(1) Terrain. Analysis of all manmade and natural terrain features is critical for planning a defense in built-up terrain. A defending commander has the advantage of occupying the terrain on which the fight will occur, and therefore usually knows it better than the attacker.

The defender must make best use of the battlespace that he controls: ground level (streets and parks), above ground (buildings), and below ground (subways and sewers). Defenders must be aware that built-up areas are normally interdependent. A reservoir supplying water to a city may be many miles away, electricity can come from remote power stations, and the control points for these services and others, such as communications lines, may be vulnerable.

.3

(a) K: Key Terrain. Primary examples of key terrain in the urban environment are ports, airfields, power grids, communication nodes, bridges, government building complexes, or parks. We identify key terrain so that we can better select our defensive positions and also better determine the enemy's objectives.

1 **Villages.** Villages often dominate high-speed avenues of approach. If the buildings are constructed of stone, brick, or concrete, providing good protection against both direct and indirect fires, a formidable defense can be constructed

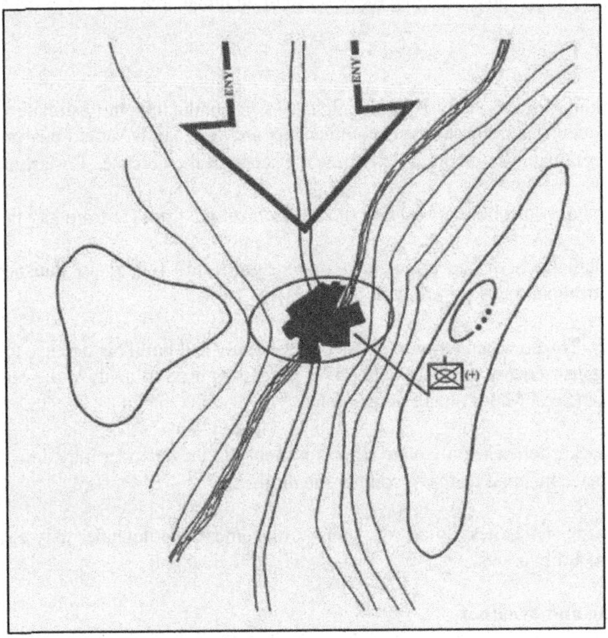

(Figure 3-1). Company-sized battle positions can be established in these small villages to block approaches into the larger urban area.

Figure 3-1. Village Defense (Chokepoint)

2 **Strip Areas.** Strip areas are built along roads or down valleys between towns and villages. They afford the defender the same advantages as villages. If effective fields of fire are available, a unit can occupy a few key positions within the strip to create an effective defense. (Figures 3-2A and 3-2B).

3 **Towns and Cities.** When a small town or city is positioned on a key enemy avenue of approach, a force can defend from the urban area while other forces

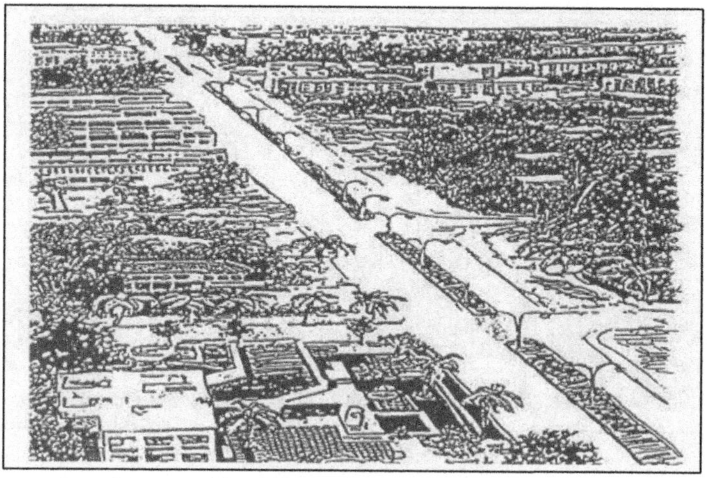

deny the enemy the ability to bypass. The town or city can be made into a strongpoint defense. (Figures 3-3A and 3-3B on page 3-6).

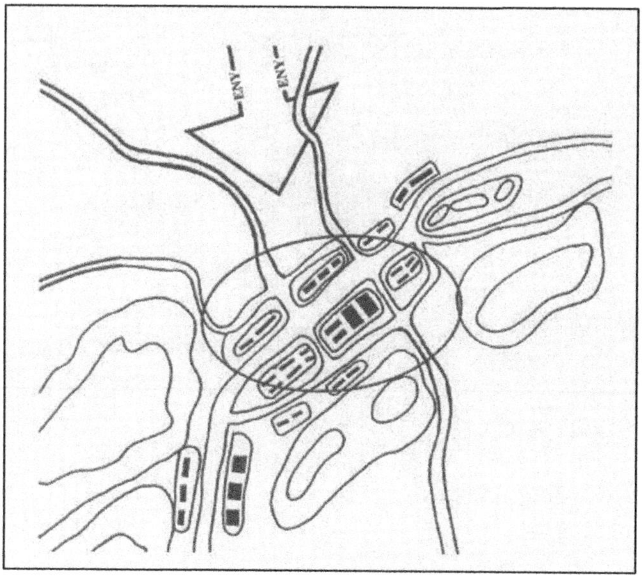

Figure 3-2A. Strip Area

.3

Figure 3-2B. Strip Area Defense

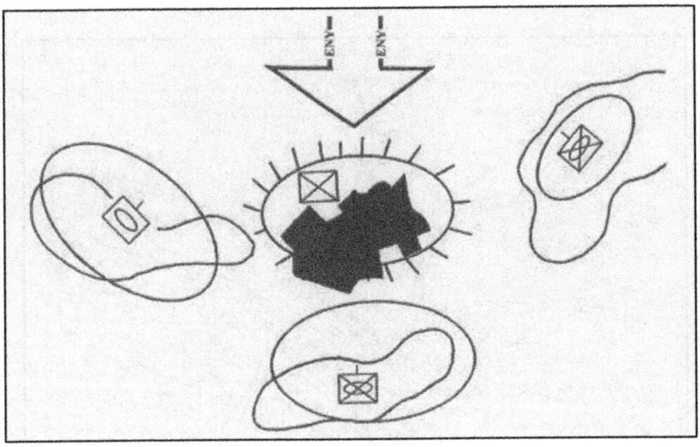

Figure 3-3A. Towns and Cities

Figure 3-3B. Defense of a Town or City

4 **Large Built-Up Areas.** Large buildings that are close together may require more forces and smaller defensive sectors than natural open terrain. The density of buildings, rubble, and street patterns will dictate the frontage of the unit.

(b) O: Observation and Fields of Fire. Attackers must generally advance by crossing streets and open areas between buildings where they are exposed to fires from concealed positions.

1 **Weapons and Range.** The defender must position weapons to obtain maximum effect and mutual support. FACs and FOs should be placed well above street level for increased observation. Fires and final protective fires should be preregistered on the most likely approaches.

2 **Limited Visibility.** The commander can expect the attacker to use periods of limited visibility to sustain or gain momentum. (See Appendix H for more detailed discussion.) The commander considers the following during periods of limited visibility:

- Unoccupied areas that can be observed and covered by fire during daylight may have to be occupied or patrolled at night.
- Remote sensors should be emplaced in dead space.
- Mines, noise-making devices, tanglefoot, and OPs should be positioned on avenues of approach to provide early warning.
- Artificial illumination to include street lamps, stadium lights, and flares should be integrated into the overall defense.
- When defenses are probed, indirect-fire weapons, grenade launchers, and hand grenades can be used to avoid disclosure of defensive position locations.

(c) C: Cover and Concealment. Although walls, floors, and ceilings can be used as protective cover, Marines should always improve these positions by using sandbags, rubble, etc. A defender can reduce his exposure by establishing routes between positions using:

- prepared breaches through buildings

.3

- reconnoitered and marked underground systems
- trenches and sewage systems
- concealment offered by smoke and darkness to cross open areas.

(d) O: Obstacles. A city is an obstacle because it canalizes and impedes an attack. Avenues of approach should be blocked by obstacles and covered by fire. (See Paragraphs 3404.d., 3404.f., 3405.e., and Appendix F.)

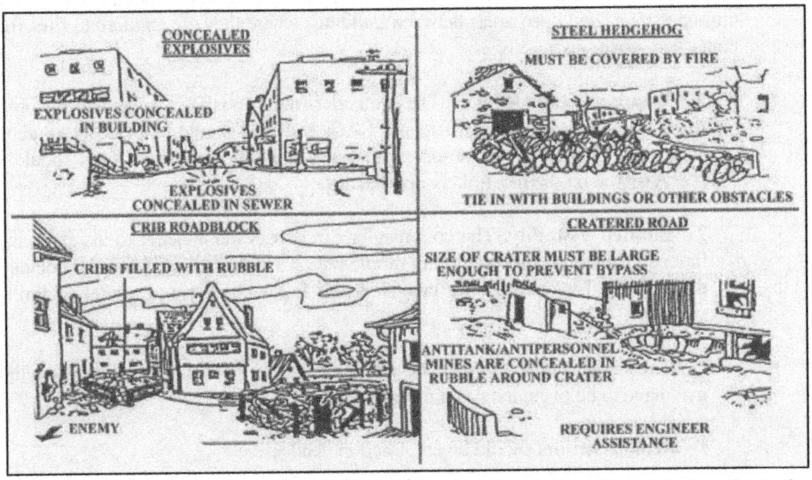

(e) A: Avenues of Approach. Avenues of approach (surface and subsurface) should be denied to the enemy in keeping with the overall defensive plan (Figure 3-4).

(2) Weather. The effects of weather as discussed in Chapter 2, Section II, pertain equally to defensive operations.

D. Troops and Support Available. The defender of a built-up area has a terrain advantage and can resist the attacker with much smaller forces. The defender stands a greater chance of success when he fights with combined arms and task organizes his forces. The discussion of troops and support available in Chapter 2, Section II, also pertains to defensive operations.

Figure 3-4. Denial of Avenues of Approach

e. Time Available. The commander establishes priorities of work, which depend on the time available. Many tasks can be accomplished at the same time, but priorities for work are accomplished according to the commander's order. An example priority-of-work sequence follows:

- Establish security and communications
- Assign sectors of responsibility, final protective fires
- Clear fields of fire
- Select and prepare initial fighting positions
- Establish and mark routes between positions (including alternate and supplementary routes)
- Emplace obstacles and mines
- Improve fighting positions.

3103. Organizing for a Defense

a. Types of Defense. The defense of a built-up area should be organized around key terrain features that preserve the integrity of the defense and provide the defender ease of movement. There are two types of defense: position and mobile. (See MCWP 3-1, *Ground Combat Operations* [under development] for details on defensive operations.) Most defenses will include a combination of position and mobile defense. The type of defense chosen is predicated on commander's intent and METT-T.

(1) Position Defense. The type of defense in which the bulk of the defending force is disposed in selected tactical localities where the decisive battle is to be fought. Principal reliance is placed on the ability of the forces in the defended localities to maintain their positions and to control the terrain between them. The reserve is used to add depth, to block, or restore the battle position by counterattack. (Joint Pub 1-02.) The position defense is usually characterized by defending key terrain. In the urban environment, this equates to the physical occupation of key public buildings, avenues of approach, transportation centers, industrial parks, etc. This defense focuses combat power to repel the attacker's advance and/or penetration while retaining the terrain. This type of defense is common for battalion-sized units and below.

The commander conducting a position defense organizes forces utilizes the same fundamentals that apply to any defense. Forces can be assigned sectors to defend and/or battle positions or strongpoints based on the commander's analysis of METT-T. Reserves are identified and located to respond quickly to exploit success.

.3

(2) Mobile Defense. Defense of an area or position in which maneuver is used with organization of fire and utilization of terrain to seize the initiative from the enemy. (Joint Pub 1-02.) The mobile defense is characterized by its focus on the enemy, not on terrain. Terrain is still important. However, this defense is organized with a mobility capability to exploit enemy weakness. In order to conduct the mobile defense, the defender must have enough forces to defend the built-up area, and to rapidly conduct an enveloping or flanking attack.

As in the position defense, the sector, battle position, strongpoint, or any combinations thereof are also used in shaping the battlespace by canalizing the enemy into a designated location where he can be destroyed.

b. Defensive Options

(1) Defense Outside the City. The commander may defend from outside the city. This option is often selected when the terrain surrounding an urban area offers an advantage to a defender and when his defending force is of sufficient strength (See Cherbourg paragraph in Chapter 1). The presence of a large, friendly, civilian population, or buildings of historic significance or specific cultural value, would also contribute to deciding on an outside defense. A situation in which the built-up area has a preponderance of building structures that offer little protection to defenders or terrain that favors the attacker.

A commander organizes his defenses around strongpoints selected on the outer edges of the urban area when these may be suitable structures for siting weapons to gain maximum firepower range. The organization of an outer-city defense is also heavily influenced by the number and type of weapons with which the hostile force is armed. A defense against a predominantly infantry threat would differ from that planned to defeat an armored threat. The decision to defend or concentrate assets solely on the outer edge of a city are weighed against the enemy's ability to mass fires, infiltrate, or bypass strongpoints.

(2) Defense Inside the City. The commander may decide that his best defense is to defend within the city. This option is usually chosen when hostile forces possess strong, accurate, long-range fires and the defender wishes to minimize their effectiveness (See Quang Tri City example in Chapter 1). The strength of the defending force relative to the attacker may also lead to this option as an economy-of-force measure.

Structures on the outer edge of an urban area that obstruct fields of fire or that are likely to be used by an attacker should be prepared for demolition or demolished ahead of time. All defensive positions should be supplemented or covered by direct-fire weapons. Indirect fires should be planned for most likely avenues of approach. Additionally, close air support should be planned and integrated with the defense.

Security forces are placed outside the defended area to gain early enemy contact, inflict maximum casualties, and attempt to deceive the enemy as to the location of the friendly positions.

(3) Defense of a Key Sector. This option is usually employed when only a portion of an urban area be held or when strength is insufficient to defend an entire area (See the Japanese defense of Manila in Chapter 1). This option is often employed in situations where the mission requires clearing and holding a key sector, or sectors, in order to facilitate other operations.

(4) Entrapment and Ambush. This option is preferred when the defender lacks requisite strength to man organized, in-depth, defensive positions (See the highly successful Egyptian defense of Suez City example in Chapter 1). It requires the preparation of a series of blocking and/or ambush positions along major avenues of approach through an urban area. The defending force uses mines and obstacles to block alternate or possible detour routes in order to canalize the attack force into prepared ambush sites. The defender employs security forces forward of the main battle area and establishes inner or outer defensive strongpoints with extensive obstacles designed to assist in canalize the enemy into kill zones.

The entrapment and ambush option requires detailed planning and rehearsal by all echelons. Alternate plans should be prepared in the event the ambush is only partially successful or if the enemy fails to enter prepared ambush routes. Alternate ambush positions or counterattacks are planned.

Ambush planning in an urban environment does not differ significantly from planning in other environments. However, the urban environment offers an abundance of kill zones that afford the enemy little or no opportunity to escape. Ambush positions are established in selected buildings in the kill zone which offer optimum fields of fire for all weapons. Indirect fires and close air support should be planned to obtain maximum effect in the kill zone. The kill zone should also be covered with targets for indirect fire and CAS. Entrances into buildings not used as firing points should be blocked to entrap hostile forces in the kill zones.

(5) Defense in Depth. Defense in depth should be incorporated into any defensive plan. Units should be given enough terrain to allow for primary, supplementary, and alternate fighting positions as well as successive battle positions or strongpoints. This allows enough room for multiple ambush/entrapment kill zones and the positioning of a reserve.

c. Defensive Organization. The defensive battlespace is divided into three areas: the *security area*, *main battle area*, and *rear area*. Depending on the mission of the unit, forces defending a built-up area may have missions in any one or more of these areas. (Figure 3-5 on page 3-12).

.3

(1) Security Area. The defensive battle begins with a combined-arms force conducting security operations well forward of the main battle area. Security area missions can include screen , guard, and cover. Based on their assigned mission, security forces use all

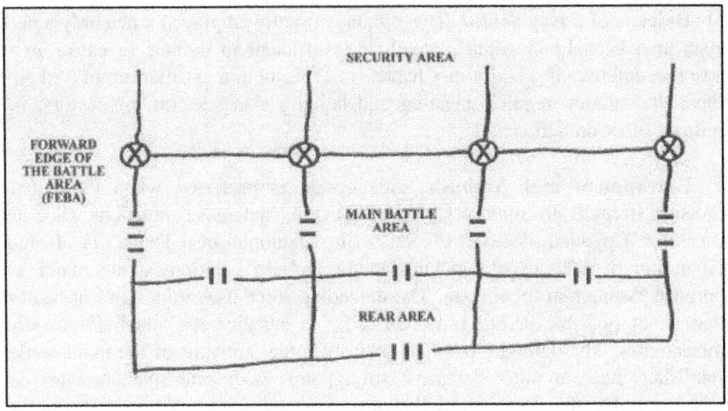

available combat power to gain time, destroy the enemy, or slow his momentum. The initial battle is fought using aviation assets and indirect fire weapons.

(a) Value of Security Area Operations. Security area operations complement to the defense by:

- Alerting the defense to the strength, location, and general direction of the enemy's main and supporting attacks
- Delaying enemy first-echelon detachments
- Initiating early engagement of enemy forces
- Deceiving the enemy as to the true location of the defender's main forces and main effort.

Figure 3-5. Organization of the Battlefield

(b) Withdrawal of Security Area Forces. The urban environment may complicate battle handover from the security force to the main battle area force. However, this transition must be accomplished smoothly to prevent the enemy from gaining momentum. The withdrawal of the security area forces must not result in an easing of pressure on the enemy.

(2) Main Battle Area. The decisive battle is fought in the main battle area. However, the

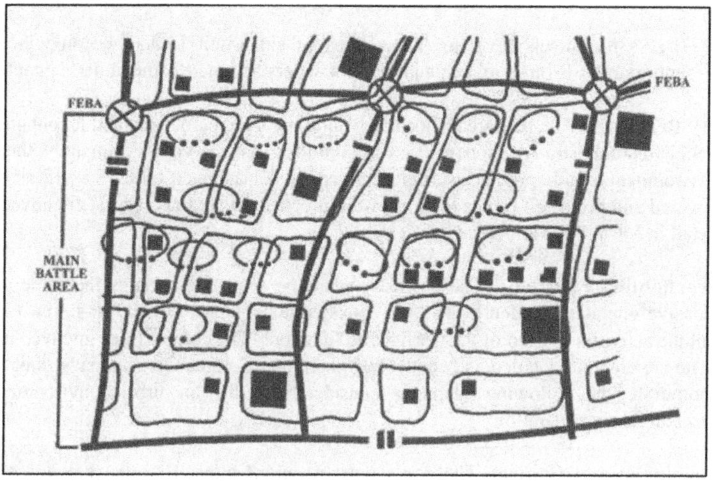

commander may deploy units on the forward edges of the city or in battle positions in depth. The defensive scheme should include forces to defend along the flanks. The commander normally employs security forces to the front to provide early warning and to deny the enemy intelligence on the unit's defensive dispositions (counterreconnaissance) (Figure 3-6).

(a) The size and location of battle positions depend on METT-T and the type of defense prepared. A sophisticated defense will often include a combination of both position and mobile defense but will always be planned in depth.

(b) A unit assigned battle positions on the forward edge of a city or town should:

- Provide early warning of the enemy's advance
- Engage the enemy at long range
- Deceive the enemy as to the true location of the defense.

Figure 3-6. Main Battle Area

(c) The defender employs all available fires to destroy and suppress the direct-fire weapons that support the ground attack. Depending on the concept of operations, weapons may engage targets at their maximum effective range or engage simultaneously for an ambush.

(d) As the attack develops, the actions of small-unit leaders become increasingly important. It is imperitive that all leaders understand their commander's intent.

(3) Rear Area. The rear area is located behind the main battle area and it contains the CSS and administrative support. Protection of these units is vital. Fortunately, the urban environment usually provides sufficient cover and concealment for CSS assets as well as covered and protected routes to the forward areas. Rear area operations are covered in detail in MCWP 3-41.1, *Rear Area Operations*.

3104. Warfighting Functions. The defensive planning sequence remains the same as in any other defensive operation. Defense planning considerations, based on METT-T, must take into account all the activities in each of the warfighting functions. The complexities involved in defense of a built-up area require detailed and centralized planning. Commanders and subordinate leaders must incorporate the following planning considerations for an urban environment when conducting a defensive operation.

a. Command and Control. The commander positions himself so that he can control the action. In an urban environment, this can be more difficult because of the close nature of fighting, prolific obstacles, poor visibility, and difficulty in communication.

Graphic control measures oriented on prominent terrain features are also used in planning and conducting combat in built-up areas. Prominent streets are ideal for use as phase lines and boundries. These and other control measures assist the commander in controlling manuever and fires throughout the battlespace.

COCs should be located in secure locations. (These can be inside or underground when possible.) (See Chapter 2, Section IV, Paragraph 2402.)

Commanders consider the effects of built-up areas on communications when they allocate time to establish communications. LOS limitations affect both visual and radio communications. Wire laid at street level is easily damaged by rubble and vehicle traffic. Also, the noise of urban combat is much louder than in other areas, making sound signals difficult to hear. Therefore, the time needed to establish an effective communications system may be greater than in more conventional terrain.

Wire is the primary means of communication for controlling the defense of a city and for enforcing security. However, wire can be compromised if interdicted by the enemy. Radio communication in built-up areas is normally degraded by structures and a high concentration of electrical power lines. Nonetheless, radio is an alternate means of communication. Messengers can be used as another means. Visual signals may also be used but are often not effective because of the screening effects of buildings, walls, and so forth. Signals must be planned, widely disseminated, and understood by all assigned and attached units. Commanders consider the following techniques when planning for communications:

(1) If possible, lay wire through buildings for maximum protection.

(2) Use existing telephone systems. (Remember, telephones are not secure.)

(3) Emplace radios and retransmission sites in buildings on top floors or atop buildings that are higher than surrounding structures. Ensure that antennas are placed on the side of buildings where the radio signal can be directed at friendly units. This will enhance LOS radio communications.

(4) Use messengers at all levels because they are the most reliable and secure means of communications.

b. Intelligence. The discussion in Chapter 2 pertains equally to a defensive operation. Intelligence gathering for defensive operations is not limited to studying only the enemy. Commanders must emphasize obtaining and using all intelligence. Commanders must emphasize the utility of the urban infrastructure to enhance his intelligence gathering and dissemination effort. The defender usually has the advantage of being familiar with the intricacies of the urban terrain that he is defending. For example, the:

- The sanitation or public works department will have knowledge of the subterranean sewer lines
- Civil engineers will have extensive knowledge of water and power distribution nodes
- Law enforcement representatives may have access to lists of people sympathetic to the attacker
- Local journalists may agree to use local sources to assist in intelligence gathering effort.

c. Manuever

(1) Positioning of Forces. Finding fighting positions in urban areas that provide both good fields of fire and cover is often difficult. The forward edges of a town usually offer the best fields of fire but can be easily targeted by enemy observation and supporting fire. These areas often contain residential buildings constructed of light material. Factories, civic buildings, and other heavy structures that provide adequate cover and are more suitable for a defense are more likely to be found deeper within the town, but they have limited fields of fire.

Because the forward edge of a town is the obvious position for the defender, it should generally be avoided. However, the defender may choose to use the edge of the town if

.3

the terrain limits the enemy's ability for engagement or if strongly constructed buildings that give defenders protection are available. The mission of such a force is to provide early warning of the enemy's advance, to engage the enemy at long range, and to deceive the enemy as to the true location of the defense. This force should withdraw in time to avoid decisive engagement. If there is limited observation from the forward edge, a force should be positioned on more favorable terrain to gain better observation and to engage the enemy at long range.

Reconnaissance by all defending elements should help select routes for use by defenders moving between positions. Movement is crucial in fighting in built-up areas. Early selection and marking of routes adds to the defender's advantages.

The commander covers probable landing and drop zones such as parks, stadiums, or large rooftops and heliports with obstacles or fire to prevent air assault. (See Appendix F.) In a large built-up area, a unit is given a sector to defend and normally establishes a series of defensive positions. Although mutual support between positions should be maintained, built-up terrain often allows the enemy to infiltrate between positions. Therefore, the defender must identify the following:

- Positions that allow suppressive fires on infiltrating routes
- Covered and concealed routes, such as subways and sewers, for friendly forces to use to move between positions
- Structures that dominate large areas
- Areas such as parks, boulevards, rivers, highways, and railroads where antiarmor weapons have longer fields of fire
- Firing positions for mortars
- Command locations that offer cover, concealment, and ease of command and control
- Protected storage areas for supplies.

Buildings that enhance the general plan of defense are selected. Mutual support between these positions is vital to prevent the attacker from maneuvering and outflanking the defensive position. Buildings chosen for occupation as defensive positions should:

- Offer good protection
- Have strong floors to keep the structure from collapsing under the weight of debris

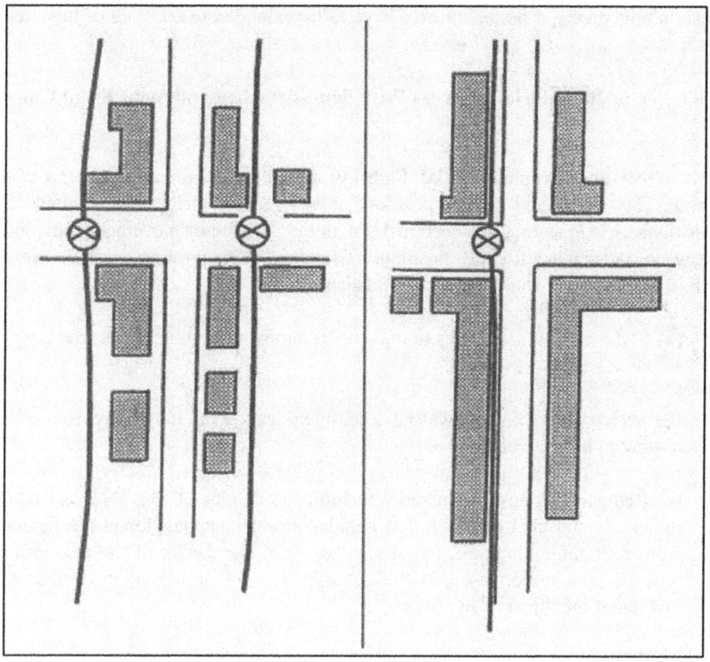

- Have thick walls
- Be constructed of nonflammable materials
- Be strategically located (corner buildings and prominent structures)
- Be adjacent to streets, alleys, vacant lots, and park sites (these buildings usually provide better fields of fire and are more easily tied in with other buildings)
- Provide for 360-degree protection and fields of fire
- Be stocked with adequate supplies as positions may become isolated and cut off for extended periods.

(2) Assign Sectors of Responsibility. Boundaries define sectors of responsibility. Sectors include areas where units may fire and maneuver without interference or coordination with other units. Responsibility for primary avenues of approach should never be split among units. In areas of semi-detached construction, where observation and movement are less restricted, boundaries should be established parallel to alleys or streets so that both sides of a street will be in a single sector. Where buildings present a solid

.3

front along streets, a boundary may have to be extended to one side of the street (Figure 3-7).

Figure 3-7. Sector Boundaries in Areas With Semi-detached and Solid-Front Construction

(3) Select and Prepare Initial Fighting Positions. Commanders select positions in depth. The unit should prepare positions as soon as they arrive and continue improving positions as long as they are occupied. Enemy infiltration or movement sometimes occurs between and behind friendly positions. Therefore, each position must be organized for all-around defense. The defender should also:

(a) Make minimum changes to the outside appearance of buildings where positions are located.

(b) Screen or block windows and other openings to keep the enemy from seeing in and tossing in hand grenades.

(c) Remove combustible material to limit the danger of fire. Fires are dangerous to defenders and create smoke that could conceal attacking forces. Defenders should store firefighting equipment (water, sand, etc.). The danger of fire also influences the type of ammunition used in the defense. Tracers or incendiary rounds should be avoided if the threat of fire exists.

(d) Turn off electricity and gas. Both propane and natural gas are explosive. Natural gas is also poisonous and is not filtered out by a protective mask. Propane gas, although not poisonous, is heavier than air. If it leaks into an enclosed area, it displaces the oxygen and can cause suffocation. Gas mains and electricity should be shut off at the main distribution nodes.

(e) Battle positions should be located so as not to establish a pattern. The unit should avoid obvious firing locations like church steeples.

(f) The unit should rehearse movements between positions.

(g) Camouflage positions.

(h) Reinforce positions with available materials, such as beds, furniture, etc.

(i) Block stairwells and doors with wire or other material to prevent or delay enemy movement. Create holes between floors and rooms to allow for covered movement within a building.

(j) Prepare range cards, fire plans, and sector sketches.

(k) Emplace machine guns in basements windows, where the guns can provide grazing fires. When basements are not used, access to them should be sealed to prevent enemy entry.

(l) Stockpile V (ammunition) and class VIII (medical supplies) items.

(4) Clear Fields of Fire. Commanders may need to rubble certain buildings and structures to provide greater protection and fields of fire. If the ceiling of a lower story room can support the weight of the rubble, collapsing the top floor of a building before the battle starts may afford better protection from indirect fires. Rubbling an entire building can increase the fields of fire and create an obstacle to enemy movement. Rubbling buildings too soon (or rubbling too many) may give away defensive locations and may destroy cover from direct fire. Rubbled buildings may also interfere with planned routes of withdrawal or counterattack.

(5) Improve Fighting Positions. When time permits, all positions, including supplementary and alternate positions, should be reinforced with sandbags and provided

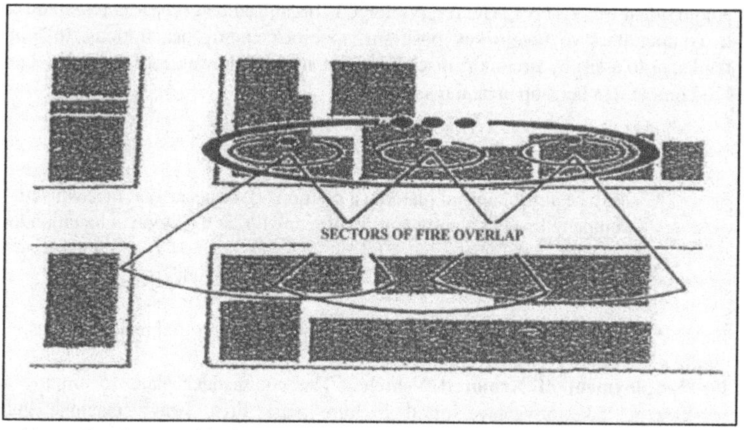

with overhead cover. Obtain support from attached engineers. (See Appendix D).

(6) Employment of Infantry

(a) The infantry battalion is normally assigned a sector to defend. However, depending on METT-T, it may be assigned to defend from a battle position or strongpoint. The battalion will usually assign its companies to either a battle position, a strongpoint, or a sector. These battle positions, strongpoints, or sectors are placed along avenues of approach to block or restrict the enemy's movement.

Depending on the type of built-up area, a company may be employed on the forward edge of the flanks of the area. This forces the enemy to deploy early

.3

without decisive engagement because it deceives the enemy as to the true location of the main defense. Other companies may then be employed in a series of battle positions and/or strongpoints in the center of the city or town. In all cases, mutual support between positions is vital. Companies and platoons should have designated alternate and supplementary positions.

(b) Once the rifle platoon commander has received his battle position or sector, he then selects the positions for his squads and crew-served weapons. Squad positions within the built-up area may be separated by rooms within buildings or be deployed in different buildings. Squad positions must be mutually supporting and allow for overlapping sectors of fire (Figure 3-8).

Figure 3-8. Sectors of Fire

(7) Employment of the Reserve. The commander's defensive plan always considers the employment of a reserve. The reserve force is organized and prepared to exploit success, to counterattack to regain key positions, to block enemy penetrations, to protect the flanks, or to assist by fire in the disengagement and withdrawal from endangered positions. For combat in a built-up area, a reserve force:

- Must be as mobile as possible
- May be a mechanized platoon, a platoon (-), or a squad at mechanized company level or a company, a company (-), or a platoon at battalion level
- May be supported by tanks or other armored vehicles
- Should be centrally located to facilitate manuever to threatened areas.

(8) Employment of Armored Vehicles. The commander plans to employ armored vehicles to take advantage of their long-range fires, heavy machine gun fires, self-generating smoke, and mobility. However, built-up areas restrict the mobility of armored vehicles and make them vulnerable to the antiarmor weapons of the enemy. Both the LAV and AAV are lightly armored and can be penetrated by heavy machine gun and antiarmor fires. Armored vehicles provide the commander with a mobile force to respond quickly to enemy threats on different avenues of approach.

(a) When armored vehicles are employed in the defense of a city, infantry should be positioned to provide them with security against close antitank fires and to detect targets for the armored vehicles. Armored vehicles should be assigned *primary*, *alternate*, and *supplementary* positions, as well as *primary* and *alternate* sectors.

(b) Armored vehicles should be located on likely avenues of approach to take advantage of their long-range fires. They may be:

- Positioned on the edge of the city in mutually supporting positions
- Positioned on key terrain on the flanks of towns and villages
- Used to cover barricades and obstacles by fire
- Placed with the reserve.

(c) Armored vehicles are normally employed as a platoon. However, sections of armored vehicles may be employed with rifle platoons or squads. This provides the armored vehicles with the close security of the infantry.

(9) Employment of Helicopterborne Forces

(a) Helicopterborne Task Force in the Defense. A helicopterborne force can defend against an infantry-heavy threat by utilizing its mobility to achieve a maneuver advantage. This force can be key in the execution of a mobile defense in both the security area and main battle area.

(b) Security Area. The helicopterborne force may be able to conduct security force operations for a larger force. Helicopters can position combat power and combat support quickly during rapidly changing situations.

(c) Main Battle Area. The mobility advantage that the helicopterborne force has over enemy infantry-heavy units may allow it to defend in greater depth in a large city. The helicopterborne force focuses on the destruction of advancing enemy forces and fights a series of battles in depth, it can attack the enemy from the front, flanks, and rear. Battle positions are selected and prepared throughout the main battle area along likely avenues of approach. Primary and alternate LZs and pick-up zones should be selected.

d. Fires. Fire support in the defense is as important as it is in the offense and has many of the same characteristics. The proximity of buildings to targets, minimum range restrictions, and repositioning requirements are all factors in the utilization of fire support in the defense. Additionally indirect fires are planned on top of and immediately around defensive positions for close support. Some of the indirect fire support considerations for the defense include:

(1) Historically, artillery fire has been used in some unique and innovative ways in an urban environment. Artillery fire support may be used in the direct- or indirect-fire mode. Artillery fire should be used to:

- Suppress and blind enemy overwatch elements

.3

- Disrupt or destroy an assault
- Provide counterfire
- Support counterattacks
- Provide direct fire when necessary
- Provide illumination in hours of darkness.

(2) Mortars at the infantry battalion and rifle company level are employed to maximize the effect of their high-angle fires. Mortars are one of the most effective indirect fire weapons in the urban environment. They should be used to engage:

- Enemy overwatch positions
- Enemy infantry before they seize a foothold
- Targets on rooftops
- Enemy reinforcements within range.

(3) Final protective fires are planned to stop dismounted assaults in front of the defensive positions. Fires within the city are planned along likely routes of advance to delay, disrupt, and destroy the enemy as he attempts to deepen a penetration.

(4) Commanders establish priorities of fire on the basis of enemy avenues of approach and threat systems that present the greatest danger to the defense. For example, during the attacker's initial advance, tanks, IFVs, and overwatching elements are the greatest threat to the defense. ATGMs should concentrate on destroying tanks first, then IFVs. Indirect fires should suppress and destroy enemy ATGMs and overwatch positions and/or elements. If enemy formations secure a foothold, priority is shifted to the destruction of enemy forces within the penetration.

(5) As the enemy attack progresses into the city, fires are increased to separate enemy infantry from its supporting armored vehicles. Friendly artillery concentrates on attacking infantry, conducting counterfire missions, and destroying reinforcements.

(6) Ample consideration must also be given to the rubbling effects of supporting arms. These effects often create massive rubbling of structures and buildings, making obstacles to both friendly and enemy movement.

(7) Mortars may be initially positioned forward in support of security forces. After their withdrawal, mortars are repositioned where they can support the in the main battle area.

(8) The commander assigns final protective fires and machine gun final protective lines (FPLs). Machine gun positions provide interlocking grazing fire and mutual support between adjacent units. FPLs are fired on planned signals. Proposed FPLs must be "walked out" to determine the extent of grazing fire available and to locate dead space. Dead space can be covered by:

- Rifle and light machine gun fire
- Grenade launchers
- Mines and boobytraps
- Indirect fires.

(9) Long-range antiarmor weapons are employed in the forward areas of the main battle area and the security area. Antiarmor weapons must be carefully positioned to prevent terrain and obstacles from reducing their range. Dismounted and manpacked antiarmor weapons should be positioned in buildings and along routes where engagement distances are reduced but where surprise and ambush are ideal.

e. Logistics. Just as in the offense, the defense utilizes a greater amount of class V and class VIII supplies in the urban environment. The defender needs to plan for multiple routes and means of supply. These can include:

- Using the city's existing transportation infrastructure to move supplies
- Prepositioning supplies in the defense in depth
- Using helicopter assets for rapid resupply
- Establishing and protecting supply routes
- Plan for utilization of existing repair and maintenance facilities
- Identify potable water sources

f. Force Protection

(1) Employment of Engineers. Engineers are employed at the battalion level or attached to companies and integrated down to squad level. Commanders must consider and assign the priority of work for engineer tasks that enhance survivability, mobility, and countermobility. Tasks that engineers can accomplish in the defense of a built-up area include:

- Constructing obstacles and rubbling for countermobility

.3

- Clearing fields of fire
- Laying mines
- Preparing routes to the rear
- Preparing fighting positions.

(a) Employment of Obstacles. Obstacles are used to shape the battlespace. (See FM 5-102, *Countermobility*, and FM 20-32, *Mine/Countermine Operations*, for details.) The four types of obstacles are:

- **Disrupting.** Disrupting obstacles are used to disrupt the enemy's timetable and force him into conducting a time-costly bypass or breach, or cause his march columns to become staggered or "telescoped."
- **Turning.** Turning obstacles are used to canalize the enemy into engagement areas or kill zones where combined-arms fires can be concentrated and flanking shots are increased.
- **Fixing.** Fixing obstacles increase the time the enemy can be fired on in an engagement or target area and to cause increased confusion.
- **Blocking.** Blocking obstacles prevent or delay the enemy from moving past a certain point.

(b) Obstacle Belts. In defensive obstacle planning, obstacles may be emplaced in a series of belts. This enhances the overall effectiveness of the obstacle plan with successive belts designed for specific purposes. The following example lays out a three-belt obstacle plan.

1 **First Obstacle Belt.** The first obstacle belt is enplaced at the nearest buildings across from and parallel to the main defensive position. This belt consists of wire and improvised barriers (including inside buildings, in subterranean avenues of approach, and outside in open areas), danger areas, and dead space. These barriers and obstacles should be heavily boobytrapped and covered by fires. This belt is designed to impede enemy movement, breaks up and disorganize attack formations, and inflicts casualties.

2 **Second Obstacle Belt.** The second obstacle belt is placed between the first belt and the main defensive position buildings, but out of hand grenade range. It is designed to impede movement, canalize the enemy into the best fields of fire, break up attack formations, and inflict casualties. This belt is not meant to stop the enemy permanently. It consists mainly of wire obstacles, improvised barriers, road

craters, and minefields. It should be heavily boobytrapped. Triple-strand concertina wire is placed along the machine gun FPL to delay the enemy in the kill zone.

3 **Third Obstacle Belt.** The third obstacle belt is the defensive position's denial belt. It consists of wire obstacles placed around and through the defensive buildings, close-in minefields, and subterranean accesses. It is designed to impede the enemy's ability to gain a foothold in the defensive area. It should be heavily boobytrapped, and integrated with mines.

(c) Emplace Obstacles and Mines. While principles for employing mines and obstacles do not change in the defense of a urban area, techniques for their employment change. For example, burying and concealing mines in streets is usually not done. To save time and resources in preparing the defense, commanders must emphasize using all available materials (automobiles, railcars, rubble) to create obstacles. Civilian construction equipment and materials should be utilized. Obstacles must be integrated into buildings and rubble areas to increase their effectiveness and to canalize the enemy. The family of scatterable mines (FASCAM) may be effective on the outskirts of a city or in parks and open areas. (See Appendix F for further discussion on obstacles, mines, and demolitions.)

(2) Air Defense. Air defense (Stinger) weapons are positioned on rooftops and parking garages in order to obtain LOS coverage for the units they are protecting. Stinger teams can be assigned the mission of protecting specific positions or placed in general support (GS) of the battalion.

(3) Fire Hazards. All cities are vulnerable to destruction by fire, especially those with many wooden structures. The defender's detailed knowledge of the terrain permits him to avoid areas that are likely to be fire hazards. The defender can deliberately set fires:

- To disrupt and disorganize the attackers
- To canalize the attackers into more favorable engagement areas
- To obscure the attackers' observation.

.3

Section II

Infantry Battalion Defense of a Built-Up Area

This section addresses some of the missions a task-organized and/or reinforced infantry battalion could be assigned. These missions are normally conducted as part of a larger joint operation.

3201. Defense in Sector. An infantry battalion is normally given the mission of defending a sector in a city (Figure 3-9). The interior buildings in the battalion sector form battle positions for the battalion's fixed defense. The size of a sector varies, but may cover the outlying structures and can be used to provide early warning and to delay the enemy.

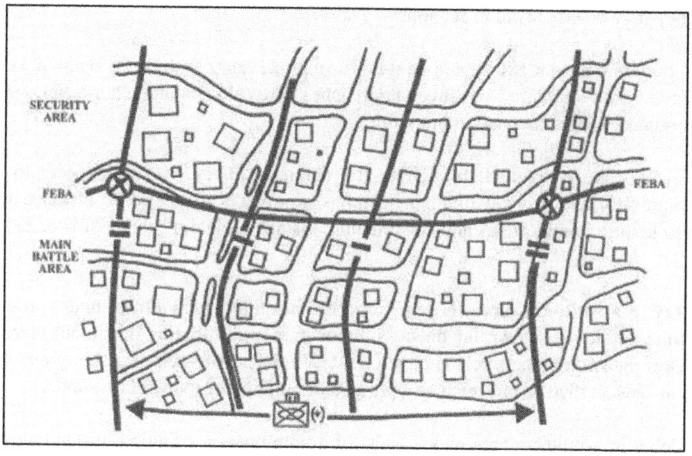

Figure 3-9. Defense of a Built-Up Sector

a. Defensive operations begin with the reconnoitering of the assigned sector to learn the terrain and to orient toward the enemy's expected advance. The main defensive positions should be formed on key terrain features, such as prominent buildings, road junctions, and positions which offer good fields of fire. Long-range antiarmor weapon systems should be deployed to engage the enemy as early as possible.

b. Depending on the situation, a reinforced rifle company may be used as a security force. The security force would operate forward of the FEBA and, depending on the mission, cause the enemy to deploy without without becoming decisively engaged. This can be done through maximum use of ambushes and obstacles while using covered and concealed routes. These routes are used for disengagement. Covered and concealed routes can be artificially made by making holes through buildings. The security force also attempts to detect the location of the enemy's main attack. Once the security force withdraws to the main battle area, it can be used as a reserve to reinforce or counterattack.

c. Based on METT-T considerations, the defense in sector (behind the FEBA in the main battle area) may consist of either company sectors, strongpoints, or battle positions. Strongpoints located on or covering decisive terrain are extremely effective in the defense.

d. Antiarmor vehicles and weapons should be used to engage enemy armored vehicles and/or cover obstacles with fire on likely avenues of approach.

e. Tanks are best used to engage enemy tanks, cover obstacles by fire, and engage in counterattacks. They should be employed in platoons where possible. However, in congested areas they may be employed in sections.

f. Artillery and mortar fire should be used to suppress and blind enemy overwatch elements, to engage enemy infantry on street approaches, provide counterbattery fire, and support counterattacks with direct and indirect fire.

g. Engineers are integrated throughout the defense. They provide support to forces in strongpoints to help prepare fighting positions. Engineers may also be attached to security forces to help in laying mines and constructing obstacles, clearing fields of fire, and preparing routes.

3202. Delay in a Built-Up Area. A delay is an operation in which a force under pressure trades space for time by slowing down the enemy's momentum and inflicting maximum damage on the enemy without becoming decisively engaged (MCWP 3-1). The delay can be oriented either on the enemy or on specified terrain such as a prominent building or industrial complex.

a. A delay in a built-up area may consist of a combination of ambushes and alternating or successive battle positions (Figures 3-10A and 3-10B on page 3-28).

(1) Ambushes are normally planned on observed obstacles. They require decentralized control and execution. Ambushes can be combined with limited-objective attacks on the enemy's flanks. They are most effective at the edge of open spaces, parks, wide streets, and so on. These attacks may be executed by dismounted infantry supported by tanks, LAVs/AAVs, or HMMWV-mounted machine guns and antitank vehicles.

(2) Battle positions should be placed where heavy weapons, antiarmor weapons, and machine guns will have the best fields of fire. Such locations are normally found at major street intersections, parks, and the edge of open residential areas. Battle positions should be carefully and deliberately prepared, reinforced by obstacles, and supported by indirect fire weapons. Battle positions are designed to inflict maximum casualties on the enemy and cause him to deploy for a deliberate attack.

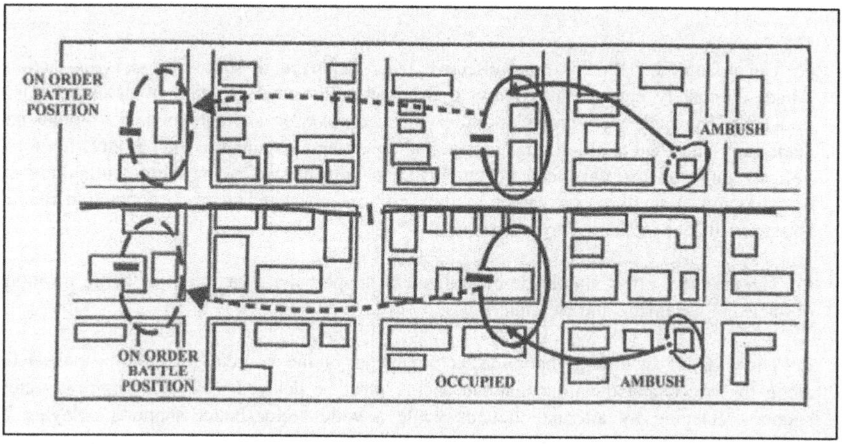

b. Tanks, LAVs/AAVs, and antiarmor weapons should prepare primary and alternate positions to improve their surivability and flexibility.

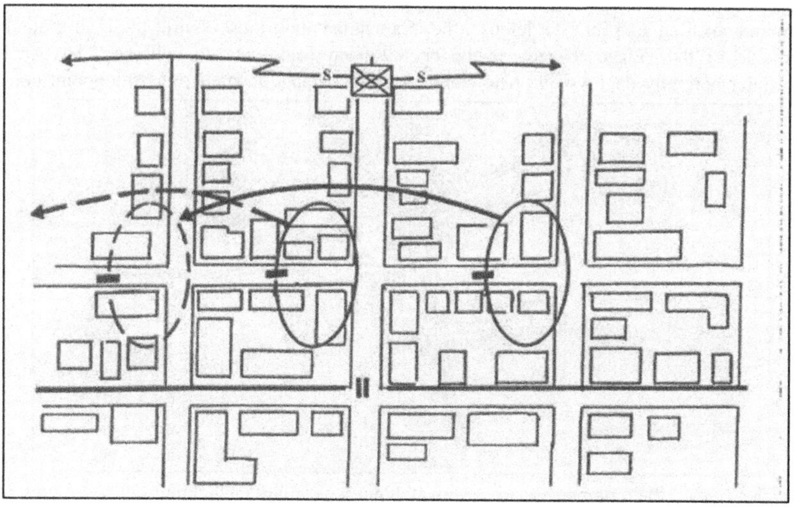

Figure 3-10A. Delay in a Built-Up Area (Successive Bounds)

Figure 3-10B. Delay in a Built-Up Area (Alternate Bounds)

c. The infantry battalion is most effective when deployed in successive delaying actions. Forces alternately engage the enemy to the maximum extent possible without becoming decisively engaged, then move on order in successive or alternate bounds. Smoke and demolitions are used to aid disengagement and cover the movement of the withdrawing force. Security units on the flank can be employed to prevent the enemy from outflanking the delaying unit. A small reserve can be used to react to unexpected enemy action and to conduct continued attacks on the enemy's flank.

d. The engineer effort should be centralized to support the preparation of battle positions, obstacle emplacement, and clearing routes.

e. The width of the infantry battalion sector depends on the nature of the buildings, obstacles along the street, and the time that the enemy must be delayed. Normally, narrow sectors supports delaying by alternate bounds while a wide sector better supports delaying by successive bounds.

3203. Defense of a Village. An infantry battalion assigned a defensive sector that includes a village may decide to incorporate the village as a strongpoint. This use is most common where the village stands astride a high-speed avenue of approach or lies between two difficult obstacles. To incorporate such an area into its defense, the infantry battalion must control the high ground on either side of the village to prevent the enemy from firing into the village. The battalion commander normally deploys one rifle company in the village to prepare a strongpoint defense

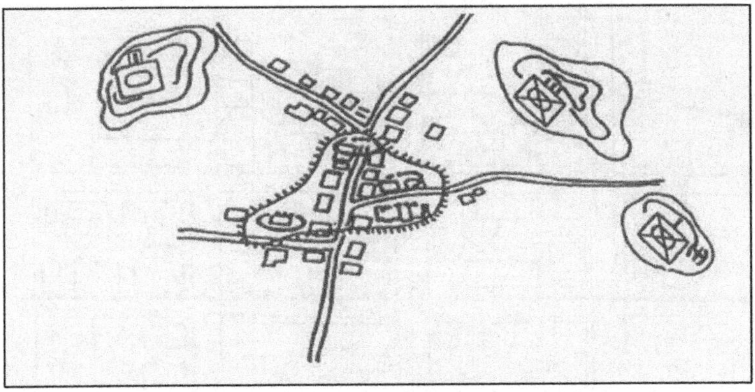

while the other rifle companies are overwatching the village, blocking bypass routes, and protecting the strongpoint's flanks. The strongpoint defense should provide a location where the enemy can be stopped and counterattacks can be launched (Figure 3-11).

Figure 3-11. Defense of a Village

a. Armored Vehicles and antiarmor weapons should be employed where the maneuver room is the greatest. This is usually on key terrain to the flanks of the village. As the security force withdraws and rifle companies become engaged, armored vehicles and weapons support by fire.

b. Although the infantry battalion's disposition should prevent large enemy forces from threatening the rear and flanks of the village, the danger of small-unit enemy infiltration requires the village be prepared for all-around defense.

c. Engineers should be assigned to the rifle company in the village strongpoint to provide continuous engineer support in preparation for the enemy's attack. The priority of engineer support should go to the strongpoint.

.3

Section III

Rifle Company Operations

This section addresses some of the missions the infantry battalion could assign to the rifle company. For the purpose of this publication, the rifle company is a task-organized and/or reinforced unit. Assigned missions are normally part of a larger battalion operation.

3301. Defense of a Village. Once the commander has completed his reconnaissance of the village and the surrounding terrain, he develops his plan for the defense (Figure 3-12). The commander must decide whether to defend on the leading edge of the village or within the village.

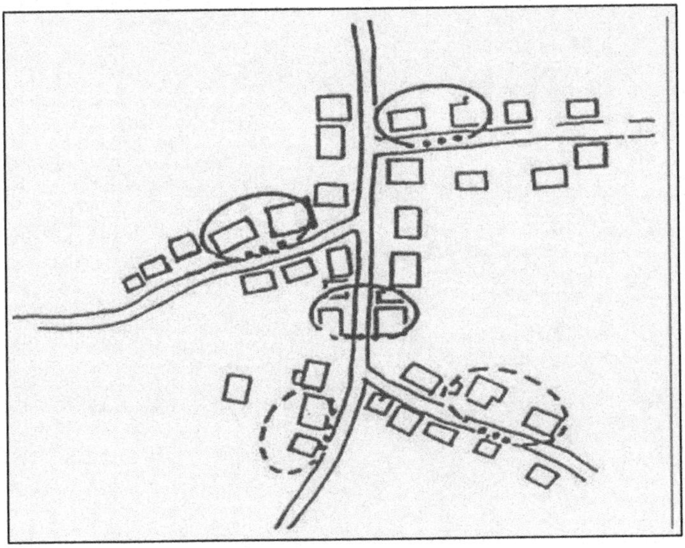

Figure 3-12. Defense of a Village

a. Several factors influence his decision. First, he must know the type of enemy that he will defend against. If the threat is mainly dismounted infantry, the greatest danger is in allowing them to gain a foothold in the town. If the threat is armor or motorized infantry, the greatest danger is that the company's defensive positions can be destroyed by direct fire. The commander must also consider the terrain forward and to the flanks of the village from which the enemy can direct fires against his positions.

b. The commander places platoons in a small group of buildings where each platoon leader establishes mutually supporting squad-sized positions. This positioning increases the area that the rifle platoon can control and hampers the enemy's ability to isolate or bypass a platoon. A rifle platoon can be responsible for the road through a village. The rest of the rifle company is positioned to provide all-around security and defense in depth.

c. Mortar positions must be protected from direct fire and allow for overhead clearance.

d. Combat vehicles should be assigned primary, alternate, and supplementary positions as well as primary and secondary sectors of fire. They should be positioned in defilade behind rubble and walls or inside buildings for cover, concealment, and movement into and out of the area.

e. If the company is reinforced with antiarmor weapons, the reinforcements should be positioned on high ground in and around the town to attain good fields of fire.

f. If the company is reinforced with tanks, the commander could place the tanks along the leading edge complement the other antiarmor weapons. The tank platoon leader selects exact firing positions and assign sectors of fire. If attacked by enemy infantry, the tanks should move to alternate positions protected by infantry. The alternate positions should allow the tanks to engage to the front and flanks with as little movement as possible. Tanks can be used as a mobile reserve after withdrawal from the leading edge of the town.

g. Obstacles are easily constructed in a built-up area. Commanders must stop enemy vehicles without interfering with his own movement in the village. Command detonating cratering charges at key street locations may be used to hamper enemy mobility. Mines should be laid on the outskirts of the town and along routes that the company will not use.

h. Engineers assist rifle platoons to prepare defensive positions. Engineers use explosives to make firing ports, mouseholes, and demolition obstacles. They also assist in the emplacement and recording of the minefields.

i. Ammunition expenditure is increased when fighting in an urban environment. Ammunition should be stockpiled at each squad position.

Other factors that the company commander may consider are:

- Resupply
- MEDEVAC
- Communications
- Firefighting

.3

- Sleep and alert plans
- Security
- Limited visibility
- Control of noncombatants.

j. The company should install a wire net and develop a plan for pyrotechnic signals to ensure adequate communications. Backup wire should be laid in case primary lines are cut by vehicles, fires, or the enemy. The commander also plans for the use of messengers .

3302. Delay in a Built-Up Area. A rifle company may be assigned the mission to delay in a built-up area.(Figure 3-13). To perform this mission, a rifle company can destroy enemy reconnaissance elements forward of the outskirts of the town, prevent their penetration into the built-up areas, and gain and maintain contact with the enemy to determine the strength and

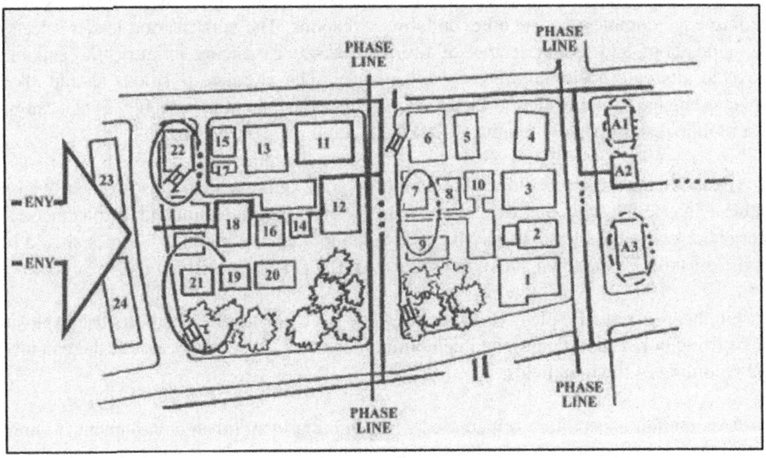

location of the main attack.

Figure 3-13. Delay in a Built-Up Area

a. Platoons delay by using patrols, OPs, and ambushes. Following each action, a platoon will disengage and withdraw.

b. The company's sector should be prepared with obstacles to enhance the delay. Engineers should prepare obstacles on anticipated enemy main routes of advance.

c. Antiarmor weapons may be positioned on the outskirts of the town to destroy the enemy at maximum range. They should be located in defilade positions or in prepared shelters. They fire at visible targets and then fall back or proceed to alternate positions.

d. Tanks in support may be used to engage enemy tanks, provide reinforcing fires, aid the disengagement of platoons, and cover obstacles by fire.

3303. Defense of a City Block. The rifle company may be assigned to defend a city block. The company actions should be coordinated with those of the security forces charged with delaying to the front of the company's position. The defense should take advantage of the protection of buildings that dominate the roads.

a. A well-organized company defense:

- Stops the attack of the enemy on the roads by using obstacles and enfilade fire
- Destroys the enemy by ambush and direct fire from prepared positions
- Ejects the enemy from footholds or fixes them for the battalion's counterattack.

b. The defensive forces should prepare ambushes on the avenues of approach, cover the obstacles by fire, and build a defense inside the buildings (Appendix D). The reserve can be tasked:

- To reinforce the fires of the defense
- To react to a danger on the flank
- To counterattack to throw the enemy from a foothold.

c. Engineers should be employed in GS to construct obstacles, prepare access routes, and assist in preparing defensive positions.

d. Tanks attached to the rifle company may be used to direct-fire support, engage enemy tanks, and support a counterattack.

3304. Defense of a Traffic Circle or Key Intersection. A rifle company may be assigned the mission to defend a key traffic circle or road junction (Figure 3-14).

a. The commander analyzes enemy avenues of approach and buildings that dominate those avenues. He plans to focus all possible firepower on the traffic circle and its approaches. He also plans for the all-around defense of the buildings that dominate the traffic circle to prevent encirclement. The commander prepares as many covered and concealed routes between these

buildings as possible. These routes make it easier to mass or shift fires and to execute counterattacks.

Figure 3-14. Defense of a Traffic Circle

b. Obstacles can be used to deny the enemy the use of the traffic circle. In this case, obstacle planning should consider whether friendly forces require access to the traffic circle. Antiarmor weapons can be employed across the traffic circle if designated fields of fire exist. Tanks can be used to engage enemy tanks and provide direct-fire for counterattacks.

Section IV

Rifle Platoon Operations

This section discusses some missions the rifle company might assign to a rifle platoon.

3401. Defense of a Strongpoint. A rifle platoon may be given the mission to provide or conduct the strongpoint defense of a building, part of a building, or a group of small buildings (Figure 3-15). The rifle platoon's defense is normally integrated into the rifle company's defense. The rifle platoon must keep the enemy from gaining a foothold in its assigned buildings. It makes the best use of its weapons and supporting fires, organizes all-around defense, and counterattacks or calls for a rifle company counterattack to eject an enemy that has a foothold. The platoon leader analyzes his defensive sector to recommend to the company commander the best use of obstacles and supporting fires. (See Appendices B and D.)

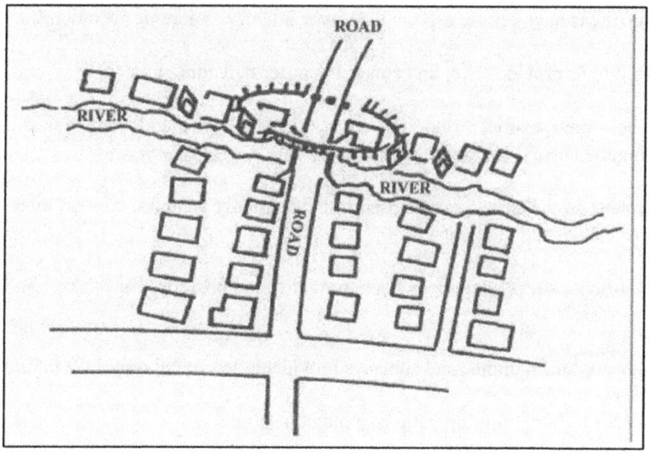

Figure 3-15. Defense of a Strongpoint

a. The rifle platoon should be organized into a series of firing positions located to cover avenues of approach, cover obstacles, and provide mutual support. Snipers may be located on the upper floors of the buildings. Unengaged units should be ready to counterattack, fight fires, or reinforce other elements of the rifle platoon.

b. Depending on the length of the mission, the rifle platoon should have available:

- Pioneer equipment (axes, shovels, hammers, picket pounders)

- Barrier material (barbed wire, sandbags)
- Munitions (especially grenades)
- Food and water
- Medical supplies
- Firefighting equipment.

3402. Defense Against Armor. Urbanized terrain is well suited to an infantry's defense against mechanized infantry and armored forces. Armored forces will usually avoid built-up areas but may be forced to pass through them.

a. Built-up areas have certain aspects that favor infantry antiarmor operations.

(1) Buildings provide cover and concealment for defending infantry.

(2) The streets restrict armor maneuver, fields of fire, and communications, thereby reducing the enemy's ability to reinforce.

(3) Rubble in the streets can be used to block enemy vehicles, conceal mines, and cover and conceal defending infantry.

(4) Rooftops, alleys, and upper floors provide good positions for flanking and overhead fire.

(5) Sewers, storm drains, and subways provide underground routes for infantry forces.

b. To prepare for antiarmor operations, a platoon should:

(1) Choose a Good Engagement Area. The best way for infantrymen to engage tanks is one at a time so that they can destroy one tank without being vulnerable to the fires of another. Good engagement areas include narrow streets, turns in the road, "T" intersections, bridges, tunnels, split-level roads, and rubbled areas. Less obvious locations can also be created by using demolitions or mines to create obstacles.

(2) Select Good Weapons Positions. The best weapons positions are places where the tank is weakest (tank deadspace) and the infantry is most protected. Forcing tanks to button up greatly reduces their ability to see and fire on close-in street and overhead targets. Figure 3-16 (on page 3-38) shows close-in tank weapon vision and engagement limitations (tank buttoned up) at street and overhead levels.

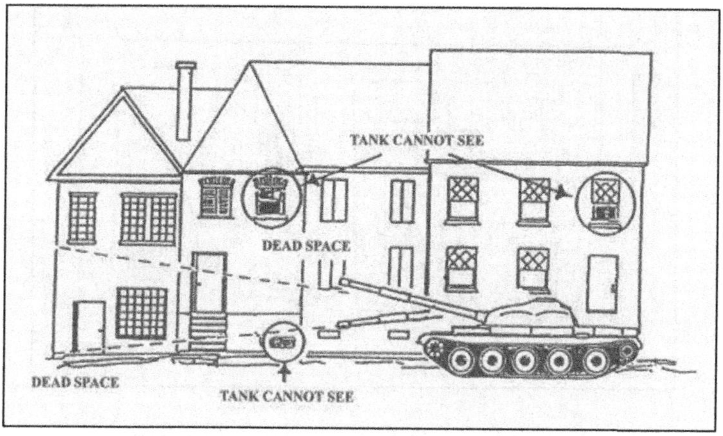

Figure 3-16. Close-in Tank Limitations at Street and Overhead Levels

(3) Assign Target Reference Points and Select Method of Engagement. After selecting the weapons positions, target reference points should be assigned to ensure adequate coverage and control fires. The target reference points should be clearly visible through the gunner's sights and should be relatively resistant to battle damage (for example, large buildings or bridge abutments, but not trees or cars). The three types of antiarmor engagement techniques are frontal, crossfire, and depth. Frontal fire is the least preferred because it exposes the gunner and it attacks the tank where the armor is the thickest. (For more information on target engagement techniques, see MCWP 3-15.5 *Antiarmor Operations.*)

Infantry forces should fire at tanks on the flanks and rear from ground level or at the top of tanks from an elevated position. A suitable antiarmor defense might be set up as shown in Figure 3-17. For a safe engagement from an elevated position, Marines should allow the tank to approach to a range three times the elevation of the weapons. However, engagement at longer ranges risks counterfire because the attacking weapon's position will not be in the tank's overhead dead space.

(4) Coordinate Target Engagement. Tanks are more vulnerable to infantry operations when they are buttoned up. The first task is to compel the tanks to button up by using direct and indirect fire. The next task is to coordinate the fires of the antitank weapons so that all targets in the engagement area are attacked simultaneously (Figure 3-18).

Figure 3-17. A Rifle Platoon's Antiarmor Defense

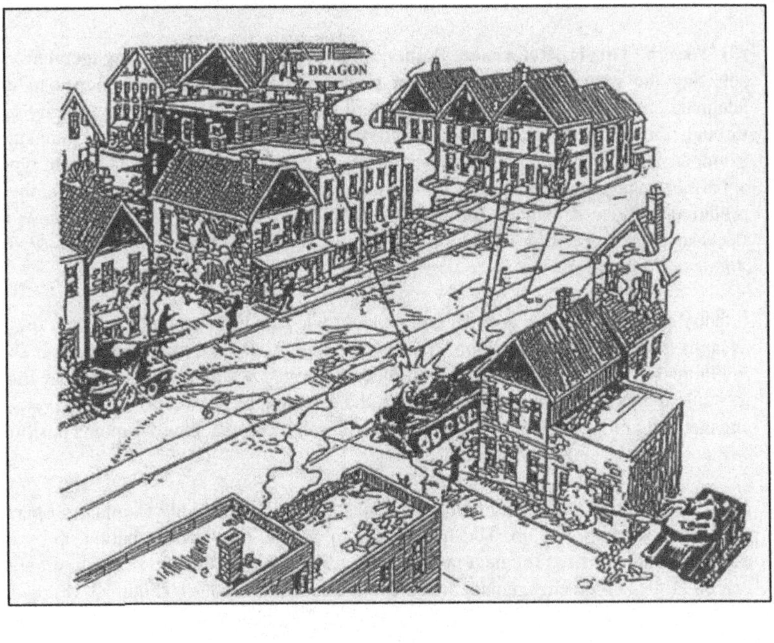

Figure 3-18. Coordinated Antiarmor Ambush

c. Infantry often accompany armored vehicles in built-up areas to provide them with close-in protection. Antiarmor weapons should be supported by an effective antipersonnel defense.

d. At a planned signal (for example, the detonation of a mine), all targets in the engagement area are attcked at the same time. If that is not possible, then the most dangerous target should be engaged first. Although tanks present the greatest threat, BMPs are also dangerous because their infantry can dismount and destroy friendly antiarmor positions. Rubble and mines should be used to reduce target mobility so that a target's time in the engagement area is increased.

3403. Conduct of an Armored Ambush. A rifle company can use an attached tank platoon to conduct an armored ambush (Figure 3-19). The tank platoon should be reinforced with infantry.

Figure 3-19. Armored Ambush

a. An armored ambush requires maneuver on obstacle-free roads. Obstacles outside the ambush area can be used to canalize and delay the enemy.

b. Ambushing tanks should hide approximately 1,000 meters from the expected enemy avenue of approach. A security post, located at a chokepoint, observes and reports the approach, speed, security posture, and activity of the enemy. When the enemy is reported at a trigger point or target reference point, the tank platoon leader orders the ambush.

c. The tanks move quickly from their hiding positions to firing positions, taking advantage of all available concealment. They attempt to fire flank shots on the approaching enemy at an average range of 300 to 400 meters. Firing at these ranges does not expose tanks to enemy infantry. After the engagement, the tank breaks contact and moves to a rally point with close security provided by infantry. They may then move to a new ambush site.

Chapter 4

Combat Support

Combat support units provide critical combat functions in conjunction with combat arms units to secure victory (MCRP 5-2A).

4001. Introduction. Success against a determined enemy is achieved in part by generating decisive combat power through the application of *combined arms*. This is especially true in MOUT. The MAGTF commander must effectively integrate all the capabilities of the command element (CE), GCE, ACE, and CSSE to win the battle in the shortest amount of time with the fewest casualties. The MOUT environment requires special considerations pertaining to combat support forces and capabilities.

Section I

Fire Support

4101. Indirect and Direct Fire Support

a. General. MOUT operations present unique fire support considerations. On urbanized terrain, buildings provide excellent cover and concealment to the enemy while limiting friendly observation and targeting efforts. Targets are generally exposed for brief periods of time and are often in close proximity to friendly forces. Observers will experience difficulty in finding OPs with adequate fields of observation. Terrain masking by tall buildings may restrict the delivery of indirect fires. Collateral damage and rubble effects must be considered during ammunition selection. The importance of effective communications, ROE, control measures, and procedures to prevent fratricide is magnified.

b. Offensive Support. In the offense, fire support plans should include fires to isolate the objective area, support the assault, and support the clearing action. Fires are delivered to isolate and fix the enemy and deny him the use of avenues of approach into and out of the built-up area. Most fires are normally planned and executed at the GCE or higher echelons. Fires are employed to rupture the enemy's established defenses and screen friendly maneuver, in order to maintain the momentum of the attack. Fire support is also allocated to units involved in clearing operations. Fire support plans should incorporate the employment of aerial observers and UAVs to compensate for restrictions to observation and to assist in the delivery of deep fires. Procedures for designating the forward line of own troops (FLOT), marking targets, shifting fires, and communicating in the urban environment should also be considered.

c. Defensive Support. In the defense, fire support plans address fires to disrupt and slow the enemy attack. Fires are delivered at maximum ranges along avenues of approach to separate armor and infantry forces, to canalize the enemy into killing zones, and to deny the enemy use of key facilities in the city, such as communications and transportation facilities. Most fires are planned and executed at the GCE or higher echelon. Defensive fires planned at lower echelons support fighting within the built-up area. Defensive fire support plans identify fire support coordination procedures necessary to execute the delivery of fire support.

4102. Artillery

a. Employment of Artillery. The mission of artillery does not change in MOUT. However, the method of employment is tailored to the unique requirements posed by the urban battlespace. MCWP 3-16.3, *Marine Artillery Support,* provides a detailed discussion on the employment of artillery. Considerations for MOUT include:

(1) Organization for Combat. Standard command relationships for tactical missions are used to organize for combat. During the initial phases of offensive MOUT, particularly isolation of the objective area and the initial assault to secure a foothold, control of artillery is more centralized than in other types of offensive operations. Centralized control provides better responsiveness in shifting and massing fires. As the attack progresses into the clearance phase, control of artillery becomes more decentralized with direct support (DS) artillery providing direct and indirect fires. In the defense, control of artillery should be sufficiently centralized to enable the rapid massing of fires.

(2) Movement and Positioning. Key considerations for the positioning of artillery in MOUT include:

- Mobility restrictions
- Limited availability of firing positions
- Masking of fires
- Security
- Enemy counterbattery fires

(a) Offense. Artillery is initially positioned outside or in the outskirts of the city. Artillery may displace to suitable firing positions within the city as the attack progresses. Positions should be selected that minimize masking, provide several routes for escape, and afford as much cover and concealment as possible. Potential firing positions can be found in parks, racetracks, sports fields, school yards, stadiums, parking lots, and similar areas. Special techniques for the emplacement of howitzers may be required if the ground is not suitable for normal emplacement. Some artillery positions may be maintained outside the city (see appendix B). Artillery generally makes few displacements in MOUT because it requires extensive route reconnaissance and security. If and when artillery is moved, it should be given priority of movement during displacements. Movement is conducted during night or periods of reduced visibility.

(b) Defense. Artillery is positioned on the friendly outskirts of the built-up area. Artillery displaces along predetermined routes to avoid traffic congestion.

(3) Delivery of Fires. The commander, in coordination with the fire support coordinator (FSC), considers the following when planning fire support in a built-up area:

(a) Target acquisition may be more difficult because of the increased cover and concealment afforded by the terrain. Ground observation is limited; therefore, FOs should be placed near tops of buildings.

(b) For targets in defilade, fire may be adjusted laterally so that the rounds impact on the street perpendicular to the observer's position. Airburst rounds are more suitable for this adjustment. Adjustments may have to be made by sound. When rounds impact on a perpendicular street, they are adjusted for range. When the range is correct, a lateral shift is made onto the target and the observer fires for effect.

(c) Special consideration must be given to shell and fuze combinations when effects of munitions are influenced by buildings. These include:

- Careful use of variable time is required to avoid premature arming.
- HE ammunition with impact or delay fuzes may create unwanted rubble.
- The close proximity of enemy and friendly forces requires careful coordination.
- White phosphorus (WP) may create unwanted fires and smoke.
- Fuze delay should be used to penetrate fortifications.
- Illumination rounds can be effective. However, friendly positions should remain in shadows and enemy positions should be highlighted. Tall buildings may mask the effects of illumination rounds.
- HE/variable time, HE/time fuze, and improved conventional munitions are effective for clearing enemy positions, observers, and antennas from rooftops.
- Swirling winds may degrade smoke operations.
- FASCAM may be used to impede enemy movements. FASCAM effectiveness is reduced when delivered on a hard surface.

(4) Targeting is difficult on urbanized terrain because the enemy has many covered and concealed positions and movement lanes. The enemy may be on rooftops and in buildings and may use sewer and subway systems. Targets should be planned on rooftops to clear enemy FOs, communications, and radar equipment. Targets should also be planned on major roads, at road intersections, and on known or likely enemy fortifications. Employing artillery in the direct-fire mode to destroy fortifications should be considered. Also, restrictive fire support coordination measures (such as a restrictive fire area or no fire area) may be imposed to protect civilians and critical installations.

(5) The 155-mm howitzer can be effective in neutralizing concrete targets by direct fire. A 155-mm round can penetrate 36 inches of concrete at ranges up to 2,200 meters. Restrictions may be placed on the type of ammunition used in order to reduce rubbling on friendly avenues of approach.

(6) FOs must be able to determine the location and size of dead space. Dead space is the area in which indirect fires cannot reach the street level because of buildings. This area is a safe haven for the enemy because he is protected from indirect fires. For low-angle artillery, the dead space is about five times the height of the building. For mortars and high-angle artillery, the dead space is about one-half the height of the building.

(7) Aerial observers and UAVs are effective in providing observation behind buildings immediately to the front of friendly forces. They are extremely helpful in adjusting fire. Aerial observers can also relay calls for fire when communications are degraded because of power lines or building masking.

(8) Weapons locating radars are effective in locating enemy artillery and mortar targets in an urban environment because of their high percentage of high-angle fires. However, LOS must be maintained in order to exploit this advantage. Radars should not be placed in the midst of a city due to masking. They are best located on the outskirts of the city.

(9) The use of airburst fires is an effective means of clearing snipers and manpads from rooftops. HE shells with delay fuzes may be effective against enemy troops in the upper floors of buildings, however, because of the overhead cover provided by the building, such shells have little effect on the enemy in the lower floors. (The planning and employment of artillery in offensive and defensive operations is also addressed in Chapters 2 and 3.)

b. Local Security. Artillery crews and howitzers require close security when used in the direct-fire mode. Artillery is particularly vulnerable when units are employed in isolated firing positions.

c. Communications and Information Systems (CIS). CIS must be able to satisfy the unique C2 requirements for artillery in MOUT. CIS should provide the commander with the ability to collect, process, analyze, and exchange information. Requirements such as massing fires and call for fire must be carefully planned and adapted to meet the requirements of

MOUT. For example, small units conducting independent operations must have the communications capabilities and maps to allow them to call for fire.

4103. Mortars

a. Employment of Mortars. Mortars are often the primary indirect-fire weapon for forward units in the assault or defense of a built-up area. Their high rate of fire, short minimum range, and high trajectory give mortar units the ability to fire in the tight confines of the city. Mortars can provide obscuration, neutralization, suppression, or illumination fires.

(1) Organization for Combat. Normally, 81-mm mortars are employed in GS of the battalion. The 60-mm mortar is normally used at the discretion of the rifle company commander.

(2) Movement and Positioning. 81-mm mortars employed in GS normally occupy a position within or near the infantry battalion reserve. Key considerations for the selection of positions include:

- The minimum range of the weapon
- Mask and overhead clearance
- Terrain suitable for setting base plates
- Dispersion, and accessibility.

Special techniques can be used to position and lay the 81-mm mortar. These techniques include placing buffers under base plates, using curbs and sandbags to anchor or brace mortars, and filling cans with dirt or using sandbags to emplace aiming posts. Fixed objects may be used as distant aiming points. When the depth of the defensive position is shallow, or when suitable firing positions are not available, mortars may have to be positioned behind the reserve. Positioning of mortars behind buildings greatly enhances survivability. Displacement is often executed by section. Mortars are usually not placed on top of buildings because the location lacks cover and concealment and recoil/concussion could collapse roofs.

(3) Delivery of Fires. Mortar fires are effective in the offense as well as in the defense. The indirect fires are extremely responsive. Mortars are well suited for combat in built-up areas because of their high rate of fire, steep angle of fall, and short minimum range. Mortar fires can be used to inhibit enemy fires and movement, allowing friendly forces to maneuver to a position of advantage. Effectively integrating mortar fires with dismounted maneuver is key to successful combat in a built-up area.

.3

(a) HE Ammunition. HE is used more than any other type of ammunition. Although mortar fires are often targeted against roads and other open areas, the natural dispersion of indirect fires will result in many hits on buildings. Leaders must use care when planning mortar fires to minimize collateral damage. Considerations when using HE include:

- HE ammunition is particularly effective when used on lightly built structures within cities. However, it does not perform well against the reinforced concrete found in larger urban areas.

- Point-detonating fuzes should be used. The use of proximity fuzes should be avoided because the nature of built-up areas causes proximity fuzes to function prematurely. However, proximity fuzes are useful in attacking targets such as OPs on tops of buildings.

- During World War II, the Middle East conflicts, and most recently in Bosnia, mortar HE fires have been used extensively to deny the use of streets, parks, and plazas to enemy and civilian personnel.

(b) Incendiary Fires. Employment of these munitions is planned in detail to prevent the burning of buildings and surrounding terrain that may create obstacles to friendly forces.

(c) Illumination. In the offense, illumination rounds should burst above the objective to put enemy troops in the light. Buildings reduce the effectiveness of the illumination by creating shadows. If the illumination is behind the objective, the enemy troops would be in the shadows rather than in the light. In the defense, illumination is planned to burst behind friendly forces to put them in the shadows and place the enemy forces in the light. Continuous illumination requires close coordination between the FO and the fire direction center (FDC) to produce the proper effect.

(d) Other Considerations. When planning the use of mortars, commanders must consider the following:

- FOs should be positioned on tops of buildings so target acquisition and adjustments in fire can best be accomplished.

- Commanders must understand ammunition effects to correctly estimate the number of volleys needed for the specific target coverage. Also, the effects of using white or red phosphorus may create unwanted smoke screens or limit visibility, which could interfere with the tactical plan.

- Mortar sections should plan to provide their own security.

- Commanders must give consideration to where, when, and how mortars are to displace in order to maintain immediate indirect fire support. Combat in built-up areas may adversely affects the ability of mortars to displace because of rubbling.

b. Communications. An increased use of wire, messenger, and visual signals will be required. However, wire should be the primary means of communication between the FOs and the FDC because elements are close to each other. Also, frequency-modulated (FM) radio transmissions in built-up areas are likely to be erratic. Remote siting of antennas to upper floors or roofs may improve communications and enhance operator survivability. The use of radio retransmissions can also be effective. Both in the offense and defense, commanders consider using existing civilian systems to supplement the unit's capability.

c. Magnetic Interference. Anything that uses or generates electricity generates its own magnetic field. In an urban environment, all magnetic instruments are affected by surrounding structural steel, electrical cables, and automobiles. Minimum distance restrictions for the use of the M-2 aiming circle will be difficult to overcome. However, an azimuth may be obtained from a distant aiming point.

4104. Naval Surface Fire Support (NSFS) and Naval Gunfire (NGF)

a. General. When NSFS ships are available, naval surface fires and NGF can provide some fire support.

b. Employment of Naval Surface Fire Support. The urban environment severely limits NSFS, and NGF in particular, due to its flat trajectory. NSFS can be used to supplement existing fire support but should not be planned as the primary fire support asset for MOUT. NSFS is most effective when firing on targets to isolate the battlefield and engage targets located on the seaward side of a city. Only when the ROE allows extensive collateral damage can the full effect of NSFS be brought to bear on units occupying a city.

(1) Organization for Combat. NSFS is best employed under centralized control because of the limited use of NSFS within the confines of the city.

(2) Positioning. Many of the fires delivered by NSFS ships require a careful orientation of the gun target line and may limit the size of a fire support area.

(3) Delivery of Fires. NGF can deliver a variety of ammunition in MOUT. NGF can deliver HE, incendiary, and illuminating projectiles. The NGF high-capacity projectile is very effective in producing fragmenting and penetration effects. Offensive and defensive fire support plans can incorporate NSFS fires on bridges, roads, reinforcements, artillery positions, and command and control centers. NSFS can be effectively delivered on targets along the outskirts of the city to rupture established defenses. These fires are particularly effective on buildings and on targets with a linear axis such as streets and airfields. Because of its flat trajectory, NGF employment is limited on targets within the city. Firing reduced charges at high angles does enable NGF to hit more targets but will reduce its range and accuracy.

4105. Aviation

a. Employment of Aviation. Both fixed- and rotary-wing aircraft can provide fire support in MOUT. The aviation combat element (ACE) provides the MAGTF commander with a variety of options in ordnance delivery. Aerial delivered ordnance is particularly effective against hard targets. Cluster bomb units, rockets, cannons, laser-guided bombs, and electro-optically guided missiles can be used against a variety of targets. However, the presence of civilians, the requirement to preserve key facilities, and the residual effects caused by rubble and collateral damage must be taken into consideration when planning for and using aviation in an urban environment. (See MCWP 3-24 *Assault Support Helicopter Tactical Manual* (under development), MCWP 3-23.1 *Close Air Support* (under development),and NWP 55-3-AH1, *AH1 Tactical Manual,* for greater detail on MOUT considerations for aviation.)

(1) Fixed-Wing Aircraft

(a) Offensive/Defensive Operations. During offensive operations, fixed-wing aircraft can isolate, interdict, and reduce enemy strongpoints throughout the battlespace. During defensive operations, they attrite enemy forces as they approach, attack forces attempting to bypass, or canalize the enemy to take avenues of approach that are advantageous to the defender.

(b) Other Considerations

1 Fixed-Wing Attack of a Built-Up Area. Air strikes can reduce or destroy vital supplies and facilities supporting the enemy. The resulting shock, concussion, and psychological effects of such attacks can reduce the efficiency and fighting spirit of enemy forces.

2 Rubble and Debris. The rubble and debris resulting from air attacks may increase the defender's cover and create obstacles to the movement of friendly forces.

3 Proximity of Friendly Forces. The proximity of opposing forces to friendly forces increases the potential for fratricide. This may require the use of precision guided munitions (PGMs) or the temporary disengagement of friendly forces in contact.

4 Locations of Indigenous Civilians or Key Facilities. The use of aerial delivered ordnance may be restricted by the ROE because of the presence of civilians or the requirement to preserve key facilities within a city.

5 **Limited Ground Observation.** Limited ground observation may require the use of forward air controllers (airborne) (FAC(A)).

(2) Rotary-Wing Aircraft. Attack helicopters provide a distinct advantage to the commander in MOUT. They can rapidly maneuver within the built-up area to attack a variety of targets. Urban structures offer ideal cover for attack helicopters. The ability to deliver PGMs, rockets, and 20 mm cannon fire makes attack helicopters valuable assets in reducing strongpoints, breaking up enemy concentrations, and providing suppressive fires for attacking forces. Attack helicopters also provide a highly manueverable antiarmor capability.

(a) Attack Helicopter Fire Support During Offensive/Defensive Operations

1 Provide close air support in support of the commander's scheme of maneuver

2 Provide suppression and/or destruction of enemy forces within buildings and strongpoints

3 Conduct counterattacks

4 Isolate the built-up area or objectives within the built-up area

5 Conduct armed reconnaissance and patrols.

6 Conduct rapid concentration of fires.

4106. Armored Vehicles

a. Employment of Armored Vehicles. Tanks, LAVs, and AAVs are primarily mobility assets that function best in an unrestricted, mobile environment. However, these vehicles can play an important role inside the city by using their firepower in support of offensive and defensive operations. LAVs and AAVs highly are susceptible to armor-piercing rounds and manpacked antiarmor weapons. All vehicles are susceptible to top-attack munitions. Some of the fire support roles armored vehicles may provide include:

- Suppression and/or destruction of enemy forces within buildings and strongpoints (tanks/LAVs/AAVs)
- Provide machinegun suppression fires
- Reserve or counterattack suppression fires (tanks and/or AAVs)
- Creating entry/exit points in buildings
- Isolation of the built-up area or objectives within the built-up area (tanks/LAVs/AAVs)

.3

- Artillery-towing vehicle (LAVs/AAVs)

- Antiarmor fires

- Breaching obstacles in a direct fire mode.

b. Special Consideration. The existence of waterways, canals, and rivers in and around urban areas can provide avenues of ingress into designated objective areas. If used in conjunction with supporting fires, smoke, and/or deception operations, AAVs can capitalize on their amphibious capability and make use of waterways previously viewed as obstacles.

4107. Antiarmor Weapons. Antiarmor weapons have more utility in the defense than in the offense. Their primary task is defeating enemy armor. Secondary tasks may involve blasting holes in structures and fortified positions. As with other weapons systems, antiarmor weapons are normally employed in a decentralized fashion at the small unit level. Enemy armor should be engaged in predetermined locations that afford the most advantage to the defender. Weapon positions should be selected that provide maximum range fires. TOWs and Dragons may be fired from an enclosed area if it is a sturdy ventilated structure of appropriate size. For a more detailed discussion of antiarmor weapons, see MCWP 3-15.5, *Antiarmor Operations*.

4108. Employment of Snipers

a. General. Snipers are highly trained in field skills and marksmanship and can deliver long-range precision fire at selected targets from concealed positions. The method by which snipers are employed in urban warfare will be governed by the nature of the terrain/weather, number of snipers available, and whether the enemy employs snipers. A sniper can play an important role in MOUT. Employment of snipers influences the action by:

- Creating adverse psychological affects on the enemy by negating feelings of security

- Inflicting casualties on enemy leadership

- Enabling the infantry to seem to cover a large area, regardless of whether the terrain is physically occupied or not

- Detailing enemy positions and activity through the use of advanced optical equipment and observation techniques

- Controlling fire support

- Conducting reconnaissance in conjunction with their sniper role

- Disrupting enemy movement, observation, and infiltration, and negating the possibility of surprise.

b. Offensive Employment. Snipers should be used to gain and maintain contact with the enemy. This enhances security and prevents surprise by keeping constant and unrelenting pressure on the enemy. Some of the considerations for sniper employment include:

(1) As the "eyes" of a commander, a sniper increases the commander's flexibility by gathering and transmitting information on fire support targets.

(2) Snipers attack those targets that affect the enemy's ability to wage battle and those that will cause the maximum amount of confusion on the battlefield. Potential targets for a sniper are:

- Officers and noncommissioned officers
- Enemy snipers and marksmen
- Engineer personnel
- Personnel manning crew-served weapons
- Communications personnel and equipment
- Fire support observers and/or equipment
- Commanders of armored vehicles.

c. Combat on Urbanized Terrain

(1) Snipers can be employed to operate for extended periods of time from hidden positions to dominate and establish a "no-man's land," screen flanks, protect the rear, and deny the enemy access to certain areas or avenues of approach. Snipers can operate with the covering element to deliver accurate fires in support of search elements. Optical devices enhances their ability to detect movement and engage it instantly.

(2) Snipers can also support (by fire) infantry movement across streets. They can provide immediate precision fire on enemy machine gun nests and enemy snipers. The best countersniper weapon is the sniper. Snipers are assigned the following supporting missions:

- Delaying and inflicting casualties on the advancing enemy
- Observing and reporting potential targets
- Covering (by fire) avenues of approach and obstacles

- Harassing the enemy and causing him to deploy prematurely
- Directing supporting arms fire.

The urban environment provides an abundance of secure firing positions that are not highly accessible to countersniping.

d. Defensive Employment. In defensive operations, the sniper is best employed forward of the FEBA to provide early warning of enemy approach, disrupt it, and if possible, cause the enemy to deploy prematurely. Snipers are employed to:

- Prevent enemy infiltration
- Detect and destroy infiltrators
- Protect unit at the FEBA from surprise attacks
- Protect patrols from ambush
- Screen the flanks and rear of the defensive positions.

More detailed information regarding the employment of snipers can be found in Marine Corps Warfighting Publication (MCWP) 3-15.3, *Scout Sniping* (under development).

Section II

Other Combat Support

4201. Engineers

a. General. The nature of the urban battlespace requires extensive use of engineers. In both the offense and defense, the proper employment of engineers can be decisive. Engineers must also be prepared to repair and maintain the internal support facilities of the city in order to support further operations. MAGTF commanders plan to use all available engineer assets, including engineers in the GCE, ACE, and CSSE, as well as any naval construction forces (NCFs).

b. Engineer Organization. Engineers are almost always task organized. The compartmentalized, closed-in nature of the city, which limits observation, mobility, and communications, will dictate the best organization for the mission. Teams should be formed to perform specific tasks for small units and must be positioned for quick employment. In MOUT, engineer support is key to maintaining the mometum of the attack. Engineer teams will usually move with the units they are supporting. Two-man demolition teams may be assigned to support a rifle platoon or squad to breach obstacles or to neutralize booby traps and improvised explosive devices (IEDs).

Commanders must carefully analyze mission requirements, the alignment of enemy forces, the condition of the battlespace, and available assets, to ensure an organization that will provide optimum engineer support. Most of the actual engineer work is done at the small unit level.

c. Engineer Tasks

(1) Offense. Urban combat is by nature a very slow, deliberate affair. Even if the enemy has not prepared the area for a defense, it will take time to secure the city. A well-constructed urban defense will slow or stop any attack unless impediments to movement can be quickly removed or bypassed. Properly organized and positioned engineers can reduce delays by performing a variety of tasks which include:

(a) Engineer reconnaissance to determine the level of effort required to penetrate outer and inner defenses, clear obstacles, and identify bypass routes.

(b) Bridge repair and construction on critical avenues of approach. Bridge construction may also open additional avenues of approach that the enemy is not prepared to defend.

(c) Breaching obstacles both outside and inside the city. This includes breaching minefields and neutralizing booby traps and IEDs.

(d) Assault breaching into buildings.

(e) Reducing enemy strongpoints using demolitions and heavy equipment.

(f) Preparing helicopter landing zones (HLZs).

(g) Assist in the preparation of defensive positions in the event of a counterattack.

(2) Defense. An urban area can be made nearly impregnable with proper organization of the battlespace and skillful preparation by engineers. Obstacles in outlying areas can delay/disrupt the enemy and canalize him into engagement areas. The streets and alleys inside a city can also be use to canalize enemy forces into engagement areas and kill zones. In preparing the defense of an urban area, engineers perform several critical tasks to include:

(a) Engineer preparation of the urban battlespace. This should begin well outside the city with complex obstacles in support of security forces. Selective demolition of bridges and roads and other countermobility measures can force the enemy into taking avenues of approach that favor the defense.

(b) Constructing obstacles forward of and between defensive positions to block, disrupt, fix, and turn enemy attacks. A majority of the obstacles will be directed at interfering with the movement of personnel and vehicles, but some effort must be directed at preventing vertical and subterranean movement. Material designed especially for obstacles and barricades, such as barbed wire and tetrahedron(s), is necessary, but improvised barriers can be just as effective and are readily available. Vehicles and building rubble are the most common improvised barriers.

(c) Enhancing mobility by opening and/or marking preplanned routes into attack positions for counterattacks or alternate positions to protect a flank.

(d) Providing technical advice on which buildings are more suitable for use as strongpoints or which rooftops are strong enough for helicopter landings can be vital. Any reinforcement of building roofs should be supervised by the engineers.

(e) Preparing tank positions that offer them cover and concealment.

(f) Assisting infantry units in clearing clear fields of fire, especially for longer range weapons like the TOW.

(g) Playing a potential role in the maintenance of critical urban services such as power, water, and communications. This function may also be performed by naval construction units.

(h) Assisting in the preparation and maintenance of railroads, airfields, and ports. This function may also be performed by naval construction units.

4202. Assault Support Helicopters

a. Combat Assault Transportation. The mobility provided by helicopters can be a significant advantage in the urban environment. In many modern cities, office buildings may have helipads on their roofs. Other buildings, such as parking garages, are usually strong enough to support the weight of a helicopter. The delivery of Marines onto a building can also be accomplished by rappelling, fastroping, or jumping out of the helicopter as it hovers just above the roof. The rooftops should be inspected to ensure that no obstacles exist, such as electrical wires, telephone poles, antennas, or mines and wire. The use of helicopters in this role is usually predicated on the antiair threat. Helicopterborne forces are highly vulnerable to small arms fire, AAA and manpads.

b. Small-Scale Helicopterborne Assaults. Small units may be helo-lifted onto the rooftop of a key building. Depending on the construction of the roof, rappelling or fastroping Marines from the helicopter may be more advantageous than landing them on the rooftop. Fastroping provides a rapid and safer means of insertion than jumping from a low hover. With practice, Marines can accomplish a fastrope insertion with minimum exposure.

c. Large-Scale Helicopterborne Assaults. For large-scale assaults, rooftop landings are not practical; however, open spaces (parks, parking lots, sports arenas) within the built-up area may be used. Depending on METT-T, helicopterborne forces may assault into an LZ in one wave or scheduled waves.

d. Mobility of Forces and Supplies. In the urban environment, movement of forces by helicopters may become a major requirement. Units engaged in house-to-house fighting normally suffer more casualties than units fighting in open terrain. These casualties must be evacuated and replaced quickly with new forces. At the same time, roads are likely to be crowded with resupply and evacuation vehicles and may also be blocked with craters or rubble. Helicopters provide a means of overcoming these obstacles.

e. Helicopterborne Assault Techniques. Helicopter assaults into enemy-held territory are extremely difficult (Figure 4-1 on page 4-16). One technique is to fly nap-of-the-earth down a broad street or commercial ribbon while attack helicopters and assault helicopter door gunners suppress buildings on either side of the street. Artillery preparations can be incorporated into the helicopter assault plan to suppress rooftop and building antiair threat. Feints and demonstrations in the form of false insertions can deceive the enemy as to the intended objective. Helicopters often deliver forces to the last covered position, short of the fighting, then return without exposure to enemy direct fire. Similar flight techniques can be used for transportation of supplies and casualties.

Figure 4-1. Helicopterborne Assault of a Built-Up Area

f. Combat Roles. Below are some of the roles that assault support helicopters may play in urban warfare:

- Conduct insertion and extraction of forces (See MCRP 3-11.7A, *Helicopter insertions and Extractions* [under development])
- Provide command/control and aerial retransmission
- Support CSS operations
- Provide air reconnaissance.

4203. Employment of Reconnaissance Forces. Reconnaissance units traditionally make maximum use of rural camouflage and terrain. Area, zone, and route reconnaissance are conducted from well-concealed vantage points through clandestine insertion by foot, vehicle, rubber boat, or helicopter. These methods of insertion may still apply in village and strip areas for collection of information, sensor employment, conducting OP radio relays, or a combination of missions. In major urban areas, undetected insertions are difficult to accomplish because of the high density civilian population.

The use of reconnaissance assets must be used wisely. Reconnaissance assets must not be risked indiscriminately if the information can be collected by other means. Consideration should be given to flank and rearward employment of ground reconnaissance units, in addition to close (0 to about 2,000 meters), distant (approximately 2,000 - 20,000 meters), and deep (from about 20,000 meters and beyond, plus preassault) missions. Reconnaissance of bypassed pockets of resistance, avenues of approach, canals, bridges, subways, and sewer systems (possibly in conjunction with combat engineers) is feasible. Observation, sniper, and listening posts manned by reconnaissance personnel are also missions that reconnaisance personnel perform. The limited combatant diving capability of division reconnaissance may be employed to advantage in many coastal cities.

4204. Military Police (MP). MPs can play a significant role in the urban environment. They may conduct battlefield circulation control, area security, enemy prisoner of war (EPW) and law and order operations. These operations require continuous coordination with MAGTF units as well as with host nation military and civilian police.

a. MP units support area damage control operations. With the increased possibility of rubbling, MP units report on and block off affected areas and reroute movement to alternate road networks.

b. When conducting EPW operations, MPs collect prisoners as far forward as possible and transport them to the rear. They operate temporary collecting points and holding areas to briefly retain EPWs and civilian internees. EPW operations are important in built-up areas because the rate of capture can be higher than normal.

c. Commanders must realize that MP support may not always be available and that infantry units may have to assume certain MP functions. These functions include:

(1) Providing convoy escort and security for lines of communications.

(2) Controling roads, waterways, and railroad terminals that are critical chokepoints in the main supply routes.

(3) Providing security of critical sites and facilities, including communications centers, government buildings, water and electrical supply facilities, command and control nodes, airfields, storage facilities, and other essential areas.

(4) Provide refugee control in close cooperation with host nation civil authorities.

(5) Collect and escort EPWs.

4205. Communications

a. General. Manmade structures create many challenges for CIS in an urban environment. Other CIS challenges include extensive commercial and communications networks and public service radio nets. Commercial communication networks are commposed of miles of

.3

underground protected cable which connect and control telephones and computer networks. Public service radio nets, (including police, fire, civil defense, and taxi) compete with existing communication systems. To communicate on an effective, continuous basis in the urban environment, commanders must minimize communication limitations imposed by this environment and maximize the advantage of existing civil communications capabilities.

b. Radio. Manmade structures will decrease the effectiveness of VHF radio communications. The physical locations of radios are critical for effective communications. Some specific locations to avoid include powerlines, underground areas, and steel bridges.

Very high frequency (VHF) radios provide the primary communication capability at the battalion level. It is imperative that these radios be employed as effectively as possible. Communicators can use the following techniques to enhance VHF communications equipment:

- Relocation of VHF radios to overcome phase cancellation due to numerous dead spots
- Use of the higher end (50 - 75 MHz) of the available frequency spectrum
- Use of retransmission facilities to improve the reliability of communications
- Use of radio remotes.
- Use of field expedient antennas.

In setting up retransmission stations, antennas should be camouflaged to blend in with urban surroundings and to prevent them from being targeted by enemy fires. Antennas can be concealed by such objects as water towers and steeples.

In larger cities with many tall buildings, it may be necessary to construct a terrain profile of the area of operations. A communications plateau can be deduced by calculating the average height of the structures in a zone. Operation of antennas above the median line of this plateau will result in better wave propagation and aid in retransmission.

c. Wire Communications. Wire may be used extensively in urban environments. It offers two primary advantages over radio communications:

- It is not susceptible to interference and absorption suffered by radio systems.
- Security of the overall system is greater because wire systems are essentially point-to-point.

Maximum use should be made of existing facilities for concealment. Sewers, subway tunnels, cable stairwells, elevator shafts, intact buildings, and pole lines offer ideal concealment for military wire systems. Precautions must be taken to lessen the possibility of vehicles tearing lines and disrupting communications. Existing telephone poles and tunnels can be used to keep wire above or below the street level.

Maximum use of existing commercial facilities should be made to supplement or replace military CIS. Telephone exchanges are linked together by underground cable systems and constitute a key resource in the urban communications system. These facilities are normally windowless buildings constructed of steel-reinforced concrete and provide ideal CPs and communications centers. The physical location of these exchanges, along with circuit diagrams and telephone directories, should be determined as soon as possible. Action should be initiated to secure these facilities early. Other facilities that contain extensive internal/external telephone systems may also be used as command centers. These include most municipal buildings, department stores, local police stations, and large business offices.

d. Messengers. Messengers provide a secure and flexible means of communication, particularly in an electronic warfare (EW) environment. However, careful selection of messenger routes is necessary to avoid pockets of enemy resistance. Disorientation in an urban environment occurs more often and more easily than in a rural environment. Marking facilities, transportation networks, and city streets on overlays often facilitates messenger service. Selected primary and alternate routes can often interface with designated logistics/resupply routes. Early restoration of street signs and recognizable landmarks are is essential.

e. Telephone Networks. Captured civil or military communications facilities can be utilized to augment organic communications capabilities. A civilian phone system can provide a reliable means of communication. If a telephone encryption device is not available, codes and authentication tables should be used.

Telephone technology differs from place to place, but there are universal concepts and practices in their operation. The most important concept to the military planner is network hierarchy.

Network hierarchy is a concept of organization by which a telephone company can rapidly and expeditiously allocate calls through a particular group of offices. As an ancillary function, it represents a control feature by which traffic can be rerouted to avoid congested areas or damaged facilities. This network hierarchy could play an important role in the planning and execution of the scheme of maneuver and fire support.

4206. Employment of Sensors

a. Sensor systems (seismic, infrared, acoustic, or magnetic) are normally employed in a rural context but also have traditional applications in village and strip area terrain. Use of sensors in urban areas is largely undocumented. In major urban areas, the capability for emplacement by air over select areas is reduced. Emplacement in, under, or around urban structures may also reduce microphone efficiency. Tactical employment will be degraded by such obstacles as concrete and steel structures, high-density population, and traffic.

b. Training in camouflage techniques for microphones and sensor terminals in an urban environment should be conducted. Operators should receive sufficient training to distinguish background-generated clutter from enemy foot, vehicular, and mechanized traffic. Use of and reliance on sensors may be seriously reduced during offensive operations in major urban areas. However, greater use of sensors will occur in subterranean defensive operations. Sensors should continue to be employed as far forward as possible and be centrally controlled.

c. Sensor employment should be coordinated with counterintelligence (CI) efforts for placement, readouts, and recovery. Sensors can also play a role in rear areas to augment rear security operations. Sensors, in themselves, are only component elements of an integrated intelligence system. Their employment is limited only by the imagination.

4207. Employment of Air Defense. Basic air defense doctrine does not change when units operate on urbanized terrain. The employment principles of mix, mass, mobility, and integration all apply to the employment of air defense assets. The commander must consider the following when developing his air defense plan:

a. Enemy air targets such as principal lines of communication and road and rail networks and bridges are often found in and around built-up areas.

b. Good firing positions may be difficult to find and occupy for long-range air defense missile systems in the built-up areas.

c. Movement between positions is normally restricted in built-up areas.

d. Long-range systems can provide air defense cover from positions on or outside of the edge of the city.

e. Radar masking and degraded communications reduce air defense warning time for all units. Air defense control measures must be adjusted to permit responsive air defense within this reduced warning environment.

f. Positioning of air defense weapons in built-up areas may be limited to more open areas without masking, such as parks, fields, and rail yards.

g. Stinger teams provide protection for infantry battalions the same as in any operation. When employed within the built-up area, rooftops normally offer the best firing positions.

h. Heavy machine guns emplaced on rooftops can also provide additional air defense.

Chapter 5

Combat Service Support and Legal Aspects of Combat

During combat in built-up areas, the terrain and the nature of operations create unique demands on the CSS system. Meeting these demands will require innovative techniques and in-depth planning.

Section I

Combat Service Support

5101. Introduction. CSS is "the essential capabilities, functions, activities, and tasks necessary to sustain all elements of operating forces in theater at all levels of war" (Joint Pub 1-02). The CSS organization throughout the MAGTF, down to battalion level, must provide a responsive CSS system that can perform all functions and tasks associated with meeting identified CSS requirements. This responsibility does not change during urban operations. Forward support for combat forces continues to be the basic concept governing CSS operations. No significant changes in doctrine or organization are required. However, the characteristics and nature of urban combat do affect how CSS is provided.

5102. Combat Service Support Resupply, Maintenance/Repair, and Replacement

a. Resupply of Ammunition. Combat in built-up areas is characterized by extremely high ammunition expenditure rates. Not only do individual Marines expend more ammunition, but they also use greater quantities of munitions such as smoke, concussion, and fragmentation grenades; AT4s; claymore mines; demolitions; and so on. The ammunition consumption rate for the first day of combat in a built-up area can be up to four times the normal rate. Even though this rate decreases during succeeding days, consumption remains high. Leaders should plan to meet these high consumption rates. The plan must include how ammunition and demolitions are to be moved forward to the companies. AAVs should be designated for the movement of ammunition if rubble or glass prevents wheeled-vehicle traffic. Marines may need to organize into carrying elements if streets are blocked by rubble.

b. Resupply of Fuel. The amount of bulk fuel needed by forces during combat in built-up areas is greatly reduced. Combat vehicles normally use less fuel in built-up areas because they travel shorter distances and perform less cross-country traveling. Engineer equipment and power generation equipment may use more fuel, but requirements are small. A unit may not use much fuel daily, but when it does need fuel, a problem exists in delivering bulk fuel to the vehicle. In open terrain, a vehicle that has run out of fuel can be recovered later. In built-up areas, the same vehicle is probably going to be lost quickly. Leaders should plan for and provide the means of moving limited amounts of bulk fuel forward to combat units.

.3

c. Maintenance and Repair. Maintenance contact teams must operate well forward to support units fighting in built-up areas. Although some maintenance operations may be consolidated in civilian facilities, many vehicles will have to be repaired near fighting positions. Battle damage assessment (BDA) and repair procedures allow mechanics to analyze, repair, and return damaged vehicles to a serviceable condition. Other considerations:

(1) Combat in built-up areas generates a high demand for tires.

(2) The dust and rough handling that are characteristic of combat in built-up areas also place great strains on communications and night vision devices.

(3) The unit armorers and their small-arms repair kits provide only limited maintenance. S-4s should plan for increased weapons maintenance demands.

d. Replacements. Units conducting combat in built-up areas must expect high casualty rates. Casualty reports must be prepared scrupulously and forwarded via the S-1 to the battalion personnel officer located at the consolidated administration center (CONAD). Other considerations:

(1) The S-4s must plan to expedite the evacuation of wounded from the built-up area. Location of battalion aid stations and evacuation routes must be planned and disseminated to the lowest level. Higher casualty rates should be expected and may require the stockpiling of medical supplies and augmentation of additional HSS personnel.

(2) The personnel officer quickly processes replacements and coordinates their movement forward via the S-1 officer.

(3) Proper accountability of personnel at all levels is required. Timely and accurate personnel accountability and strength reporting support decisionmaking and initiate the replacement cycle. Leaders maintain accountability through the use of battle rosters and by establishing procedures for periodic reporting of numbers to higher commands. During combat, strength reports are provided on request or as significant changes occur.

5103. Critical Classes of Supply. Requirements for supply and resupply will vary from those encountered during operations in other tactical environments. The critical classes of supply listed below will assume greater importance during combat inside a built-up area.

a. Class I (Rations). The process of ordering and moving rations to forward units is complicated by the dispersed nature of combat in built-up areas and its increased caloric demands on Marines. Hot meals should be provided when practical.

(1) Combat in built-up areas not only causes great stress on Marines, but also requires great physical exertion. This combination of stress and exertion quickly causes dehydration. Unless potable water is continuously provided, Marines will seek local sources, which are usually contaminated by petroleum, oils, and lubricants (POL) runoff, sewage, bacteria, or unburied corpses. Marines who are not provided sufficient quantities of potable water become casualties as a result of drinking from contaminated sources or from dehydration. Waterborne contaminants can quickly render entire units combat ineffective.

(2) Water and other liquid supplements such as coffee, tea, or soup that must be forwarded to exposed positions may need to be backpacked at night.

b. Class II (General Supplies). Combat in built-up areas increases wear and tear on combat uniforms and footgear. Supply officers should increase on-hand stocks of uniforms, boots, and individual combat equipment such as protective masks and armored vests. Nuclear, biological, and chemical (NBC) protective suits become damaged quickly when worn in the urban environment. Extra stocks of these and protective mask filters should be kept on hand. (For further information on NBC considerations, see Appendix G.) Limited amounts of other Class II and IV (barrier materials) items may be available locally. These should be gathered and used if authorized and practical. Local shops may provide such items as hand tools, nails, bolts, chains, and light construction equipment, which are useful in preparing a defense or reducing enemy-held positions. The unit's organic wire communications net may be augmented with locally obtained telephone wire and electrical wire.

c. Class III (POL). Bulk fuel may have to be brought forward from fuel tankers by using 5-gallon cans. One man can carry a fuel can long distances, even over rubble, if it is lashed to a pack frame. Supplies of bulk Class III items and some prepackaged POL may be available at local gas stations and garages. These may be contaminated or of poor quality. The S-4s should coordinate to have fuel tests performed.

d. Class IV (Barrier Materials). If a unit is defending a built-up area, the requirements for Class IV materials are less than in other areas. This class of supply is probably the most available locally. After coordinating the effort with higher headquarters, S-4s and supporting engineer officers gather materials for use in strengthening the defense. Cargo trucks, wreckers, or recovery vehicles from maintenance platoons or engineer units can be used to load and move barrier materials. The defense of a built-up area may require concertina wire and/or barbed wire to restrict the enemy infantry's movements. Barriers can be built of abandoned cars and buses, which are dragged into position, turned on their sides, and chained together through the axles.

e. Class V (Munitions). Combat in built-up areas causes ammunition to be expended at extremely high rates. Commanders should plan for early resupply of explosives, grenades, and ammunition for small arms, direct fire, and indirect fire.

(1) In the defense, the S-4 should prestage as much ammunition as practical in dispersed storage areas. These storage areas should be protected, and they should be easily accessible to the forward defensive positions. In the offense, attacking forces should not be overburdened with excessive ammunition. Mobile distribution points may be set up as low as at company level.

(2) Leaders should plan to continuously deliver ammunition to the leading units as they advance. Ammunition may be carried by armored vehicles close behind the advancing forces or by designated carrying elements. Modern ammunition, particularly missiles, is characterized by extensive amounts of packing material. S-4s must remove the ammunition depot overpack before the ammunition is transported forward. Resupply by helicopter (prepackaged slingloads) may be feasible.

(3) Removing the overpack from large amounts of ammunition can be a time-consuming process. It may require being augmented by available Marines. If carrying elements are used to move ammunition forward, a Marine can carry about 75 to 90 pounds by using a pack frame. Bulky and heavier loads can be carried by lashing them to litters and using teams of two to four men. Loads of up to 400 pounds can be carried moderate distances using four-man teams.

Note: DO NOT use medical corpsmen to carry ammunition forward as described above—it is a violation of the Geneva Accords.

f. Class VIII (Medical Supplies). Because of the decentralized nature of combat in built-up areas, medical supplies should be dispersed throughout the unit, not just consolidated with the aid station and the individual corpsman. Marines should carry additional bandages, and units should have additional splints and stretchers.

5104. Health Service Support. The regimental and battalion surgeons are responsible for planning and executing HSS within their respective units. The most critical functions during combat in built-up areas include preventive medicine, trauma treatment, and evacuation. In addition, there should be a plan for the treatment and evacuation of NBC-related casualties that could occur in combat in built-up areas.

a. Combat in built-up areas exposes Marines not only to combat wounds, but also to the diseases endemic to the AO. Commanders must enforce preventive measures against the spread of infectious diseases. The unit surgeon advises the commander on how best to implement the use of prophylactics.

b. Corpsmen should be placed at strongpoints, at battle positions, and in units likely to be fighting in somewhat isolated positions.

c. Battalion aid stations must be placed farther forward than in operations in more open terrain. Protection offered by urban structures will permit this forward location, but the limited range of observation will require that personnel involved in evacuation be aware of their surroundings and the threat.

d. Corpsmen attached to rifle platoons are trained in the treatment of traumatic injuries; however, they can quickly become overwhelmed by the number of casualties needing care. If time and training are available before the conduct of urban operations, the commander should increase first aid training. Immediate first aid support by a buddy to an injured Marine may be the difference between life and death and will greatly assist the corpsmen. The aid station should plan to care for the increased casualties that are inherent to combat in built-up areas, as the incidence of crushing injuries, eye injuries, burns, shrapnel wounds, and fractures increases.

e. The difficulties encountered when evacuating casualties from urbanized terrain are many and require innovative techniques and procedures. The planning for evacuation on urbanized terrain must include special equipment requirements, use of litter teams, use of air MEDEVAC, use of an ambulance shuttle system, communications requirements, and techniques for locating casualties.

(1) Special equipment requirements include ropes, pulleys, skid litters, axes, crowbars, and other tools used to break through barriers.

(2) Although litter teams are labor intensive, they are required for evacuation from buildings, where casualties can occur on any level. Also, rubble in the streets, barricades, and demolition of roads will require a heavy reliance on litter teams.

(3) Communications presents one of the biggest obstacles to casualty evacuation. Urbanized terrain renders LOS radios ineffective. Also, individual Marines normally do not have access to radios. Therefore, a wounded Marine within a building may be difficult to find and evacuate. The unit SOP should contain alternate forms of communication such as colored panels or other forms of markers that can be displayed to hasten rescue when the battle is over. Also, a systematic search of the area after the battle may be required to recover casualties.

f. The use of local medical facilities, hospitals, professional medical help, and medical supplies may be available during combat in large, built-up areas. The commander must adhere to the guidelines established within the theater as to when and how these facilities can be used. If civilians are wounded in the unit's area, the commander is responsible for providing them with aid and protection without disrupting military operations. A commander cannot confiscate civilian medical supplies unless he makes provisions to provide adequate replacements if civilians are wounded.

g. The commander is responsible for the evacuation of deceased personnel to the nearest mortuary affairs collection point, whether they are U.S., allied, or enemy personnel, or civilians. (See Joint Pub 4-06, *JTTP for Mortuary Affairs in Joint Operations*.) Some general considerations for the handling of deceased personnel include:

(1) The theater commander is the approval authority for hasty burial.

(2) The deceased person's personal effects must remain with the body to assist in the identification of the body and to facilitate shipment of personal effects to the next of kin.

.3

Retention of personal items is considered looting and is, therefore, punishable by the Uniform Code of Military Justice (UCMJ).

(3) When operating under NBC conditions, the bodies of deceased personnel should be decontaminated before removal from contaminated areas to prevent further contamination and casualties.

(4) Care must be exercised when handling deceased personnel. Improper handling of deceased personnel can result in a significant decrease in unit and civilian morale.

5105. Personnel Services. Timely and accurate personnel services are just as important during combat in built-up areas as in any other operation.

a. The S-1 plans for all personnel services that support and sustain the morale and fighting spirit of the battalion. Among the most important of these services are:

- Replacement operations
- Strength accounting
- Casualty reporting.

These functions allow the commander to maintain accountability of his forces and to redistribute personnel or change taskings for subordinate units to maintain combat power and tempo.

Chapter 6

Noncombatant Considerations in Urban Operations

Noncombatants can have a significant impact on the conduct of military operations.

Section I

6101. Introduction. Commanders must be well educated in the legal aspects of military operations involving noncombatants. The urban environment contains numerous legal pitfalls in dealing with the civilian population. Adherence to the ROE and the standards of the Geneva Accords will keep the commander on firm legal ground.

6102. Noncombatant Impact. Large concentrations of civilians can greatly impede tactical operations. Noncombatants may have the following effects on military operations:

a. Mobility. Noncombatants civilians, attempting to escape the battlespace, can block military movement. Commanders plan routes to be used by civilians and seek the assistance of the civil police in refugee control.

b. Firepower. The presence of noncombatants can restrict the use of firepower. Areas may be designated no-fire areas to prevent noncombatant casualties. Other areas may be limited to small-arms fire and grenades. The control of fire missions may be complicated by the requirement for positive target identification. Detailed guidance on the use of fires in the presence of noncombatants is promulgated by the MAGTF commander. In the absence of guidance, the general rules of the law of land warfare apply.

c. Security. The presence of noncombatants increases security requirements in an urban environment to preclude:

- Noncombatants entering defensive areas
- Pilferage of equipment
- Sabotage
- Terrorism

d. Obstacle Employment. The presence and movement of Noncombatants will influence the MAGTFs commander's obstacle plan. Minefields may not be allowed on designated refugee routes or, if allowed, must be guarded until the passage of refugees is completed. The use of boobytraps may be curtailed until noncombatants have been evacuated. Commanders need to review and understand current national and international treaties concerning the employment of mines and boobytraps.

6103. Command Authority. The limits of command authority over government officials and noncombatants must be established. The host government's responsibility for its populace and territory can affect the commander's authority in civil-military matters. In less secure areas, where the host government may be only partly effective, the commander may have to assume greater responsibility for noncombatants.

6104. Source Utilization. Operations in highly populated areas may require the diversion of men, time, equipment, and supplies to accomplish humanitarian tasks. If host government agencies collapse, the impact on military resources could be substantial.

6105. Health and Welfare. The disruption of noncombatant health and sanitary services may sharply increases the risk of disease among both noncombatant and military personnel.

6106. Law and Order. Marines may have to augment civilian forces to protect life and property and to restore order. Marines may also have to secure vital government facilities for the host nation. (For more information on how to control civilians violating civil law, see FM 9-15, *Civil Disturbances*.)

6107. Public Affairs Officer and Media Relations. Free media access to units in the field is desirable; however, operational security, existing guidelines, and ROE considerations have priority. All members of the media should have an escort officer. Escort officers may be detailed from line units when public affairs personnel are unavailable. Enforcing established guidelines helps prevent negative publicity which could jeopardize the operation or national and strategic objectives.

6108. Civil Affairs Activities and Psychological Operations (PSYOP).

a. General. CA and PSYOP are mutually supporting and may have essential roles in MOUT. They can be critical force multipliers. CA and PSYOP may offer the possibility of victory without the destruction, suffering, and horror of protracted urban conflict.

(1) CA activities are activities performed by commanders, staffs, department of defense elements and units, and foreign military forces that (1) embrace the relationship between military forces and civil authorities and populations in areas where military forces are present; and (2) involve application of civil affairs functional specialty skills, in areas normally the responsibility of civilian government, which enhance conduct of civil-military operations (Joint Pub 1-02). CA activities are inherently civil-military in nature. They assist in coordinating military efforts to strengthen host government legitimacy, and help prevent or reduce violence by bridging critical gaps between civil and military sectors. MAGTF responsibilities for CA activities are determined during planning. U.S. Army CA units can be requested to support the MAGTF. The Marine Corps Civil Affairs Group (CAG), located in the Marine Corps Forces Reserve, may be activated and assigned to the MAGTF.

(a) CA activities are characterized by applications of functional specialties in areas normally the responsibility of indigenous government or civil authority. CA activities may extend to assumption of government functions required in an occupied territory during or immediately subsequent to hostilities. Refer to Joint Pub 3-57, *Doctrine for Joint Civil Affairs*.

(b) CA support to a MAGTF would normally include procuring local goods and services for military use, reducing civil interference in military operations, and assisting a MAGTF commander in fulfilling moral and legal obligations to the local government and noncombatant populace. See MCWP 3-33.1, *MAGTF Civil Affairs*, for more detailed information.

(2) Psychological operations are planned operations to convey selected information and indicators to foreign audiences to influence their emotions, motives, objective reasoning, and ultimately, their behavior. The purpose of psychological operations is to induce or reinforce foreign attitudes and behavior favorable to the originator's objectives (Joint Pub 1-02).

PSYOPs are planned and conducted by the MAGTF. PSYOPs are integrated and coordinated down to the lowest tactical level. PSYOPs can weaken enemy resistance and reduce its combat effectiveness. PSYOPs can also help prevent civilian interference with military operations.

PSYOPs may include the use of television, radio, posters, leaflets, and loud speakers, to disseminate propaganda and information. Television, including video tapes, is one of the most effective mediums for persuasion. In areas where television is not common, televisions may be distributed to public facilities and selected individuals. See MCWP 3-36.2/FM 33-1, *Psychological Operations*, for more detail.

6109. Refugee Control. The control of refugee movements in combat operations is primarily a host nation function.

The most important principle in refugee control is reliance on host government forces (military/civilian) whenever possible. U.S. forces will have limited refugee control capabilities. If a host government is incapable of performing the function, MAGTFs and MPs may be required to conduct any or all of the following refugee control measures:

- Enforcc standfast orders or prohibit other unauthorized movements

- Screen and take custody of enemy agents or other hostile personnel discovered among the refugees

- Screen refugees for suspicious activities (including detection of infiltration by guerrillas, enemy agents, and escaping members of hostile forces)

.3

- Perform traffic control on designated refugee routes
- Escort refugee columns
- Direct or divert refugee columns to secondary roads and through areas not used by military elements
- Perform police duties such as traffic control or security at refugee assembly areas, collecting points, and centers or camps
- Enforce prescribed restrictive measures such as curfews, identification and pass systems, restricted area restraining lines, no-passage lines, checkpoints and roadblocks, travel permits or manifests, prohibition on the use of vehicles, priorities of movement, and limits on personal belongings.

Section II

6201. Commander's Legal Authority and Responsibilities

a. General. Leaders at all levels of command are responsible for protecting noncombatants and minimizing collateral damage to their property. Looting, vandalism, and brutal treatment of civilians are strictly prohibited by the Geneva Accords. Individuals who commit such acts may be severely punished. Civilians, and their religions and customs, must be treated with respect. Women are especially susceptible to many forms of abuse, and special care must be taken to protect them. Urban warfare offers many opportunities for abuses of the civilian population. This section focuses on those civilian-military confrontations that are most common in the urban environment.

b. Control Measures. Commanders may enforce control measures to conduct operations, maintain security, or ensure the safety and well-being of the civilians.

(1) Curfew. A commander with the mission of defending a town could establish a curfew to maintain security or to aid in control of military traffic. However, a curfew would not be legal if imposed strictly as punishment.

(2) Evacuation. A commander can require civilians to evacuate towns or buildings if the purpose of the evacuation is to use the town or building for essential military purposes, to enhance security, or to safeguard those civilians. The commander must specify and provide security for those being evacuated. Food, clothing, and sanitary facilities should be provided at the destination.

(3) Forced Labor. The Geneva Accords prohibit the use of civilians in combat operations. However, they may be required to perform some forms of forced labor prior to and after the conduct of combat operations. Guidelines for use of civilian labor should be published. The commander may force civilians over 18 years of age to work if the work does not oblige them to take part in military operations. An example of forced labor that is permitted would be maintenance of public utilities as long as those utilities are not used in the general conduct of the war. Jobs can also include services to the local population such as care of the wounded and burial. Civilians can also be forced to help evacuate and care for military wounded, as long as doing so does not involve any physical danger. Prohibited forced labor include digging entrenchments, constructing fortifications, transporting supplies or ammunition, or acting as guards. Only civilians who are volunteers can be employed in such work.

c. Civilian Resistance Groups. Another situation that commanders might encounter is combat with a civilian resistance group. The combatant status of these groups varies with their organization, methods, and adherance to the Geneva Accords. For example:

(1) Civilians accompanying their armed forces (with an identity card authorizing them to do so) are treated as EPWs when captured. Examples would include civilian members of military aircraft crews, war correspondents, supply contractors, and members of labor units or of service organizations responsible for the welfare of the armed forces.

(2) Captured civilians of a nonoccupying territory who have taken up arms against an invading enemy without time to form regular armed forces are considered EPWs if they wear a distinctive insignia that can be seen at a distance, carry their weapons openly, and operate according to the rules and customs of warfare. Other civilians who provide assistance to such groups may not be entitled to status as combatants, depending on whether they are actually members of the resistance group. They are normally best treated as combatants until a higher authority determines their status.

(3) Armed civilian groups that do not meet the criteria of a legal resistance group or individuals caught in the act of sabotage, terrorism, or espionage are not considered legal combatants. If captured, they will be considered criminals under the provisions of the law of land warfare. They should be detained in a facility separate from EPWs and should be quickly transferred to the MPs. Reprisals, mass punishments, hostage taking, corporal punishment, pillaging, or indiscriminate destruction of property are prohibited.

(4) The law of land warfare lets the commander control the civil population under the conditions already described using his own organic resources. However, language and cultural differences between U.S. and foreign personnel make it good practice to use local authorities such as the police for such purposes. Use of the police does not relieve a commander of his responsibility to safeguard civilians in his area.

d. Protection of Property. Like civilian personnel, civilian buildings and towns normally have a protected status. As such, they are not legal targets. They lose their protected status if authorities determine that the enemy is using them for military purposes. All doubt as to whether a town or building is a military target must be resolved before it can be targeted. For example:

(1) If the enemy is using a building or a portion of the town for military purposes—such as a supply point or a strongpoint—that building or that portion of the town is a legal target. However, engagement of the target should be limited to the buildings or portion of the town being occupied by the enemy.

(2) Normally, religious, historical, and cultural objects and buildings are not legal targets. They are sometimes marked with symbols to signify their status. Medical facilities are identified under the internationally recognized Red Cross, Red Crescent, Red Lion, or Red Star of David symbols. The fact that such symbols are absent does not relieve a commander of his responsibility to protect objects he recognizes as having religious, cultural, medical, or historical value.

(3) The misuse of such objects by the enemy is grounds to disregard their protected status. A demand should be made for the enemy to stop his misuse of the protected object before any action is taken. However, if an enemy FO or sniper uses a church for an OP, a commander would be justified in destroying it immediately because of the immediate danger posed to his troops. Another example is a religious shrine that is used as a telephone switchboard; in this case a warning would be appropriate because it would take some time to dismantle the system. Attacks on these types of targets should be limited to the least amount necessary to neutralize the enemy installations.

(4) The destruction, demolition, or military use of other buildings is permitted under the law of land warfare if required by clear military necessity. Destroying a house to obtain a better field of fire would be a legal act. But destroying it as a reprisal would not. However, firing on any houses that are occupied or defended by an enemy force is usually legal.

Chapter 7

The Urban Environment and Restrictions to Operations

"The responsibility of officers . . . and the training necessary are of a very different order from their responsibilities and training in ordinary military duties. Instead of striving to generate the maximum power with the forces available, the goal is to gain decisive results with the least application of force and the consequent minimum loss of life."

USMC, 1940 (FMFRP 12-15, *Small Wars Manual*)

Section I

Military Operations Other Than War

7101. Introduction. One of the most likely missions that U.S. Marines will undertake abroad will be military operations other than war (MOOTW). These missions typically will take place in the Third World and are normally associated with urban areas. MOOTW focuses on deterring war, resolving conflict, promoting peace, and supporting civil authorities in response to domestic crises. It is important to remember that political considerations permeate at all levels. Although some types of MOOTW, such as peace enforcement, may have the same characteristics of war, they normally have more restrictive rules of engagement (ROE).

Urban environments are inherently complex, and sometimes unpredictable. Local governments may have limited functions. Services may be overburdened, leading to allocation on the basis of personal connections and corruption. Services may be provided by a mixture of formal, high-technology and informal, labor-intensive systems. Large portions of the population may be poor, young, and underemployed. Nevertheless, there may be a high degree of social organization and interaction. Word of mouth and informal organizations may be as important as mass media and formal organizations to the communication structure of the area. Indigenous populations may have experienced centuries of foreign interventions and become quite adept at innovative ways of communicating. These factors and others often combine to make MOOTW a challenging and frustrating endeavor.

7102. Types of MOOTW. MOOTW encompasses a wide variety of military operations, that may take place in an urban environment. Forward deployed MAGTFs are ideal forces to respond to contingency operations involving MOOTW. These operations include:

- Combatting Terrorism
- DOD support to counterdrug operations
- Enforcement of Sanctions/MIO
- Enforcing Exclusion Zones

- Humanitarian Assistance
- Military Support to Civil Authorities
- Nation Assistance/Support to Counterinsurgency
- NEO
- Peace Operations
- Recovery Operations
- Show of Force Operations
- Strikes and Raids.

7103. Recent Experience and Lessons Learned. Recent Marine Corps urban experiences—Lebanon (1982 - 1984); Grenada (1983); Panama (1989); and Somalia (1992 - 1994)— are examples of MOOTW. Although each operation was unique, the lessons learned are similar. For detailed information, commanders may obtain after action reports through the Marine Corps Lessons Learned System (MCLLS). Some of the lessons learned include:

(1) Demonstrate a powerful military presence. Where and when possible, use tanks, armored vehicles, heavy automatic weapons, fixed-wing aircraft, and attack helicopters.

(2) PSYOP, CI, HUMINT teams, and linguists, are essential to communications and intelligence gathering.

(3) Topographic detachment is useful to provide necessary sketches and maps.

(4) UAVs can be used for security, reconnaissance, and surveillance.

(5) Whenever entering an AO, isolate the area, patrol, and use intelligence to evaluate specific buildings or groups of buildings for search-and-clear operations. PSYOP, CI, HUMINT teams, and linguists can be used, as appropriate, to support patrols.

(6) Take advantage of the night by using night vision devices.

(7) The most effective means of dealing with snipers is with counter snipering operations.

(8) Snipers are also a tremendous asset to overwatch patrol movements, provide intelligence, and direct fire support. (See MCWP 3-15.3.)

(9) Random patrolling, traffic control points, random checkpoints/roadblocks, and ambushes disrupt enemy movements, reduce looting, seize curfew violators, and convince the population that order has been restored.

(10) Use armored vehicles to move quickly to new checkpoints and set up roadblocks or ambushes. They also increase protection for infantry. Their night vision sights also provide increased security and recognition during periods of limited visibility.

(11) Extensive patrolling provides current combat information used to maintain situational awareness. It helps to familiarize units with the AO and presents a strong military presence. (See Appendix A, Section VII)

(12) Tanks, armored vehicles, or attack helicopters increase military presence and can provide immediate, accurate gunfire.

(13) Before search and clear of a building, use loudspeakers to explain your intentions to noncombatants and enlist their cooperation.

(14) Use a minimum amount of force when dealing with noncombatants and their property. This will assist you in obtaining them to cooperate and may prevent turning them into active combatants. Assign engineers to search-and-clear forces, as necessary.

(15) MPs can instruct small unit leaders on the proper method of conducting vehicle, individual, and room search techniques. In some cases, it may be appropriate to attach MPs to search-and-clear forces. (See MCWP 3-34.1, *Military Police in Combat.*)

(16) Maintain a strong, mobile reserve.

(17) Avoid becoming predictable.

7104. Urban Terrorism. Terrorism is "the calculated use of violence or threat of violence to inculcate fear; intended to coerce or to intimidate governments or societies in the pursuit of goals that are generally political, religious, or ideological" (Joint Pub 1-02). Terrorism is a criminal act that is symbolic in nature. Its intent is to influence an audience other than the victim.

The likelihood of Marine involvement with terrorism increases in the urban environment. Terrorism in an urban environment can occur across the spectrum of conflict. Terrorists are an integral element in an insurgency and also play a major role in unconventional warfare. Terrorists can disrupt economic functions, expose a government's incompetence, eliminate opposition leaders, and elevate social anxiety. (See MCRP 3-02D, *Combating Terrorism*; and MCRP 3-02E, *The Individual's Guide for Understanding and Surviving Terrorism.*)

In countering this threat, Marines should determine whether it is internally or externally directed terrorism. Terrorism rooted externally must be severed from its roots. Against internal terrorism,

.3

Marines should attempt to penetrate the infrastructure and destroy the leadership of the terrorist groups.

The keys to defeating terrorists are awareness, education, and intelligence in order to deny, deter, delay, and detect terrorist acts. Rapid coordination among military units, military services, local police, and other government agencies is essential in denying the terrorists targets and refuge.

a. Nature of Urban Terrorism. Terrorism is generally an urban phenomenon. Incidents of terrorism are increasing in the international community, and in the United States, and pose a significant threat for U.S. military forces. Military personnel in noncombatant roles are considered terroist targets. For example, the attack on the Marine compound in Lebanon can be considered a terrorist attack instead of a military action because the Marines were officially in a noncombat activity (a multinational peacekeeping activity).

b. Terrorists in the Urban Area. Cities are target rich environments for terrorists. They offer ample cover and concealment. They contain large amounts of resources, transportation, communications, and potential recruits or sympathizers to their cause. Cities also provide the media attention essential for terrorists to accomplish their mission.

c. Combating Terrorism

(1) Counterterrorism. Counterterrorism includes the full range of offensive measures to prevent, deter, and respond to terrorism (Joint Pub 1-02). Population protection and control to combat the terrorist threat may be a major effort within the urban environment. The ability to protect the general population from attack and intimidation is essential.

(2) Antiterrorism. Antiterrorism includes defensive measures used to reduce the vulnerability of individuals and property to terrorist acts, to include limited response and containment by local military forces (Joint Pub 1-02). The extent of these measures varies based on assessments of the local threat. These include personnel awareness and knowledge of personal protection techniques. They also include crime prevention and physical security programs to "harden" the target, making installations and personnel less appealing as terrorist targets.

Section II

Restrictive Conditions

7201. Introduction. Commanders can expect to encounter restrictions on the use of force in MOOTW, especially in the urban environment. While we tend to focus on a traditional image of urban combat, MOUT extends beyond the house-to-house, close-combat effort to seize a city. Recent history provides many examples of the need for precise application of force to minimize collateral damage. MOOTW generally requires the application of force in varying degrees. Future missions in urban areas will likely follow this same pattern. While Marine forces must be prepared for more intensive urban combat, they must also train to apply only the level of force necessary to accomplish the mission. *Our tactics may have to rely more on innovation than firepower.*

Basic tenets of manuever warfare remain the same in MOOTW, but the tactics, techniques, and procedures may have to be modified to stay within established ROE and to avoid unnecessary collateral damage.

7202. Rules of Engagement. ROE are "directives issued by competent military authority which delineate the circumstances and limitations under which United States forces will initiate and/or continue combat engagement with other forces encountered" (Joint Pub 1-02). ROE for Marine forces are usually set by the JTF commander. They are based on National Command Authorities (NCA) guidance, the mission, the threat, the laws of war, and host nation constraints on force deployment.

MOOTW focuses on deterring war, resolving conflict, promoting peace, and supporting civil authorities in response to domestic crises. It is important to remember that political considerations permeate at all levels. Although some types of MOOTW, such as peace enforcement, may have the same characteristics of war, they normally have more restrictive rules of engagement (ROE). Restraint requires the careful balancing of the need for security, the conduct of operations, and the political objective. Excessive force antagonizes those parties involved, thereby damaging the legitimacy of the organization that uses it while possibly enhancing the legitimacy of the opposing party. Ensure all personnel understand the ROE and incorporate the practice into training exercises at all levels.

ROE must be fully integrated in the MAGTFs plans. As an example, table 7-1, presents the ROE used in Operation United Shield:

ROE Used for Operation United Shield
Nothing in these Rules of Engagement limits your right to take appropriate action to defend yourself and your unit.
a. You have the right to use deadly force in response to a hostile act or when there is a clear indication of hostile intent.
b. Hostile fire may be returned effectively and promptly to stop a hostile act.
c. When US forces are attacked by unarmed hostile elements, mobs and/or rioters, US forces should use the minimum force necessary under the circumstances and proportional to the threat.
d. Inside designated security zones, once a hostile act or hostile intent is demonstrated, you have the right to use minimum force to prevent armed individuals/crew-served weapons from endangering US/UNOSOM II forces. This includes deadly force.
e. Detention of civilians is authorized for security reasons or in self-defense.
Remember: 1. The United States is not at war. 2. Treat all persons with dignity and respect. 3. Use minimum force to carry out mission. 4. Always be prepared to act in self-defense.

Table 7-1. ROE Used in Operation United Shield

7203. Planning. Commanders and planners thoroughly study each situation to identify the restrictive measures necessary to prevent needless civilian casualties and damage while also protecting the lives of the Marines who will execute the mission.

a. Mission/Command Guidance. The mission statement itself may determine much about the required level of force for a given operation. A humanitarian relief operation may require only individual weapons for security and personal protection. A raid on an insurgent communication center may need more. The requirements for an NEO conducted in a permissive environment would be different from those in a nonpermissive one. The key is to conduct a thorough analysis of the mission and determine the necessary combat power and how it must be applied to accomplish the mission. Many of these missions will be accompanied by a great deal of additional guidance concerning operational restrictions. These restrictions may come from the NCA, State Department, Department of Defense (DoD), or the joint task force (JTF) commander, and other sources. Such guidance must be carefully studied and incorporated into the plan. The commander can impose restrictions on his own forces that are more stringent then those dictated by higher authority. A careful analysis of the ROE will provide key insights into other restraints that the commander may wish to impose on his own forces.

b. Intelligence/Threat. Obtaining detailed and specific intelligence on the target and the area is essential to minimize collateral damage. Information on the physical layout of an area is often as important as information about the enemy. This is especially true for an NEO where avoiding conflict is a primary concern.

c. Location. The geographic area into which operations are to be conducted will normally dictate some restrictions on operations. In an urban environment, where the city is located in a desert, a common restriction is placed on the destruction of water producing facilities. If the operation is conducted where there is a large Hindu population, special protection may be given to cattle. If the operation is on an island nation, restrictions may be placed on destroying port or airfield facilities. In an urban environment with a large amount of shanty towns, restrictions may be placed on the use of incendiary devices.

d. Restrictions on Force Employment. The *right of self-defense is inherent in all military operations*. Marines should not be placed in a situation where they are not provided the means and authority to protect themselves or others. Beyond that, the restrictions placed on the forces conducting a particular operation depend on a variety of factors. These include:

- NCA guidance
- Type of mission
- Nature of the threat
- Political environment
- Host nation attitude.

e. Lethal Fires. Most restrictions in MOOTW are normally concerned with the application of fires. For instance, a NEO in a permissive environment may not require the landing of fire support assets However, a raid may require supporting arms to be landed with detailed instructions on their use. Aviation in MOOTW is commonly restricted in the amount and type of ordnance it can deliver because of the potential for collateral damage. However, in both a permissive and nonpermissive environment, supporting arms still should be planned. The most common types of fire support assets that will be restricted in the urban environment include artillery mortars, tanks, armored vehicles, and fixed-wing attack aircraft.

f. Search and Clear Buildings. Marine forces organize to search and clear a building under restrictive conditions in the same ways as discussed in Appendix C. However, because noncombatants may be occupying rooms within the building, clearing procedures must be modified as follows:

(1) Squad organization remains the same and is reinforced as appropriate (see Appendix A), except assault elements become search elements. Search elements enter and search rooms in the same manner as assault elements enter and clear rooms, as discussed in Appendix A.

(2) ROE must be identified and known by all personnel before entering or clearing a room or building.

(3) PSYOP or CA teams can help remove noncombatants before a search or battle starts.

(4) The building or area of search should be isolated with security forces and the intent to search the building should be communicated via loudspeakers.

(5) Assault forces move to the building and begin systematically to search room by room and floor by floor. Marines should carry ammunition and other equipment as in normal urban operations. However, Marines will be restricted in the use of their weapons and other equipment as dictated by the ROE. If permitted, a riot control agent (such as CS gas) can be used to assist in the clearing of a building. Additionally, shotguns and stun grenades are more appropriate than rifles and machine guns or fragmentation grenades for this type of restricted, close-combat operation.

g. Communications and Information Systems (CIS).

(1) CIS must be able to satisfy the C2 requirements of MOOTW. CIS systems must provide the commander with the ability to rapidly collect, process, analyze, and exchange information in order to keep the potential for violence to a minimum. These systems must make available the information needed, when it is needed, whenever it is needed. When supporting arms are on standby, fire support coordination and air traffic control nets are particularly critical.

(2) CIS officers (G/S-6) must carefully analyze the mission, situation, threat, and specific characteristics of the urban environment to design reliable, flexible, and responsive CIS. These systems should include single and multichannel radios, tactical telephones, and tactical satellite communications. A detailed backup CIS plan should also be prepared. The success of CIS in MOOTW depends on designing, planning, and employing CIS that satisfies the commander's C2 requirements.

Appendix A

Organization for Urban Combat and Fundamental Combat Skills

Successful combat operations in built-up areas depend on training in fundamental skills and proper employment techniques and procedures.

Before conducting urban operations, Marines must be trained in fundamental combat skills to conduct urban operations. Training for this unique environment will reduce casualties because properly prepared Marines will be able to fight more effectively. This appendix supersedes all other fundamental combat skills taught for fighting in urbanized terrain. Urban combat will require Marines to use their best initiative and judgment in applying the techniques and procedures described in this appendix.

Section I

Organization

1. Structure. Basic infantry organization does not change when the battlefield is moved into an urban area. However, infantry units may be task organized or reinforced down to squad level in order for small units to fight and win in this violent, three-dimensional battlespace. Several new terms that address organization and employment have been added to clearly define mission assignments. Figures A-1 through A-3 show platoon through fire team organization breakdown.

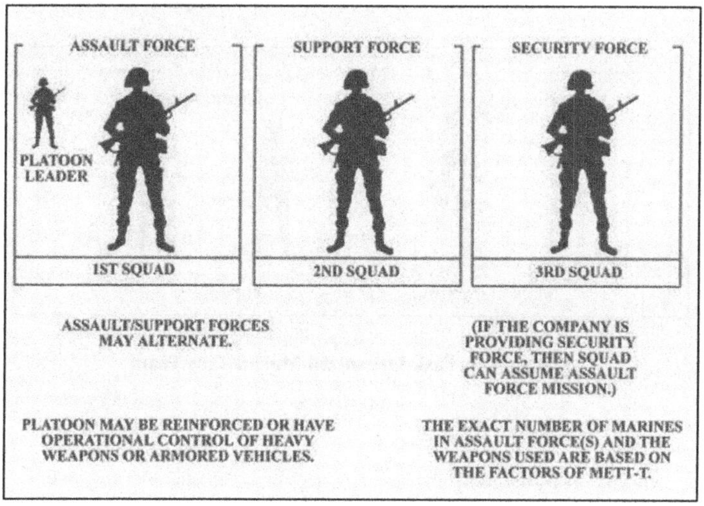

Figure A-1. Task-Organized Marine Platoon

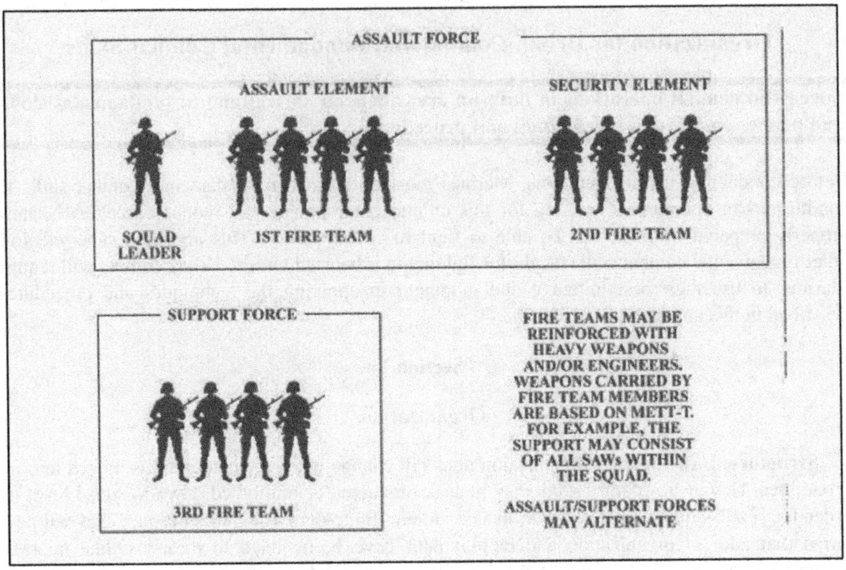

Figure A-2. Task-Organized Marine Squad

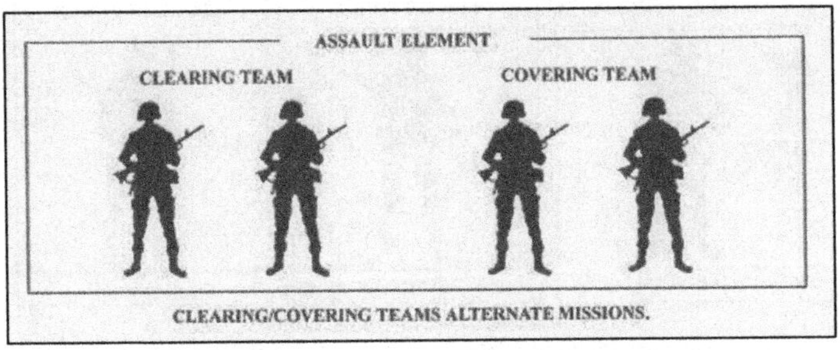

Figure A-3. Task-Organized Marine Fire Team

Section II

Weapons Handling and Firing Techniques

2. Weapons Carries. During MOUT, a Marine often finds himself very unexpectedly engaging targets at close ranges. Weapons carries provide a safe and effective method for handling the service rifle. For additional in-depth discussion of weapons handling and firing techniques, refer to MCWP 3-01X, *Field Firing of the M16A2 Rifle.*

a. Tactical Carry. The tactical carry is used when no immediate threat is present. It permits control of the rifle while moving and still allows quick engagement of the enemy. The buttstock of the rifle is placed alongside the body at approximately hip level, and the barrel is angled upward approximately 45 degrees in the general direction of the enemy (Figure A-4).

Figure A-4. Tactical Carry

b. Alert Carry. The alert carry is used when enemy contact is likely. Engagement of the enemy is faster from the alert carry than from the tactical carry. The buttstock of the rifle is placed in the shoulder with the muzzle angled down approximately 45 degrees and pointed in the likely direction of the enemy (Figure A-5).

In the alert carry, the Marine keeps both eyes open and scans for any threat with the weapon always in line with the Marine's LOS (this is called "guns 'n' eyeballs"). When a threat is spotted, the Marine quickly snaps his weapon into a firing position, looks over the rear sight aperture, and aligns the tip of the front sight post center mass on the target. Once the threat has been eliminated, the Marine returns to the alert carry and continues to scan for targets. Accuracy is the primary goal.

Figure A-5. Alert Carry

c. Ready Carry. The ready carry is employed when contact with the enemy is imminent. It allows for immediate target engagement. The buttstock of the rifle is in the shoulder with the muzzle of the rifle pointed in the direction of the enemy (Figure A-6). As in the alert carry, sight alignment and sight picture are achieved as the shot is fired. Marines should always strive for a clear tip of the front sight post and center mass hold.

Figure A-6. Ready Carry

d. Short Stocking. Rifles may be "short stocked" to increase their maneuverability in an enclosed area and to reduce possible "target indicators" for the enemy (muzzles sticking around corners, for example). The weapon is held in such a manner as to reduce the length of the weapon without sacrificing too much accuracy (Figure A-7). The stock is positioned so that the pistol grip is behind the Marine's head. The Marine may use his index finger or thumb to manipulate the trigger. The handguard of the weapon is placed against the Marine's cheek, and a firm stockweld is attained. The Marine should use the tip of the front sight post. Short stocking may be conducted from either side. However, care should be taken not to mask the ejection port cover. The short stocking technique may be used in the tactical, alert, or ready carry.

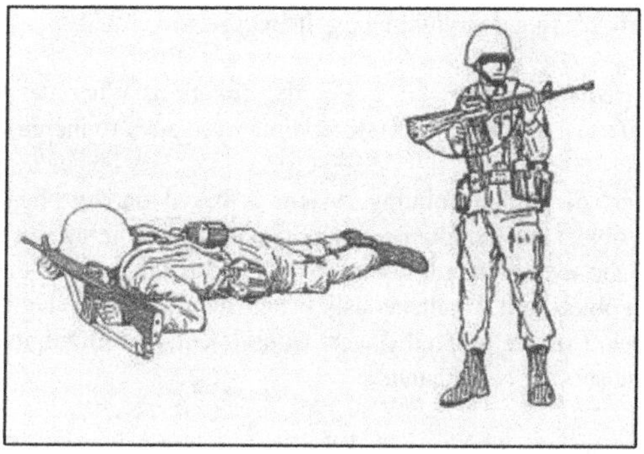

Figure A-7. Short Stocking

e. Weak-Hand Carries. Marines may carry their weapon on the weak-hand side to obtain maximum cover and fire from the weak-hand position to avoid exposing themselves. However, firing from the weak-hand position may reduce a Marine's ability to fire accurately. This may be acceptable when Marines are providing supporting fires, but in close kill-or-be-killed engagements, rapid, accurate firing from the strong-hand position is preferred.

3. Firing Techniques. The speed and uncertainty of combat require Marines to act without hesitation while achieving accurate target engagement. Several firing techniques are listed below. For an in-depth discussion of firing techniques that can be applied to MOUT operations, see MCWP 3-01X.

a. Pieing. Pieing is an effective technique for clearing dead space inside rooms and buildings to gain security of hallways, stairwells, mouseholes, and so on. It is conducted by using the ready carry position or short stocking technique. The weapon is aimed at a sector of a window, doorway, corner, or hallway and slowly moved at different angles, sectoring off the window, doorway, corner, or hallway until each sector is cleared of any threat. As soon as a hostile threat is seen, immediate, accurate fire can be placed on the threat. Figures A-11, A-34, A-35, A-38, A-47, and A-49 show examples of the pieing technique.

b. Aimed Quick Fire. The Marine's initial focus is on the target. As the rifle is brought up, the firing eye looks through or just over the rear sight aperture, and the front sight post is used to aim at the target. The Marine fires two quick shots. Focus remains on the front sight post throughout the aiming process.

Immediately after target engagement, the Marine scans for additional targets. To scan, perform the following steps:

(1) Lower the weapon to look over the sights.

(2) Place the trigger finger straight along the receiver.

(3) Scan the area for targets and assess the situation. Wherever the head moves, the muzzle moves (eyes, muzzle, target). Keep both eyes open to increase the field of view.

c. Pointing Quick Fire. The pointing system is based on the phenomenon that when a person looks at an object and simultaneously points a finger at it, the finger aligns itself on the point of focus of the eyes with no conscious effort on the part of the individual. When a Marine looks at an object and simultaneously brings his rifle to his shoulder, the rifle in effect becomes an extension of the pointed finger. Consequently, it aligns itself naturally with the object on which the shooter is focusing.

When a target appears, the Marine will keep both eyes open, concentrating intensely on a small, specific focal point near the base of the target mass. The rifle is brought simultaneously to the hollow of the shoulder. The head is held high, stock welded to the jaw. The eyes are 2 - 3 inches over the top of the sights, staring intently at the target. As soon as the rifle is brought to the shoulder, two quick shots are fired. Focus on the target is not broken during the interval between initially seeing the target and discharging the rifle.

d. Instinctive Shooting. There may be situations in which a Marine is surprised and may need to react immediately. If possible, the Marine should engage the threat by using the tip of the front sight post. However, *speed* may be more important. The Marine's weapon and body are quickly "pointed," and the target is engaged. It is important that the body be turned with the weapon in order to achieve a natural point of aim. Simply pointing the weapon will usually result in a miss. Once the first two shots have been fired and the Marine regains the initiative, the weapon should be quickly moved to the Marine's shoulder and the tip of the front post used for sighting subsequent shots.

Section III

Movement

Movement techniques within built-up areas are the first fundamental skills that Marines should master. These movement techniques should be practiced until they become habitual. To reduce exposure to enemy fire, Marines should avoid silhouetting themselves, avoid open areas, and select their next covered position before moving.

4. Crossing a Wall. Each Marine should learn the correct method of crossing a wall (Figure A-8). After reconnoitering the other side, the Marine quickly rolls over the wall, keeping a low silhouette. The speed of his movement and a low silhouette deny the enemy a good target.

Figure A-8. Crossing a Wall

5. Observation Around Corners. The area around a corner should be observed before the Marine moves beyond it. The most common mistake the Marine makes at a corner is allowing his weapon to extend beyond the corner before observing, thereby exposing his position. Using the short stocking technique (Paragraph 2.d.) reduces exposure. Additionally, Marines can use a technique called "popping the corner" to reduce exposure time (Figure A-9). The Marine will get into a prone position near the corner of a building or obstacle around which he needs to observe. The weapon is short stocked, and the muzzle is pointed in the direction the Marine is looking. This allows the Marine to engage a target, if necessary, when he observes around a corner. The Marine will crawl to the corner but not expose himself. He will raise his upper body onto his elbows. The Marine will then push his body forward with his feet and legs without moving his elbows. His upper body, with the weapon ready, will move forward. The final position will expose the weapon, the Marine's helmet, and a minimal amount of the Marine's face. The Marine's forearms will come to rest on the deck giving him a low profile, the ability to observe around the corner, and the immediate capability to engage targets with his weapon.

Figure A-9. Correct Technique for Observing Around a Corner

6. Movement Past Windows. Windows present another hazard to Marines. The most common mistake made when passing a window is exposing the head to enemy observation from inside the room. If a Marine shows his head, an enemy gunner inside the building could engage him through the window without exposing himself to return fires.

a. The correct technique for passing a window is for the Marine to stay below the window level. He makes sure he does not silhouette himself in the window; he "hugs" the side of the building. An enemy gunner inside the building would have to expose himself to covering fires to engage the Marine (Figure A-10A).

Figure A-10A. Moving Past Windows

b. The most common mistake Marines make when passing a basement window is not being aware of it. A Marine should not walk or run past a basement window because he presents a good target to an enemy gunner inside the building. When using the correct procedure for negotiating a basement window, the Marine stays close to the wall of the building and steps or jumps past the window without exposing his legs (Figure A10-B).

c. An alternate method is for the pointman to pie off the window (or other opening) as he approaches it. As the pointman moves toward the window, the second Marine (behind the pointman) steps out to provide security to the direct front. The pointman continues to cover the opening by pieing, all the while using the near edge for cover. The pointman covers and sectors off that part of the interior that can be viewed from the outside. The second Marine moves with the pointman and maintains security to the front, eventually moving past the pointman as the pointman clears the inside corners of the opening. Once the opening is cleared, the second man may assume the duties of point, or when subsequent Marines move up to secure the opening, the original pointman may reassume those duties. Figure A-11 illustrates this alternate method.

Figure A-10B. Marine Passing Basement Windows

7. Use of Doorways. Doorways should not be used as entrances or exits because they are normally covered by enemy fire. If a Marine must use a doorway as an exit, he should move quickly through it to his next position, staying as low as possible to avoid silhouetting himself (Figure A-12). Preselection of positions, speed, a low silhouette, and the use of covering fires should be emphasized when exiting doorways.

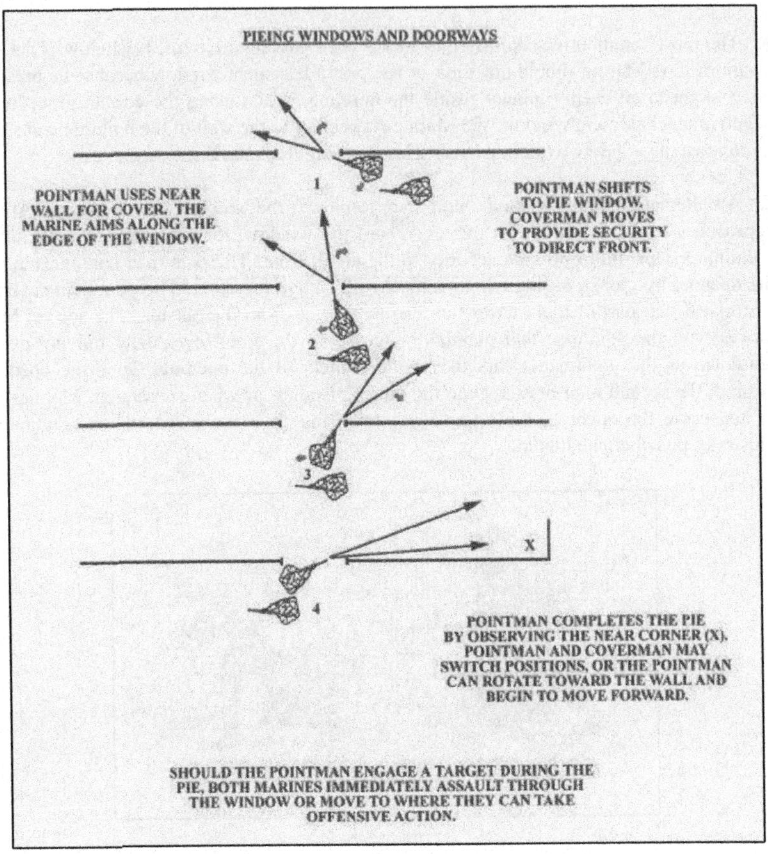

Figure A-11. Alternate Technique for Moving Past Windows

8. Movement Parallel to Buildings. Marines and small units may not always be able to use the inside of buildings as a route of advance. In such cases, they should move on the outside of the buildings (Figure A-13). Smoke, covering fires, and cover and concealment should be used to hide movement. When correctly moving along the outside of a building, the Marine hugs the side of the building, stays in the shadow, presents a low silhouette, and moves rapidly to his next position (Figure A-14). If an enemy gunner inside the building fires on a Marine, he exposes himself to fire from other squad/fire team members who are covering him. Furthermore, an enemy gunner farther down the street may have difficulty detecting and engaging the Marine.

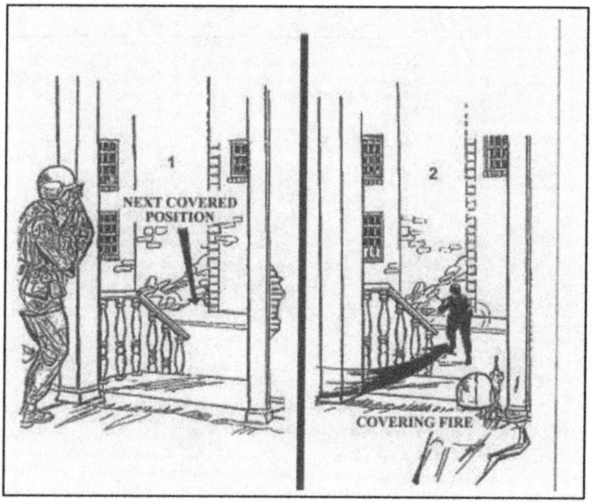

Figure A-12. Exiting a Doorway

Figure A-13. Moving Outside a Building

Figure A-14. Selection of the Next Position

9. Crossing Open Areas. Open areas such as streets, alleys, and parks should be avoided. They are natural kill zones for enemy crew-served weapons. They can be crossed safely if certain fundamentals are applied by Marines and small-unit leaders.

a. When using the correct procedure for crossing an open area, the Marine develops a plan for his own movement. (Smoke from hand grenades or smoke pots should be used to conceal the movement of all Marines.) The Marine runs the shortest distance between buildings and moves along the far building to the next position. By doing so, he reduces the amount of time during which he is exposed to enemy fire.

b. Before moving to another position, the Marine should make a visual reconnaissance and select the position that offers the best cover and concealment. At the same time, he should select the route that he will take to get to that position.

c. When moving from position to position, each Marine should be careful not to mask his supporting fires. When he reaches his next position, he should be prepared to cover the movement of other members of his assault force or element.

10. Assault Element Employment. Moving as an assault element from building to building or between buildings presents a problem because an assault element presents a large target to enemy fire. When moving from the corner of one building to another, the assault element should move across the open area in groups of two, maintaining their integrity within their respective clearing or covering team (Figure A-15). The assault element leader (fire team leader) is responsible for the secure movement of his unit. Smoke should be used to screen the assault element's movement. The covering team will move first to establish security on the far side. Their movement is protected by the fire of the clearing team. Once the covering team is in position, they provide cover as the clearing team moves across and prepares, then enters, the structure. Moving from the side of one building to the side of another presents a similar problem, and the technique of movement employed is the same (Figure A-16).

There may be situations in which it is best to move in larger groups. An example is that Marines crossing in pairs may key enemy snipers in to the movement, setting them up for good shots at the next pair to cross. Leaders should remain aware of the situation and threat and plan movements accordingly.

Figure A-15. Assault Element Movement

11. Movement Inside a Building. When moving within a building that is under attack (Figure A-17), Marines should avoid silhouetting themselves in doors and windows. If forced to use a hallway (Figure A-18), Marines should move by hugging the wall to avoid presenting targets to the enemy.

Figure A-16. Movement to an Adjacent Building

Figure A-17. Movement Within a Building Under Attack

Figure A-18. Moving Down a Hallway

Section IV

Entry Techniques

Marines should enter a building with minimum exposure. They should select an entry point before moving toward the building; avoid windows and doors; use smoke to conceal their advance to the building; use demolitions, tank rounds, and so on to make new entrances; precede the entry of the clearing team with a grenade; enter immediately after the grenade explodes; and be covered by the covering team.

12. Upper Building Levels. Clearing a building from the top down is the preferred method. Clearing or defending a building is easier from an upper story. Gravity and the building's floor plan become assets when throwing hand grenades and moving from floor to floor.

a. An enemy who is forced to the top of a building may be cornered and fight desperately or escape over the roof, but an enemy who is forced down to the ground level may withdraw from the building, thus exposing himself to friendly fires from the outside.

b. Various means such as ladders, drainpipes, vines, helicopters, or the roofs and windows of adjoining buildings may be used to reach the top floor or roof of a building. In some cases, one Marine can climb onto the shoulders of another and reach high enough to pull himself up. Another method is to attach a grappling hook to the end of a scaling rope so that a Marine can scale a wall, spring from one building to another, or gain entrance through an upstairs window.

13. Use of Ladders. Ladders offer the quickest method to gain access to the upper levels of a building (Figure A-19). Units can get ladders from local civilians or stores or obtain material to build ladders through supply channels. If required, ladders can be built with resources that are available throughout the urban area; for example, lumber can be taken from inside the walls of buildings (Figure A-20). Although ladders may not permit access to the top of some buildings, they may offer security and safety through speed.

14. Use of Grappling Hook. A suitable grappling hook and rope are selected to meet the need. The grappling hook should be sturdy, portable, easily thrown, and equipped with hooks that can hold inside a window. The scaling rope should be 5/8 of an inch to 1 inch in diameter and long enough to reach the objective window. Knots are tied in the rope at 1-foot intervals to make climbing and holding easier. Marines should follow the procedures outlined below.

a. When throwing the grappling hook, the Marine should stand as close to the building as possible (Figure A-21). If the Marine stands close to the building, he is less exposed to enemy fires and he does not have to throw the hook as far.

Figure A-19. Using Ladders To Get to Upper Levels

Figure A-20. Getting Lumber From Inside the Walls

Figure A-21. Grappling Hook Thrown at Close Range

b. Making sure there is enough rope to reach the target, the Marine should hold the hook and a few coils of rope in the throwing hand. The remainder of the rope, in loose coils, should be in the other hand. The rope should be allowed to play out freely. The throw should be a gentle, even, upward lob of the hook, with the other hand releasing the rope as it plays out.

c. Once the grappling hook is inside the window (or on the roof), the Marine should pull on the rope to obtain a good hold before beginning to climb. When using a window, the hook should be pulled to one corner to ensure a good "bite" and to reduce the Marine's exposure to lower windows during the climb.

d. The use of grappling hooks is the least preferred method for gaining entry to upper levels of buildings. Grappling hooks should be used only as a last resort and away from potential enemy positions. This method may potentially be used on adjacent buildings that offer concealed locations and a roof that connects to enemy positions.

15. Scaling Walls. When forced to scale a wall during exposure to enemy fire, all available means of concealment should be used. Smoke and diversionary measures improve the chances of a successful exposed movement. When using smoke for concealment, Marines should plan for wind direction. They should use fire, shouts, and fake movement to distract the enemy.

a. A Marine scaling an outside wall is vulnerable to enemy sniper fire. Marines moving from building to building and climbing buildings should be covered with friendly fire. Areas between buildings offer good fields of fire to the enemy. Properly positioned friendly weapons can suppress and eliminate enemy fire. The M203 grenade launcher is effective in clearing the enemy from rooms inside buildings (Figure A-22).

Figure A-22. Employment of the M203 Grenade Launcher

b. A Marine scaling a wall with a rope should avoid silhouetting himself in windows of uncleared rooms or exposing himself to enemy fires from lower windows. He should climb with his weapon slung over the firing shoulder so that he may quickly bring it to a firing position. He should clear the lower room with a hand grenade before going outside the window. Before doing this, the Marine first loosens the safety pin so that he needs only one hand to throw the grenade. The objective upper story window should not be entered before a hand grenade has been thrown in.

c. The Marine enters the objective window with a low silhouette (Figure A-23). Entry can be head first; however, a preferred method is to hook a leg over the windowsill and enter sideways, straddling the ledge.

16. Rappelling. Rappelling (Figure A-24) is an entry technique that Marines can use to descend from the rooftop of a tall building into a window. (See Army manual TC 21-24 for more information on rappelling.)

17. Entry at Lower Levels. Buildings should be cleared from the top down. However, it may be impossible to enter a building at the top; therefore, entry at the bottom or lower level may be the only course of action. When entering a building at the lower level, Marines avoid entry through windows and doors because both can be easily boobytrapped and are usually covered by enemy fire.

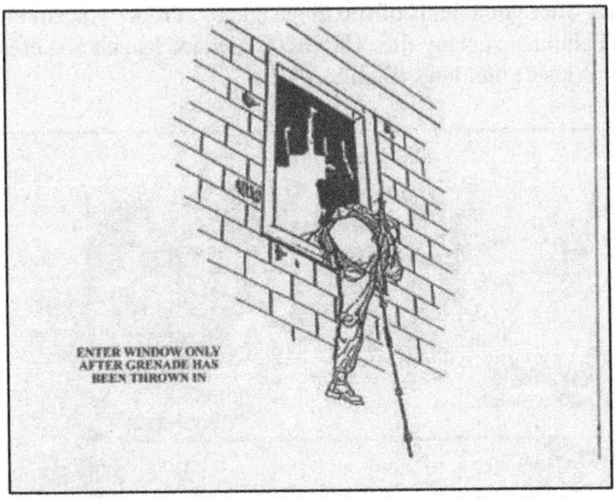

Figure A-23. Entering the Objective Window

Figure A-24. Rappelling

a. Ideally, when entering at lower levels, demolitions, artillery, tank fire, antiarmor weapons fire, or similar means are used to create a new entrance to avoid boobytraps. Quick entry is required to follow through on the effects of the blast and concussion.

b. When the only entry to a building is through a window or door, supporting fire should be directed at that location. (Specific lower level entry techniques are shown in Figure A-25.) If no supporting fire is available, AT4s can be employed instead.

c. Before entering, Marines throw a hand grenade into the new entrance to reinforce the effects of the original blast. When making a new entrance in a building, they consider the effects of the blast on the building and adjacent buildings. If there is the possibility of a fire in an adjacent building, Marines coordinate with adjacent units and obtain permission before starting the operation. In wooden-framed buildings, the blast may cause the building to collapse. In stone, brick, or cement buildings, supporting fires are aimed at the corner of the building or at weak points in the building construction.

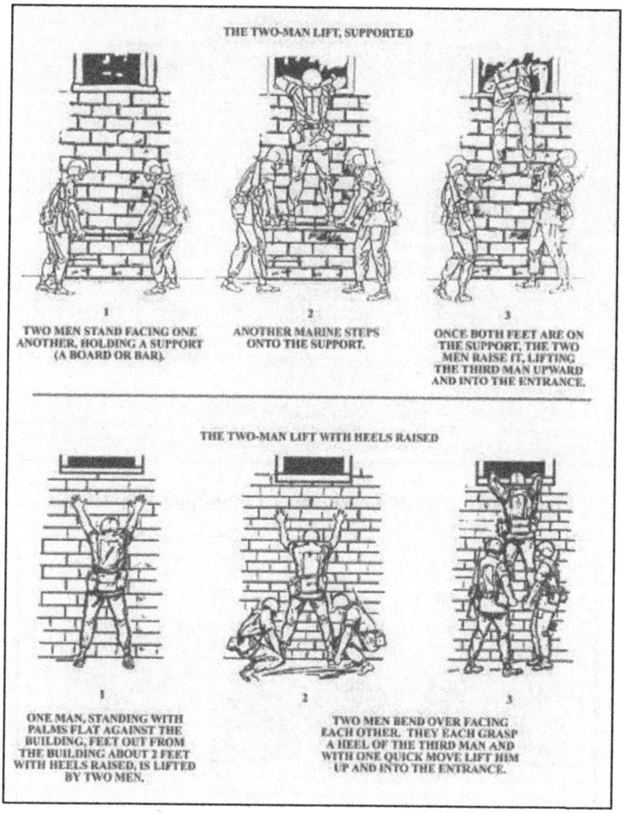

Figure A-25. Lower Level Entry Techniques

Figure A-25. Lower Level Entry Techniques (Continued)

Figure A-25. Lower Level Entry Techniques (Continued)

18. Hand Grenades. Combat in built-up areas (mainly during the attack) requires extensive use of hand grenades. The Marine should throw a grenade before entering rooms or negotiating staircases, mouseholes, and so on. This usually requires the use of both hands and employs both the overhand and underhand methods of throwing.

a. Two techniques may be used in the throwing of a grenade. The *preferred technique* involves throwing the grenade hard enough that it bounces or skips around, making it difficult to pick up. The hard-throw, skip/bounce technique may be used by Marines in training and combat. The *least preferred technique*, which involves cooking-off the grenade for two seconds then throwing it to prevent the enemy from grabbing the grenade and tossing it back, may be used by Marines, as appropriate, during actual combat but not in training.

b. Nonverbal or verbal alerts are used, as appropriate, before throwing a grenade. A visual showing (preferred) of the grenade to be thrown is made to assault element members, and a visual acknowledgment from them is received. A nonverbal alert may ensure that the enemy is surprised when the grenade is thrown. If the situation demands, a voice alert can be used, but the element of surprise may be lost. When or if a voice alert is used, the voice alert is "**FRAG OUT**;" when an enemy grenade has been identified, friendly forces shout, "**GRENADE**."

c. The construction material used in the building being cleared influences the use of grenades. In some situations, concussion grenades may be preferred over fragmentary grenades during offensive operations or when defending from hasty defensive positions. If the walls of a building are made of thin material, such as sheetrock or thin plyboard, the Marine should either lie flat on the floor with his helmet pointing toward the area of detonation or move away from any wall that might be penetrated by grenade fragments.

d. Marines should throw grenades into an opening before entering a building to eliminate any enemy that might be near the entrance (Figure A-26). Using the M203 grenade launcher is the best method for putting a grenade in an upper story window.

e. When a hand grenade must be used, the Marine throwing the grenade should stand close to the building, using it for cover. At the same time, the individual and the rest of the element should have a planned area to which they can move for safety if the grenade does not go through the window, but falls back to the ground.

f. The Marine throwing the grenade should step out far enough to lob the grenade into the upper story opening. The Marine's weapon should be kept in the nonthrowing hand so that it can be used if needed. The weapon should never be laid down outside or inside the building. Once the grenade has been thrown into the opening, assaulting forces should move swiftly to enter the building. This technique should be employed only when the window has already been broken. Otherwise, the chances are high that the thrown grenade will bounce off of the window and fall back onto the ground without going into the room.

Figure A-26. Hand Grenade Thrown Through Window

WARNING

After throwing the grenade, the Marine should immediately announce, "FRAG OUT," to indicate that a grenade has been thrown. He then takes cover because the grenade may bounce back or be thrown back, or the enemy may fire at him.

g. If Marines must enter the building by using the stairs, they first look for boobytraps. Then they throw a grenade through the stairwell door, let it detonate, and move quickly inside. They can use the staircase for cover.

h. The best way to enter a building is to breach the exterior wall. Again, a grenade should be thrown through the hole while using all available cover, such as the lower corner of the building (Figure A-27).

i. Another way to enter a room is to blast mouseholes with demolitions. In moving from room to room through mouseholes, Marines should use grenades to clear the room as they do in moving through open doorways.

j. Although buildings are best cleared from the top down, this is not always possible. While clearing the bottom floor of a building, Marines may encounter stairs that must also be cleared. Once again, grenades play an important role. Before climbing the stairs from the bottom floor, Marines should first inspect for boobytraps then toss a grenade to the head of the stairs, if appropriate (Figure A-28). The use of a grenade in these situations is based on stair/building construction and building layout. *Throwing grenades up staircases is extremely* *dangerous, and extreme caution and common sense should be used.* In most situations, throwing a grenade up a staircase is not recommended. (Again, the grenade is being used in the initial building entry or clearing of the bottom floor of a building having a staircase.) If a grenade is thrown too hard or the staircase is too steep, the grenade can roll back down the staircase. For this reason, if a grenade is thrown, it should be thrown underhand to reduce the risk of it bouncing back and rolling down the stairs. A nonverbal or, as appropriate, voice alert is used when throwing the grenade. Once the first grenade has detonated, another grenade should be thrown over and behind the staircase banister and into the hallway, destroying any enemy hiding to the rear.

Figure A-27. Entering Through a Breached Hole

Figure A-28. Tossing a Grenade Up a Stairway

k. After the stairs have been cleared, assaulting forces move to the top floor and clear it. Upon clearing the top floor, forces move downstairs to clear the center and bottom floors and to continue with the mission.

Note: Because large quantities of hand grenades are used when clearing buildings, a continuous supply must be available to forces having this mission within a built-up area.

Section V

Clearing Techniques

This section will discuss basic clearing techniques; however, the fundamentals may have to be modified depending on the various situations and building/room layouts that Marines may encounter. Caution should be taken when using a stacked type of position as personnel may be concentrating in a danger area where they could be exposed to fires penetrating walls or explosive effects from boobytraps or grenades thrown by the enemy. *Marines should always remain aware of the protective, or lack of protective, capabilities of the materials comprising the walls, floors, ceilings, and doors within the areas they are clearing.*

19. Clearing a Room, Door Closed or Open, From Stacked Positions. The following discussion and series of figures will focus on the two-man clearing team in the conduct of clearing a room. The covering team is positioned by the assault element leader to provide security as the clearing team begins its procedures to enter and clear a room.

a. Shooter Number Two is behind Shooter Number One (stacked position). This positioning is the same if the door is already open or nonexistent (Figure A-29).

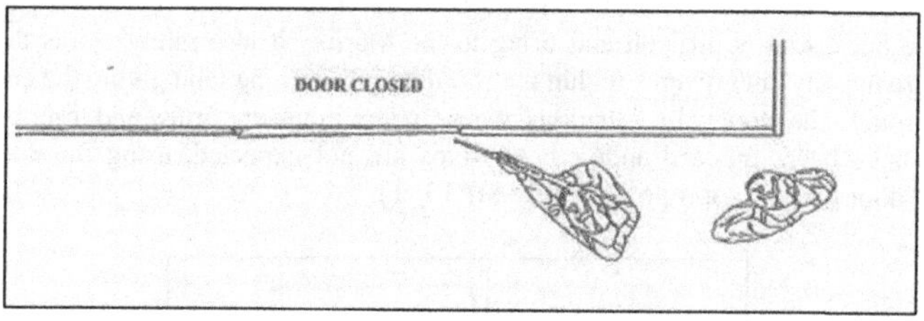

Figure A-29. Clearing a Room, Door Closed

b. Shooter Number One determines the direction in which the door opens then quickly informs Shooter Number Two and forces the door open.

When a door is the only means of entering a room, Marines should beware of fire from enemy soldiers within the room and of boobytraps. When opening a door, Marines should not expose themselves to fires through the door. The two-man clearing team forces the door open by directing a short burst of automatic fire through the door around the latch (Figure A-30) and then kicking it open. Another method of opening the door is to use a shotgun with 12-gauge slugs to shoot the hinges and/or the latch. When the doorknob is locked and/or deadbolt locks are present, shooting the hinges of the door is recommended. The door may then be opened by kicking or using a battering ram.

Figure A-30. Shooting the Door Open

Marines can also use an ax or demolitions if they are available. As a last resort, Marines can kick the door open with a front kick (Figure A-31) or mule kick. This is the least preferred technique because it is difficult and tiring to the Marine. It also rarely works the first time, thereby giving any enemy units within the room ample warning (and giving the enemy time to shoot through the door). In situations where restrictions on firing and limiting structural damage have been imposed and/or boobytraps are not expected, using the door handle to open the door may be appropriate (apply METT-T).

Figure A-31. Opening a Door With a Front Kick

c. Once the door is open, Shooter Number Two immediately throws a hand grenade into the room. Shooter Number One maintains cover for Shooter Number Two as he prepares to throw the grenade into the room. Shooter Number Two then returns to the original stacked position behind Shooter Number One (Figure A-32). If Marines have no grenades, the procedures remain the same except for throwing the grenade. *Note: When the door is already open or nonexistent, the stacked position is always used.*

WARNING

Because fragments from M67 fragmentation grenades may injure Marines outside the room, they should not be used. Marines should use MK3A2 offensive hand grenades instead. Also, the technique of cooking-off hand grenades can be deadly unless properly performed.

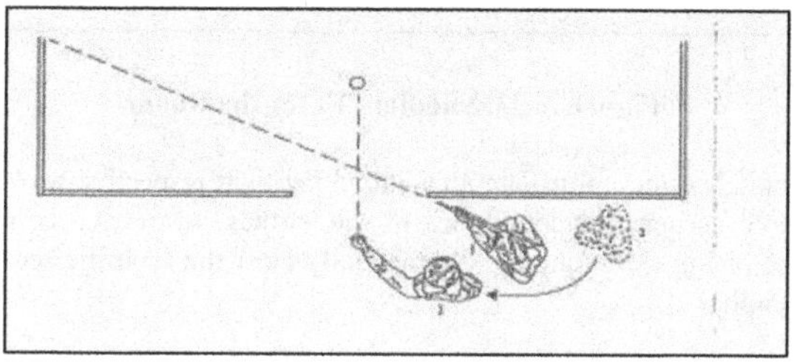

Figure A-32. Throwing a Grenade Into a Room

(1) Throwing Grenades. Two techniques are available for using grenades in room clearing. The *preferred technique* for Marines is to throw a concussion hand grenade into the room so hard that it skips and bounces, making it difficult for the enemy to pick up and throw back. The skip/bounce technique should be used by Marines during training and combat. The *least preferred technique* is to cook-off a concussion hand grenade by removing the grenade's safety pin, releasing the safety lever, counting off two seconds by thousands (one thousand and one, one thousand and two), and then throwing the grenade into the room. Cooking-off the grenade will be used only as appropriate during combat.

(2) Nonverbal and Verbal Alerts. To alert all that a grenade will be thrown, a visual showing of the grenade is made to assault element members, and a visual acknowledgment from them is received. A nonverbal alert may ensure that the enemy is surprised when the grenade is thrown. If the situation demands, a voice alert can be used, but the element of surprise may be lost. When or if a voice alert is used, the voice alert is "**FRAG OUT**;" when an enemy grenade has been identified, friendly forces shout, "**GRENADE**." This allows Marines to distinguish between warnings for outgoing and incoming grenades.

d. After the grenade explodes, Shooter Number One steps across the threshold and clears his immediate area (Figure A-33). He engages targets from the ready carry position. Shooter Number Two follows immediately behind Shooter Number One, buttonhooks, and clears his area.

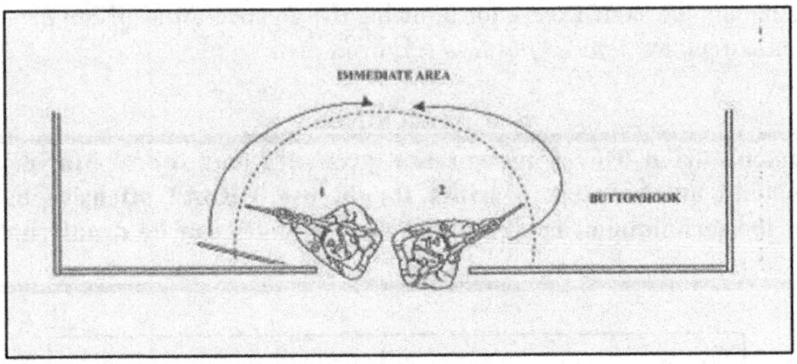

Figure A-33. Shooters Enter the Room

e. Both shooters clear the immediate area and along their respective walls, starting from the nearest respective corner and continuing to the farthest respective corner (Figure A-34). Shooters use the pieing technique to systematically clear the room by sector. (See Paragraph 3.a. of this appendix.)

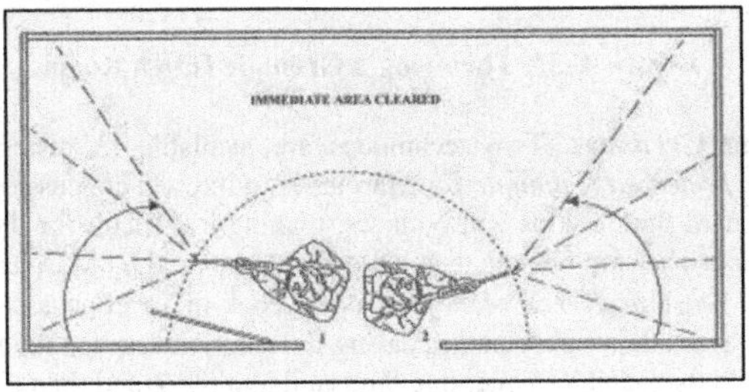

Figure A-34. Clearing a Room

f. Both shooters then establish a dominant position in the room one step away from the wall and two steps into the room and clear the room by sector, pieing to the opposite side of the room (Figure A-35). Meanwhile, the covering team (two-man team) in position outside the room being cleared provides security.

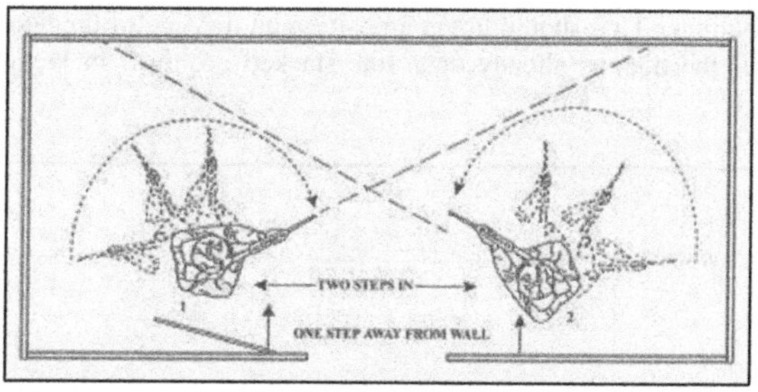

Figure A-35. Clearing a Room (Continued)

g. Another option for entering a room (situation dependent): If Shooter Number One of the clearing team enters a room without Shooter Number Two immediately following him (this could be due to room size, entering the room through a narrow opening, or a number of factors caused by the situation), then Shooter Number One will position himself inside the room to the left or right of the door and quickly scan the entire room. He then gives the command to Shooter Number Two, "**NEXT MAN IN, LEFT (RIGHT).**" Shooter Number Two shouts, "**COMING IN, LEFT (RIGHT),**" enters the room, positions himself up against the wall to the left (right) of the entrance as designated, and scans the room. Once in position, the clearing team leader can call in additional members with the "**NEXT MAN IN**" command, as the situation dictates. It is critical that all assault element members tell each other where they are to avoid fratricide.

h. When the clearing team has cleared the room and is ready to exit the room, it uses the following voice alerts. Once a room has been cleared, the clearing team yells, "**CLEAR,**" to inform the covering team. Before leaving the room and rejoining the covering team, the clearing team yells, "**COMING OUT**" and waits to hear the covering team yell, "**ALL CLEAR.**" The clearing team then executes its movement out of the room. The assault element then marks the room according to unit SOP.

When moving up or down a staircase, the appropriate team yells, "**COMING UP**" or "**COMING DOWN.**" As before, it waits for the response "**ALL CLEAR**" before executing movement and rejoining its element. When exiting a building, the clearing team or assault element yells, "**COMING OUT**" and waits for the support force or covering team to respond with "**ALL CLEAR.**"

20. Clearing a Room, Door Closed, Split Positions

a. Before opening a door, Shooter Number Two of the clearing team positions himself opposite Shooter Number One on the other side of the door and away from the wall in a safe position that allows Shooter Number One to shoot the door-opening mechanism (Figure A-36). This positioning allows Shooter Number Two to be in position to immediately move to a wall position opposite Shooter Number One. No matter what method is used to open the door, Shooter Number Two should get in a position on the side of the door opposite Shooter Number One. If the door is already open, the stacked positions in Paragraph 19.a. of this section are used.

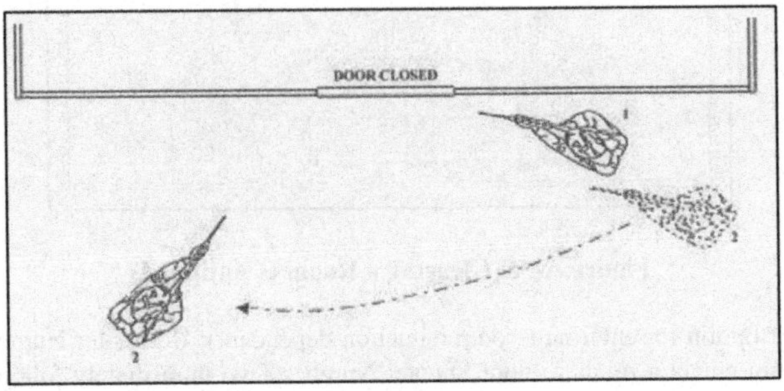

Figure A-36. Positioning To Open a Door

b. After Shooter Number One shoots the door open, Shooter Number Two immediately moves to a kneeling position against the wall (Figure A-37). Shooter Number One kicks the door open from a standing position. If the Marines have grenades, then Shooter Number Two immediately throws a grenade into the room as discussed in Paragraph 19.c. of this section. If the Marines do not have any grenades, the stand/kneel positioning of shooters places the shooters' weapons at different levels, thus allowing each shooter to engage targets inside the room while staying out of the other shooter's line of fire. The shooters scan the room for targets.

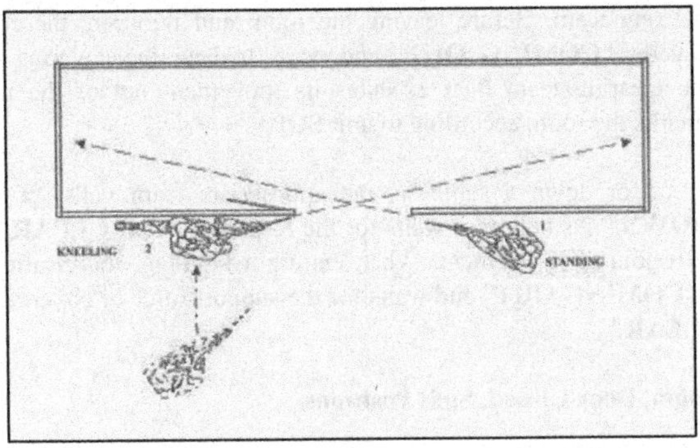

Figure A-37. Positioning To Enter a Room

c. If no grenade was thrown: Once the initial scan of the room is complete, Shooter Number One yells, "**MOVE,**" and both shooters pivot from their respective positions toward each other, pieing (see Paragraph 3.a. of this appendix) while maintaining eye-muzzle-target contact. Shooter Number Two is kneeling, and Shooter Number One is still standing (Figure A-38). Both shooters clear their respective sector of fire.

If a grenade was thrown: After the explosion, Shooter Number One yells, "**MOVE**," and both shooters pivot from their respective positions toward each other, immediately enter the room, and buttonhook (Figures A-38, A-39A, and A-39B). Shooter Number Two goes from kneeling to standing as he pivots. Both shooters clear their respective sector of fire.

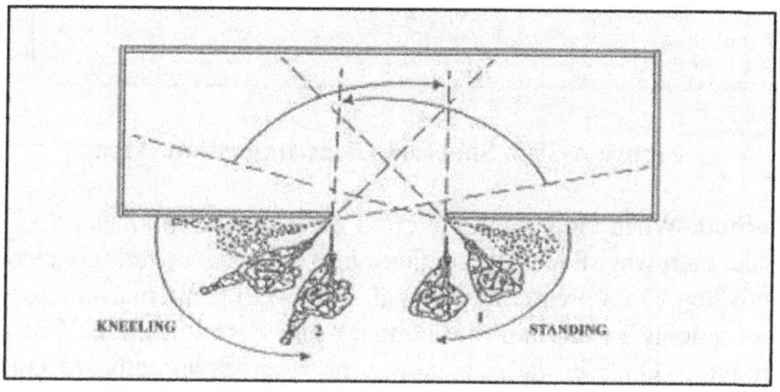

Figure A-38. Shooters Clear Sectors of Fire From Doorway

d. If no grenade was thrown: After scanning is completed, Shooter Number Two stands up and yells, "**READY**," and Shooter Number One Yells, "**MOVE**." Both shooters enter the room together (Figure A-39A).

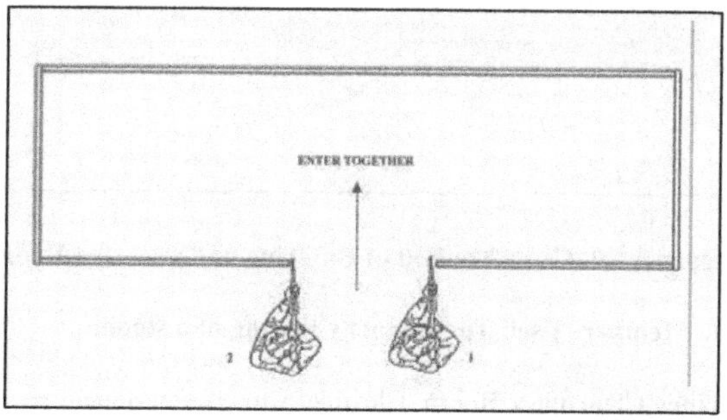

Figure A-39A. Shooters Enter the Room Together

e. Both shooters pass through the doorway together and immediately face respective corners, using the adjacent wall for cover. Both shooters buttonhook their respective areas (Figure A-39B). At this point, clearing the room is the same as discussed in Paragraphs 19.e. - 19.h.

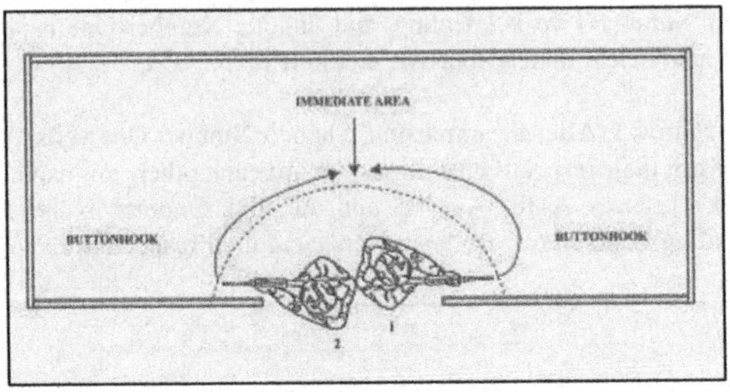

Figure A-39B. Shooters Clear Immediate Area

f. Cross Method. When employing the cross method, two Marines position themselves on either side of the entryway. Each Marine faces into the room covering the corner of the room opposite his position. On a prearranged signal, each Marine alternately enters the room. Each Marine crosses quickly to the opposite corner while covering the half of the room toward which he is moving. Once in the near corner, he assumes an outboard kneeling position to reduce his silhouette and continues to maintain coverage of his half of the room. He may change the position of his weapon to best cover the room. (Figure A-40)

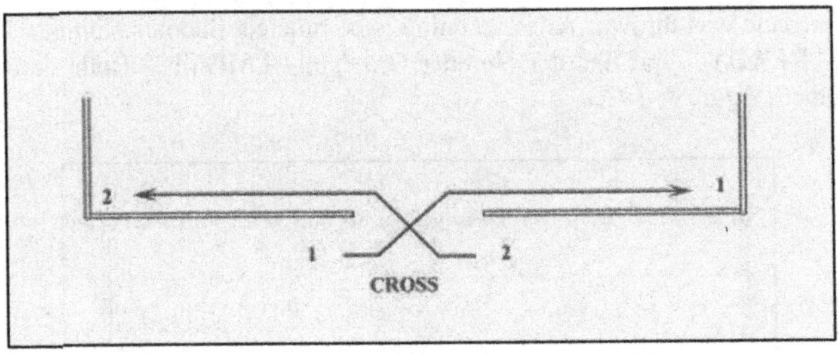

Figure A-40. Cross Method of Entering and Clearing Rooms

21. Covering Team Members Used To Support Clearing of a Room

a. Three Marines Clearing a Room. Clearing a room with three men involves the same procedures used by the aforementioned two-man clearing team, with a third Marine (from the covering team) following immediately (stacked) behind Shooters Number One and Two. Shooter Number Three moves to one side of the door and establishes a center sector of fire coverage (Figure A-41). The remaining covering team Marine provides outside security. If the split position is used, Shooters Number Two and Three are placed on the opposite side of the door.

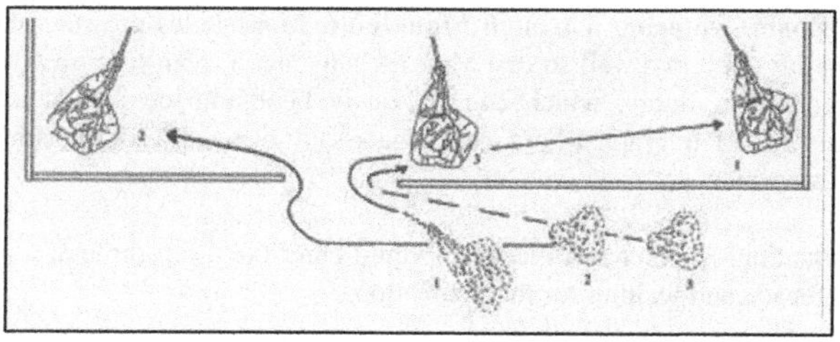

Figure A-41. Three-Marine Clearing Team (Stacked)

b. Four Marines Clearing a Room. When four Marines (entire assault element) are used to clear a room, outside security should be established by another assault element. (If outside security cannot be established, then the two- or three-Marine clearing team should be used.) The two-Marine clearing team procedures are used for initial entry into the room. Shooters Number Three and Four (split position example) stack up opposite Shooters Number One and Two. Shooters Number Three and Four follow Shooter Number Two into the room. Shooter Number Three crosses the threshold and establishes a center sector of fire coverage. Shooter Number Four follows Shooter Number Three and establishes a center sector of fire coverage opposite Shooter Number Three (Figure A-42). If the stacked position is used to enter a room, all shooters stack up on the same side of the door and follow each other in with the same positioning.

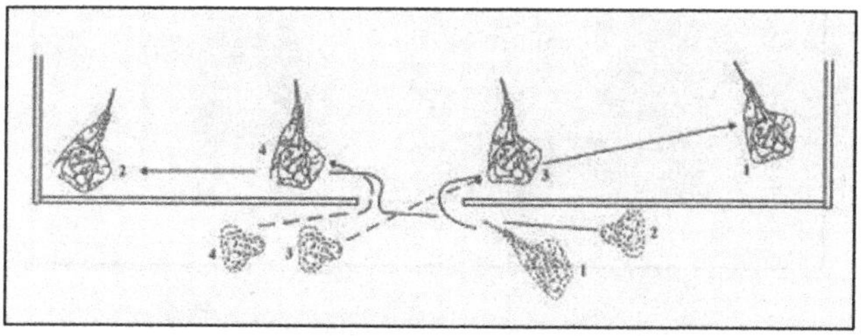

Figure A-42. Four-Marine Clearing Team (Split)

c. Single Marine Clearing a Room. This is the *least preferred* method of clearing a room. However, there may be times when this is the only method available, for example, when gaining access to an upper story window. When this is the case, the Marine gains entry to the room by shooting, kicking, ramming, or breaking a window. He then throws a grenade into the room if one is available and no restrictions apply and takes cover. Immediately after the grenade explodes, he crosses the threshold using the buttonhook technique depicted in earlier figures, scans the room for targets, and engages targets with two shots as they appear.

22. Clearing a Room, Entering Through Mousehole. Mouseholes measure about 2 feet wide and are blown or cut through a wall so that Marines can enter a room (Figure A-43). Mouseholes are safer entrances than doors, which can be easily boobytrapped and should be avoided. Procedures discussed in Paragraph 19 are used. The assault element leader positions the covering team to provide security.

a. In this situation, Shooter Number One would enter the room through a mousehole after throwing a grenade and waiting for the detonation.

b. Shooter Number One will position himself inside the room to the left or right of the mousehole and quickly scan the entire room. He then gives the command to Shooter Number Two, "**NEXT MAN IN, LEFT (RIGHT)**." Shooter Number Two shouts, "**COMING IN, LEFT (RIGHT)**," enters the room, positions himself against the wall to the left (right) of the entrance, and scans the room.

c. The clearing team continues clearing the room as discussed in Paragraph 19.

Figure A-43. Entering Through a Mousehole

23. Clearing an L-Shaped Hallway. After entering a room or building, the clearing team may encounter an L-shaped hallway that should be cleared. The following procedures are used:

a. The clearing team finds an L-shaped hallway.

b. Each shooter takes a dominant position.

c. Shooter Number Two aims his weapon to cover the dead space (Figure A-44).

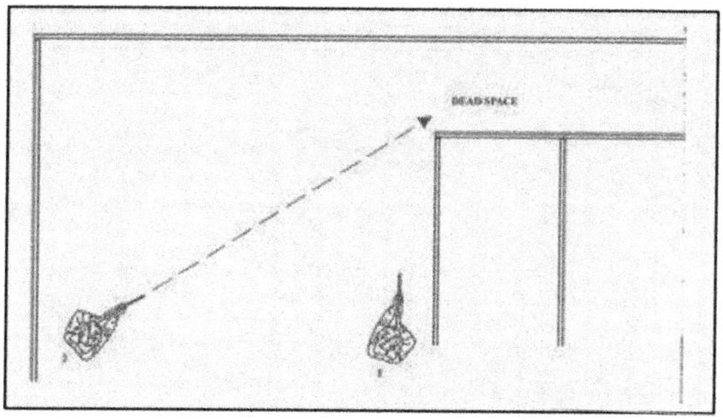

Figure A-44. Positioning To Clear an L-Shaped Hallway

d. Shooter Number Two, with his weapon covering the dead space, pies as far as possible and moves forward to increase his angle of fire further into the dead space. Shooter Number One moves accordingly with Shooter Number Two (Figure A-45).

e. Shooter Number Two stops pieing just before he reaches the position where he cannot cover the dead space. Shooter Number One stops short of the hallway and goes to a kneeling position.

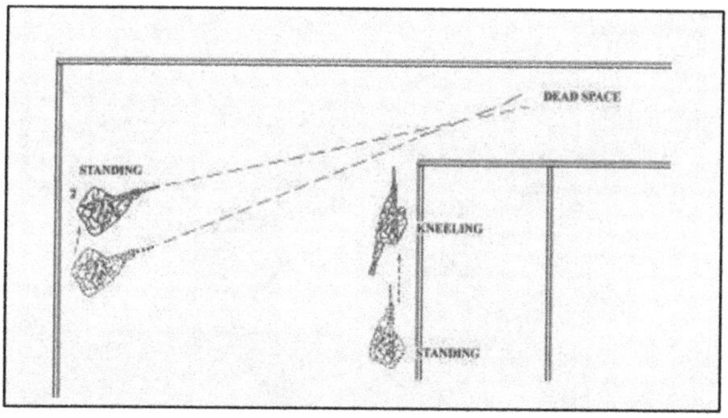

Figure A-45. Clearing an L-Shaped Hallway

f. Shooter Number One says, "**READY!**"

g. Shooter Number Two then says, "**MOVE!**" and both shooters move simultaneously to clear the hallway (Figure A-46). Shooter Number Two remains standing, and Shooter Number One remains kneeling as he pivots around the corner of the wall into the hallway.

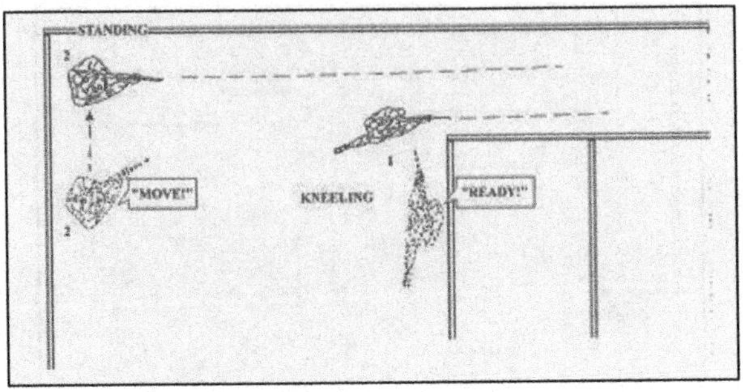

Figure A-46. Clearing an L-Shaped Hallway (Continued)

24. Clearing a T-Shaped Hallway. After entering a room or building, the clearing team may find a T-shaped hallway that should be cleared. The following procedures are used:

a. Both shooters approach the T-shaped intersection together, pieing their individual sectors while maintaining eye-muzzle-target contact (Figure A-47).

b. Both shooters move to dominant positions without entering the intersection.

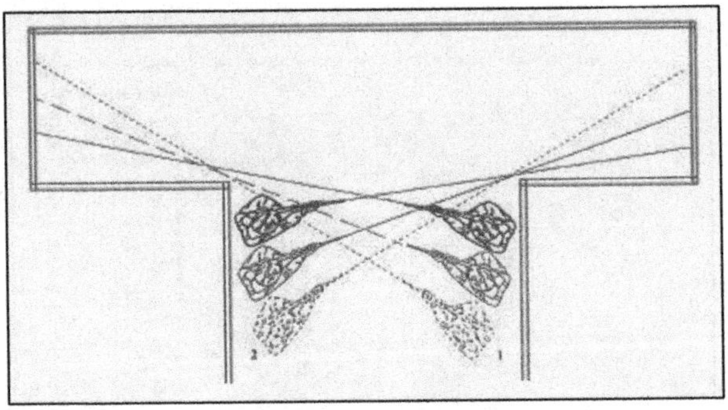

Figure A-47. Positioning To Clear a T-Shaped Hallway

c. Shooter Number One clears his sector of fire and then says, "**READY!**" (Figure A-48).

d. Shooter Number Two clears his sector of fire and then says, "**MOVE!**"

e. Both shooters buttonhook into their next positions.

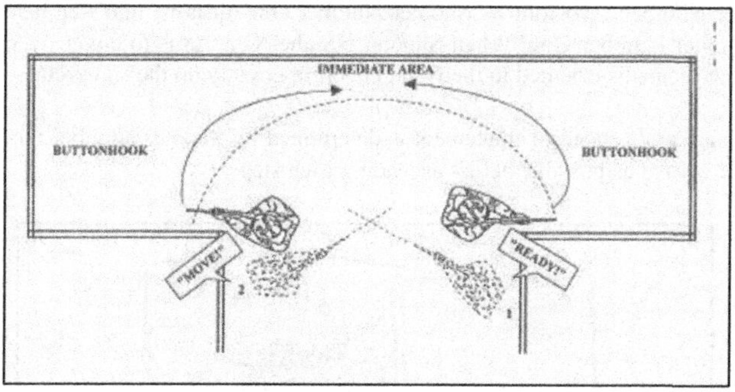

Figure A-48. Clearing a T-Shaped Hallway

f. Both shooters clear their immediate areas along their respective walls, starting from the nearest respective corner and continuing to the farthest respective corner (Figure A-49).

g. Both shooters establish dominant positions that give them control of the hallway and the doorways leading into hallways. The covering team can then be called forward to provide security while the clearing team clears one end of the hallway and associated rooms. The clearing team then clears the other end of the hallway.

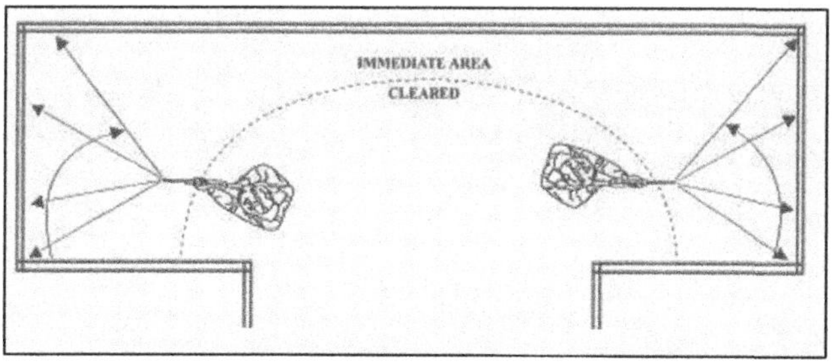

Figure A-49. Clearing a T-Shaped Hallway (Continued)

25. Clearing a Stairwell. After entering a building, the assault element may encounter stairwells. The following procedures are used:

a. Shooter Number One leads upstairs, one step ahead of Shooter Number Two (Figure A-50). Upon reaching the point just before he can be engaged from above, Shooter Number One turns around and covers overhead. From this point, Shooter Number One ascends the stairs, moving backward while covering behind and above.

b. Shooter Number Two follows Shooter Number One upstairs, one step behind and to the side of Shooter Number One. When Shooter Number One turns to cover overhead, Shooter Number Two remains oriented to the front, covering directly up the stairwell.

The clearing team's speed of movement is determined by Shooter Number One. Marines pie as much of an area as possible before ascending each step.

Figure A-50. Clearing a Stairwell

Section VI

Firing Positions

Whether a unit is attacking, defending, or conducting retrograde operations, its success or failure depends on the ability of the individual Marine to place accurate fire on the enemy while providing the least exposure to return fire. Consequently, Marines should immediately seek and properly use firing positions.

26. Hasty Firing Position. A hasty firing position is one that is normally occupied in the attack or the early stages of the defense. It is a position from which a Marine can place fire on the enemy while using available cover for protection from return fire. The Marine may occupy this position voluntarily, or he may be forced to occupy it because of enemy fire. In either case, the position lacks preparation before occupation. Some of the more common hasty firing positions in a built-up area involve firing around corners of buildings, firing from behind walls, firing from windows, firing from unprepared loopholes, and firing from the peak of a roof.

a. Corners of Buildings. The corner of a building provides cover for a hasty firing position if used properly.

(1) The shooter should be capable of both right-handed and left-handed firing of his weapon (using the "short stocking" technique) to be effective around corners. Short stocking the weapon will prevent the muzzle from protruding and keep the weapon ready to fire the instant visual contact with the enemy is made. Furthermore, it reduces the Marine's exposure as a target. (Figure A-51)

(2) A common mistake made when firing around corners is firing from the standing position. The shooter exposes himself at the height the enemy would expect a target to appear and risks exposing the entire length of his body as a target for the enemy.

Figure A-51. Firing, Using Either Hand, Around the Corner of a Building

b. Walls. When firing from behind walls, a Marine should fire around cover—not over it (Figure A-52). Marines in this situation should be able to fire from the right or left shoulder. When firing from behind cover, a Marine should remain far enough back from that cover so that the weapon does not extend beyond it. This aids in weapon retention by ensuring that no enemy on the other side of the wall can get control of the barrel.

Figure A-52. Firing Around Cover

c. Windows. In a built-up area, windows provide convenient firing ports. A Marine should avoid firing from the standing position because it exposes most of his body to return fire from the enemy and could silhouette him against a light-colored interior beyond the window. This is an obvious indicator of the firer's position, especially at night when the muzzle flash can easily be observed. When using the proper method of firing from a window (Figure A-53), the Marine is well back into the room to prevent the muzzle flash from being seen, and he is kneeling to limit exposure and avoid silhouetting himself.

Figure A-53. Firing From a Window

d. Loopholes. A Marine may fire through a hole torn in the wall and avoid windows (Figure A-54). He stays well back from the loophole so the muzzle of the weapon does not protrude beyond the wall and the muzzle flash is concealed.

Figure A-54. Firing From a Loophole

e. Roof. The peak of a roof provides a vantage point for snipers that increases their field of vision and the ranges at which they can engage targets (Figure A-55). A chimney, a smokestack, or any other object protruding from the roof of a building can reduce the size of the exposed target and should be used.

Figure A-55. Firing From the Peak of a Roof

f. No Position Available. When a Marine is subjected to enemy fire and none of the positions mentioned are available, he should try to expose as little of himself as possible. When a Marine is in an open area between buildings (a street or alley) with no cover available and he is fired upon by an enemy from one of the buildings to his front, he should lie prone as close as possible to a building on the same side of the open area as the enemy. To engage the Marine, the enemy must then lean out of the window and expose himself to return fire.

g. No Cover Available. When no cover is available, target exposure can be reduced by firing from the prone position, by firing from shadows, and by presenting no silhouette against buildings.

27. Prepared Firing Position. A prepared firing position is one built or improved to allow the shooter to engage a particular area, avenue of approach, or enemy position, while reducing his exposure to return fire. Examples of prepared positions include: barricaded windows, fortified loopholes, sniper positions, antiarmor positions, and machine gun positions.

a. The natural firing port provided by windows can be improved by barricading the window and leaving a small hole for the firer's use (Figure A-56). The barricading may be accomplished by using materials torn from the interior walls of the building or any other available material. When barricading windows, a Marine should avoid:

(1) Barricading only the windows that will be used as firing ports. The enemy will soon determine that the barricaded windows are firing positions.

(2) Neat square or rectangular holes that are easily identified by the enemy. A barricaded window should not have a neat, regular firing port. The window should keep its original shape so that the position of the firer is hard to detect. Firing from the bottom of the window gives the firer the cover advantage of the wall because the firing port is less obvious to the enemy. Sandbags are used to reinforce the wall below the window and to increase protection for the firer. All glass should be removed from the window to prevent injury to the firer. Lace curtains permit the firer to see out and prevent the enemy from seeing in. Wet blankets should be placed under weapons to reduce dust that may create a signature. Wire mesh placed over the window keeps the enemy from throwing in hand grenades.

b. Although windows are usually good firing positions, they do not always allow the firer to engage targets throughout his sector.

(1) To avoid establishing a pattern of always firing from windows, an alternate position is required such as the prepared loophole (Figure A-57). This involves cutting or blowing a small hole in the wall to allow the shooter to observe and engage targets in his sector.

(2) Sandbags are used to reinforce the walls below, around, and above the loophole. If the position is on the second floor or higher, two layers of sandbags are placed on the floor under the firer to protect him from an explosion on a lower floor. A wall of sandbags, rubble, furniture, and so on should be constructed to the rear of the position to protect the firer from explosions in the room.

Figure A-56. Window Firing Position

(3) A table, a bedstead, or other available material provides overhead cover for the position. This prevents injury to the shooter from falling debris or explosions above his position.

(4) The position should be camouflaged by knocking other holes in the wall. This makes it difficult for the enemy to determine which hole the fire is coming from. Siding material should be removed from the building in several places to make loopholes less noticeable.

Figure A-57. Prepared Loopholes

c. A chimney or other protruding structure provides a base from which a sniper position can be prepared. Part of the roofing material is removed to allow the sniper to fire around the chimney. He should stand inside the building on the beams or on a platform with only his head and shoulders above the roof (behind the chimney). Sandbags placed on the sides of the position protect the sniper's flanks.

d. When the roof has no protruding structure to provide protection, the sniper position should be prepared from underneath on the enemy side of the roof (Figure A-58). The position is reinforced with sandbags, and a small piece of roofing material should be removed to allow the sniper to engage targets in his sector. The missing piece of roofing material should be the only sign that a position exists. Other pieces of roofing should be removed to deceive the enemy as to the true sniper position. The sniper should be invisible from outside the building, and the muzzle flash should be hidden from view.

Figure A-58. Sniper Position

e. Some rules and considerations for selecting and occupying individual firing positions are:

(1) Make maximum use of available cover and concealment.

(2) Avoid firing over cover; when possible, fire around it.

(3) Avoid silhouetting against light-colored buildings, the skyline, and so on.

(4) Carefully select a new firing position before leaving an old one.

(5) Avoid setting a pattern; fire from both barricaded and unbarricaded windows.

(6) Keep exposure time to a minimum.

(7) Begin improving a hasty position immediately after occupation.

(8) Use construction material that is readily available in the built-up area for prepared positions.

(9) Remember that positions that provide cover at ground level may not provide cover on higher floors.

f. In attacking a built-up area, recoilless weapon and ATGM crews may be hampered in choosing firing positions by the backblast of their weapons. They may not have enough time to knock out walls in buildings and clear backblast areas. They should select positions that allow the backblast to escape, such as corner windows where the round fired goes out one window and the backblast escapes from another. The corner of a building can be improved with sandbags to create a firing position (Figure A-59).

Figure A-59. Corner Firing Position

g. The rifle squad is often reinforced with attached antitank weapons during an attack on and in defense of a built-up area. Therefore, the rifle squad leader should be able to choose good firing positions for the antitank weapons under his control.

h. Various principles for employing antitank weapons have universal applications, such as making maximum use of available cover, trying to achieve mutual support, and allowing for the backblast when positioning recoilless weapons, TOWs, Dragons, and AT4s.

i. Operating in a built-up area presents new considerations. Marines should select numerous alternate positions, particularly when the structure does not provide cover from small-arms fire. They should position their weapons in the shadows and within the building.

j. Recoilless weapons and ATGMs firing from the top of a building can use the chimney for cover (Figure A-60). The rear of this position should be reinforced with sandbags.

Figure A-60. Antitank Weapon Firing From a Rooftop

k. When selecting firing positions for recoilless weapons and ATGMs, Marines should make maximum use of rubble, corners of buildings, and destroyed vehicles to provide cover for the crew. Recoilless weapons and ATGMs can also be moved along rooftops to obtain a better firing angle from which to engage enemy armor. When buildings are elevated off the ground, positions can be prepared using a building for overhead cover (Figure A-61). The backblast under the building must not damage or collapse the building or injure the crew.

Note: When firing from a slope, ensure that the angle of the launcher relative to the ground or firing platform is not greater than 20 degrees. When firing within a building, ensure that the enclosure is at least 10 feet by 15 feet, is clear of debris and loose objects, and has windows, doors, or holes in the walls to allow the backblast to escape.

l. The machine gun can be emplaced almost anywhere. In the attack, windows and doors offer ready-made firing ports (Figure A-62). For this reason, the enemy normally has windows and doors under observation and fire; they should therefore be avoided. Any openings in walls that were created during the fighting may be used. When other holes are not present, small explosive charges can create loopholes (Figure A-63). Regardless of what openings are used, machine guns should be sited within the building and in the shadows.

Figure A-61. Prepared Positions Using a Building for Overhead Cover

Figure A-62. Emplacement of a Machine Gun in a Doorway

m. Upon occupying a building, Marines board up all windows and doors. By leaving small gaps between the slots, Marines can use windows and doors as good alternate firing positions.

n. Loopholes should be used extensively in the defense. They should not be constructed in any logical pattern, nor should they all be at floor or tabletop level. Varying their height and location makes them hard to pinpoint and identify. Dummy loopholes, knocked-off shingles, or holes that are cut but not intended to be used as firing positions aid in deception. Loopholes located behind shrubbery, under doorjambs, and under the eaves of a building are hard to detect.

Figure A-63. Use of a Loophole With a Machine Gun

o. Increased fields of fire can be obtained by locating the machine gun in the corner of the building or sandbagging it under a building (Figure A-64). Available materials, such as desks, overstuffed chairs, couches, and other items of furniture, should be integrated into the construction of bunkers to add both cover and concealment (Figure A-65).

Figure A-64. Sandbagged Machine Gun Emplacement Under a Building

p. Although grazing fire is desirable when employing the machine gun, it may not always be practical or possible. Where destroyed vehicles, rubble, and other obstructions restrict the fields of grazing fire, the gun can be elevated to a position from which it can deliver plunging fire over obstacles. Firing from loopholes on the second or third story may be necessary. A firing platform can be built under the roof (Figure A-66) and a loophole constructed. Again, the exact location of the position should be concealed by knocking off shingles in isolated patches over the entire roof.

Figure A-65. Corner Machine Gun Bunker

Figure A-66. Firing Platform Built Under a Roof

28. Target Acquisition. Built-up areas provide unique challenges to acquiring targets. Buildings mask movement and the effects of direct and indirect fires. Also, the rubble from destroyed buildings provides concealment and protection for attackers and defenders alike, making target acquisition difficult.

a. The techniques of patrolling and using OPs apply in the city as well as in wooded terrain. These techniques enable units to locate the enemy, to develop targets for direct and indirect fires in the defense, and to find uncovered avenues of approach in the offense.

b. Most weapons and vehicles produce recognizable signatures. These come from design features or from the environment in which the equipment is used. For example, firing a tank main gun in dry, dusty, and debris-covered streets raises a dust cloud; a tank being driven in built-up areas produces more noise than one moving through an open field; people moving through rubble on a street or in the halls of a damaged building create more noise than in a wooded area. Marines should learn to recognize signatures so they can locate and identify targets. Seeing and hearing assist in detecting and identifying signatures that lead to target location, identification, and rapid engagement. Marines should look for targets in areas where they are most likely to be employed.

c. Target acquisition should be continuous, whether halted or moving. Built-up areas provide both the attacker and defender with good cover and concealment but usually favor the defender. This makes rapid and accurate target acquisition extremely important.

d. When a unit is moving and enemy contact is likely, the unit should employ an overwatching element. This principle applies in built-up areas as it does in other kinds of terrain except that in built-up areas the overwatching element should observe both the upper floors of buildings and the street level.

e. Stealth should be used when moving in built-up areas because little distance separates attackers from defenders. Only hand and arm signals should be used until contact is made. The unit should stop periodically to listen and observe to ensure that it is not being followed or that the enemy is not moving parallel to the unit's flank for an ambush. Routes should be carefully chosen so that buildings and piles of rubble can be used to mask the unit's movement.

f. Observation duties should be clearly issued to squad members to ensure all-around security as they move. This security continues at the halt. All of the senses should be used to acquire targets, especially hearing and smelling. Marines soon recognize the sounds of vehicles and people moving through streets that are littered with rubble. The smell of fuel, cologne, or food cooking can disclose enemy positions.

g. OPs are "positions from which military observations are made, or fire directed and adjusted, and which possess appropriate communications" (Joint Pub 1-02). They are positions from which Marines can watch and listen to enemy activity in a specific sector. They warn the unit of enemy approach and should be positioned in the upper floors of buildings to give Marines a better vantage point than they would have at street level.

h. In the defense, a platoon leader positions OPs for local security as ordered by the company commander. The platoon leader selects the general location, but the squad leader establishes the OP (Figure A-67). Normally, there is at least one OP per platoon. An OP consists of two to four men and is within small-arms supporting range of the platoon. Leaders look for positions that allow good observation of the target sector. Ideally, an OP has a field of observation that overlaps those of adjacent OPs. The position selected for the OP should provide cover and concealment for units moving to and from the OP. The upper floors of houses or other buildings should be used. The squad leader should not routinely select obvious positions such as water towers or church steeples because those positions naturally attract the enemy's attention.

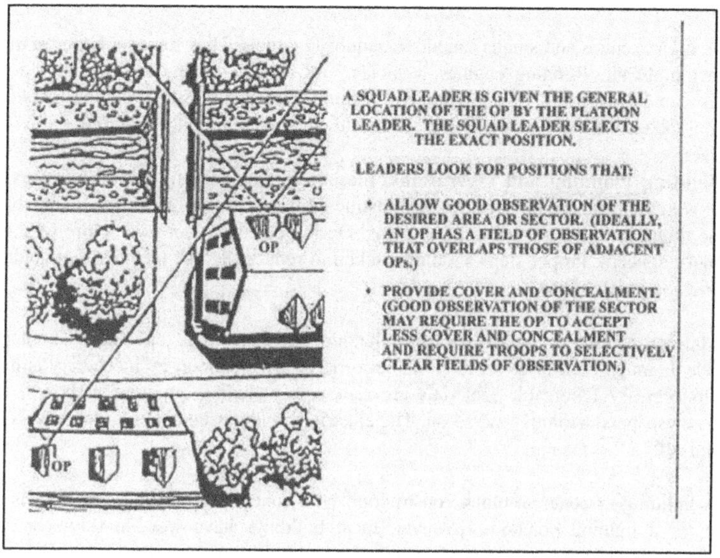

Figure A-67. Selection of an OP Location

i. The Marine should be taught how to scan a target area from OPs or from fighting positions. Use of proper scanning techniques enables observers to quickly locate and identify targets. The Marine searches quickly without optics for obvious targets, using all of his senses to detect target signatures. If no targets are found and time permits, he makes a more detailed search (using binoculars, if available) of the terrain in the assigned sector using the 50-meter method. First, he searches a strip 50 meters deep from right to left. He then searches a strip from left to right that is farther out, overlapping the first strip. This process is continued until the entire sector has been searched. In the city core or core periphery where the observer is faced with multistory buildings, the overlapping sectors may extend up rather than out.

j. Marines who man OPs and other positions should employ target acquisition devices. These devices include binoculars, image intensification devices, and thermal sights. All of these devices can enhance the unit's ability to detect and engage targets, especially at night or during periods of reduced visibility.

k. Target acquisition techniques used at night are similar to those used during the day. At night, whether using daylight optics or the unaided eye, the observer should not look directly at an object but a few degrees off to the side. The side of the eye is more sensitive to dim light. When scanning with off-center vision, the Marine moves his eyes in short, abrupt, irregular movements. At each likely target area, he pauses for a few seconds to detect any motion.

l. At night, sounds and smells can aid in acquiring targets. They transmit better in the cooler, damper night air. Running engines, vehicles, and people moving through rubble- covered streets can be heard for great distances. Odors from diesel fuel, gasoline, cooking food, burning tobacco and aftershave lotion can reveal enemy and friendly locations.

29. Firefighting Planning and Operations. Incendiary ammunition, special weapons, and the ease with which incendiary devices can be constructed from gasoline and other flammables make fire a true threat in operations in built-up areas. During defensive operations, firefighting should be a primary concern. Proper steps should be taken to reduce the risk of a fire that could make a chosen position indefensible.

a. Marines choose or create positions that do not have large openings. These positions provide as much built-in cover as possible to prevent penetration by incendiary ammunition. All unnecessary flammable materials are removed, including ammunition boxes, furniture, rugs, newspapers, curtains, and so on. The electricity and gas coming into the building should be shut off.

b. A building of concrete block construction with concrete floors and a tin roof is an ideal place for a fighting position. However, most buildings have wooden floors or subfloors, wooden rafters, and wooden inner walls that may require improvement. Inner walls are removed and replaced with blankets to resemble walls from the outside. Sand is spread 2 inches deep on floors and in attics to retard fire.

c. All available firefighting gear is prepositioned so that it can be used during combat. For the individual Marine, such gear includes entrenching tools, helmets, sand, and blankets. These items are supplemented with fire extinguishers.

d. Fire is so destructive that it can easily overwhelm personnel regardless of extraordinary precautions. Marines should plan routes of withdrawal so that evacuation can be accomplished from their fighting positions. This allows Marines to exit through areas that are free of combustible material but that still provide cover from enemy direct fire.

e. The confined space and large amounts of combustible material found in built-up areas may influence the enemy to use incendiary devices. First-aid problems that are expected include burns and smoke/flame inhalation. These can easily occur in buildings, thereby rendering the victims combat ineffective. Although there is little defense against flame inhalation and lack of oxygen, smoke inhalation may be greatly reduced by wearing the individual protective mask. Regardless of the fire hazard, defensive planning for combat in built-up areas should include location of corpsmen. Corpsmen must reach victims and their equipment and should stock extra supplies for the treatment of burns and inhalation injuries.

f. Offensive operations also require plans for firefighting because the success of the mission can easily be threatened by fire. Poorly planned use of incendiary munitions can make fires so extensive that they become obstacles to offensive operations. The enemy may use fire to cover his withdrawal and to create obstacles and barriers to the attacker.

g. Every Marine participating in an attack should be ready to deal with the problems of fire. The normal firefighting equipment available includes the entrenching tool, helmet (for carrying sand or water), and blankets (for snuffing out small fires). Fire extinguishers are normally available on vehicles.

30. Employment of Snipers. The value of the sniper to a unit operating in a built-up area depends on several factors. These factors include the type of operation, the level of conflict, and the ROE. Where ROE allow destruction, snipers may not be needed because other weapons systems available to the infantry commander have a greater destructive effect. However, snipers can contribute to the fight. Where the ROE prohibit collateral damage, snipers may be the most valuable tool the commander has. (See MCWP 3-15.3 (under development).)

a. Sniper effectiveness depends in part on the terrain. Control is decentralized by the characteristics of an urban area. To provide timely and effective support, the sniper should have a clear picture of the commander's mission and intent.

b. Where possible, snipers should be positioned in masonry buildings. These buildings should also offer long-range fields of fire and all-around observation. The sniper has an advantage because he does not have to move with or be positioned with lead elements. He may occupy a higher position to the rear or flanks of, and some distance away from, the element that he is supporting. By operating far from the other elements, a sniper avoids decisive engagement but remains close enough to kill distant targets that threaten the unit. Snipers should not be placed in obvious positions such as church steeples and rooftops because the enemy often watches these and targets them for destruction. Also, snipers should not be positioned where there is heavy traffic; these areas also invite enemy observation.

c. Snipers should operate throughout the AO, moving with and supporting the companies as necessary. Some teams may be designated to operate independently of other forces. They may be tasked to search for targets of opportunity, especially for enemy snipers. The sniper team may occupy multiple positions. A single position may not afford adequate observation for the entire team without increasing the risk of detection by the enemy. Separate positions should maintain mutual support. Alternate and supplementary positions should also be established in urban areas.

d. Snipers may be assigned the following tasks:

- Killing enemy snipers (countersniper fire)

- Killing targets of opportunity (These targets should be prioritized by the commander. Types of targets might include enemy snipers, leaders, vehicle commanders, radio men, sappers, and machine gun crews.)
- Denying enemy access to certain areas or avenues of approach (controlling key terrain)
- Providing fire support for covering barricades and other obstacles
- Maintaining surveillance of flank and rear avenues of approach (screening)
- Supporting local counterattacks with precision fire.

Section VII

Advancing/Patrolling Along City Streets

31. Advancing/Patrolling Along City Streets. During MOUT operations, skills for advancing/patrolling along city streets will be required. The platoon, squad, or fire team mission may be such that building-to-building clearing is neither required nor desired until contact with the enemy has been made. This section will discuss advancing/patrolling along urban streets. The standard tactics, techniques, and procedures for the conduct of a patrol, as outlined in MCWP 3-11.6 *Scouting and Patrolling* (draft), apply except as otherwise discussed within this section.

a. Organization. The platoon's three-squad organization or the squad's three-fire team organization is already structured to conduct an advance/patrol down an urban street. The typical patrol is organized into assault, support, security, and command and control elements. Security is divided into front, flank, and rear security responsibilities. The platoon or squad can move down the street in a double or single column.

b. Formation and Movement

(1) Double Column. The double column is the *preferred method* for moving along an urban street (Figures A-68 and A-69). The double column provides 360 degrees of security and allows for mutual support and interlocking sectors of fire. The double column is flexible but can make control more difficult. There should be at least two squads or fire teams to conduct a double column. The two squads or fire teams move down the sides of the street using the walls of the buildings for cover and concealment. The pointmen for the two columns should always stay abreast of each other and not move in front of or behind the opposite column. The Marines following in trace of the pointmen stagger themselves accordingly. The platoon or squad leaders move where they can best control their units. If a squad is conducting the advance/patrol along the street, the trail fire team is never split.

(a) Sectors of Responsibility. Every Marine moving in a column has a particular sector for whose security he is responsible. The pointmen are responsible for security to the direct front and for pieing off windows and doorways. They are also responsible for staying in line with one another.

The Marines immediately behind the pointmen are designated as the covermen. They are responsible for providing guidance to their respective pointmen from the leader of the unit and for covering the 45-degree oblique angle across from one another. The second coverman is also responsible for covering the direct front when the pointman is pieing and checking for the enemy. The third and fourth covermen in the column are responsible for the far-side flank security; one covers the upper level and one covers the lower level building windows, doors, and stairwells. Marines will alternate providing security on upper level and lower level windows, doors, and stairwells. The fourth Marine also provides rear security if he is the last man in the column or interlocks his sector of fire with that of the lead Marine in the next fire team in the column. Should the second coverman have to take the point while the pointman pies a window or door, the third coverman would automatically shift his sector of fire to the 45-degree oblique angle. If the second coverman has to assist the pointman in pieing a large opening, the third coverman would provide security to the direct front while the fourth coverman would take the 45-degree oblique angle. Marines following in the column would adjust their sectors of fire accordingly.

The two columns provide mutual support for each other and the greatest all-around security. Teamwork and coordination between the two columns are vital.

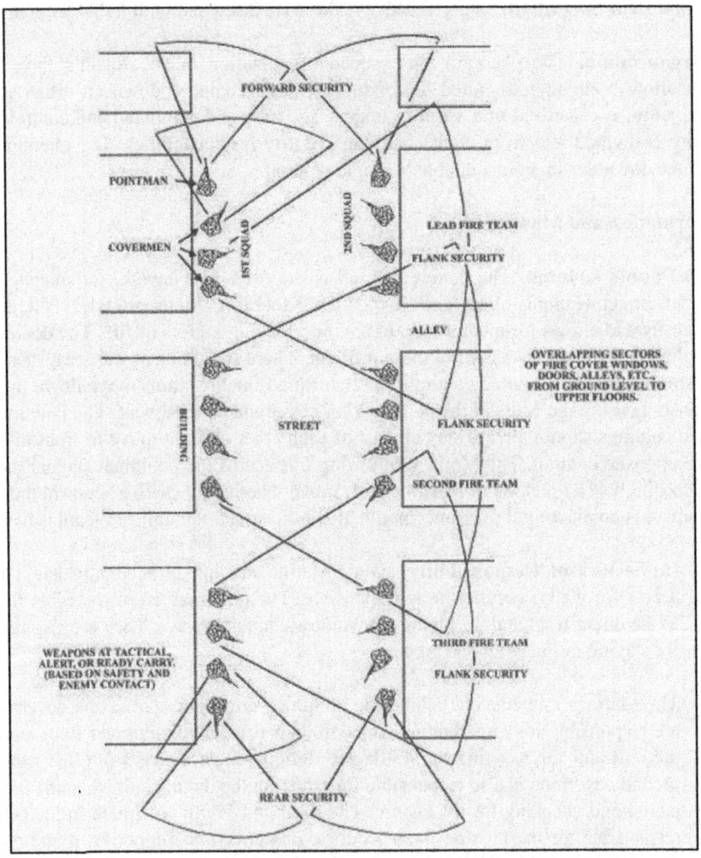

Figure A-68. Advancing/Patrolling Along an Urban Street (Squad, Double Column)

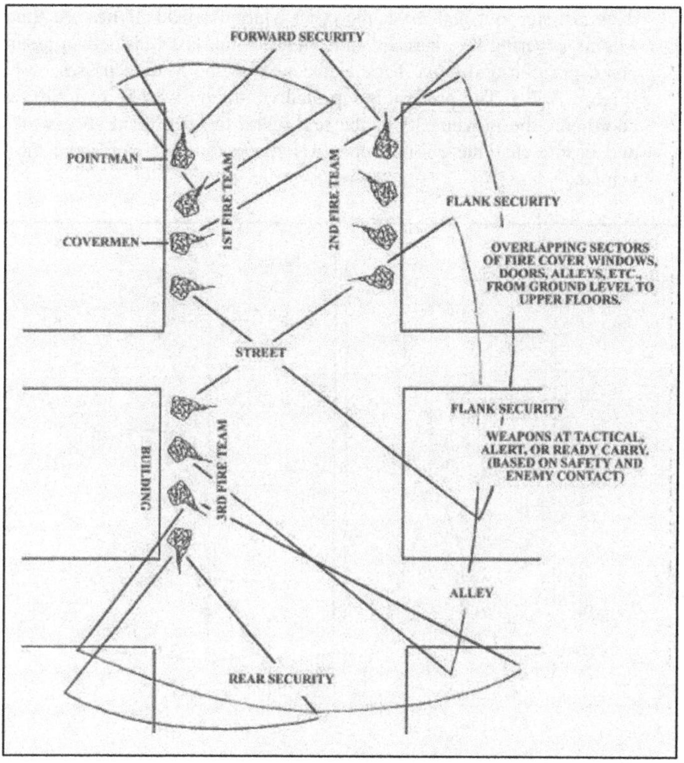

Figure A-69. Advancing/Patrolling Along an Urban Street (Fire Team, Double Column)

(b) Rear Security. The last two Marines in the squad of each column are responsible for the rear security of each of their formations. They will move in bounds in order to allow a Marine to always provide active security to the squad's rear (Figures A-70 and A-71). Two methods of bounding may be used: *alternate* or *successive bounding.*

1 **Alternate Bounding.** In this method, the two rear-security Marines alternate being the last Marine in the squad. The last Marine in the formation will have his weapon at the ready and will be in the kneeling position along the wall of the building. The second to last Marine will move in trace of the unit until he has moved approximately 10 to 15 meters. At this point, he will step out one pace from the wall. He will then assume a good kneeling firing position and announce, "**SET**." The last Marine will rise from his position, turn around, and move next to the wall to catch up with the column. It is important that the Marine moving back

does not step out and mask the other Marine's field of fire. At this point, the Marine covering the rearward movement of the last Marine can assume the rear guard position and move back to the wall as the Marine passes by his position (Figure A-70). The process is repeated continuously. The rear fire team leaders coordinate the movements of the rear guard to keep them abreast of each other and ensure that the column does not move too fast and leave the rear guard behind.

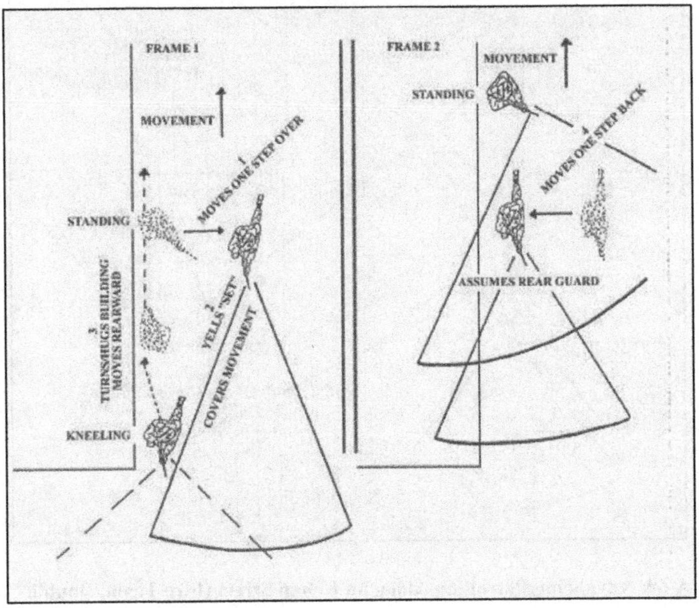

Figure A-70. Rear Security (Alternate Bounds)

2 **Successive Bounding.** In this method, the last Marine in the squad continues to be the rear guard after bounding. The same procedures are employed as in alternate bounding; however, the rear-guard Marine bounds abreast of the Marine providing cover, turns around, drops to a good kneeling position, provides rear-guard security, and announces, "**SET**." The covering Marine moves behind the rear-guard Marine and continues to move 10 to 15 meters before repeating the process (Figure A-71). The rear fire team leaders coordinate the movements of the rear guard as in alternate bounding.

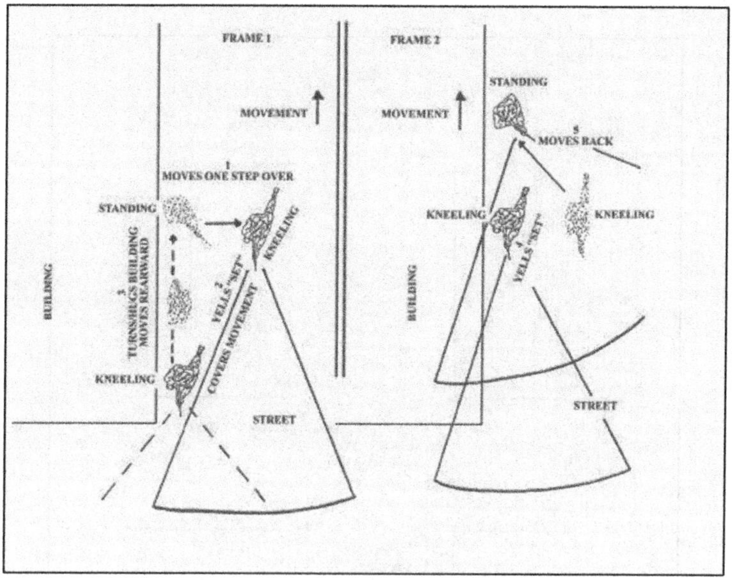

Figure A-71. Rear Security (Successive Bounds)

(2) Single Column. The single column is used to move rapidly along a street or when streets are restricted to a single column. The single column is the least preferred advance/patrol method of moving along an urban street. The same principles apply as in the double column, except the single column does not provide 360-degree security. The single column is normally used when speed is desired and contact with the enemy is considered to be unlikely.

32. Crossing a Street Intersection. Four-way intersections are very dangerous areas that should be crossed quickly. Security at an intersection is extremely important for the protection of crossing units. When moving, the pointmen notify the unit leaders, through the covermen, of an approaching intersection. Crossing the intersection will force the three elements of a unit moving along a city street to rotate positions and assume appropriate responsibilities.

a. First Fire Team To Cross. The lead (1st) fire teams in the two columns will simultaneously set security around the corners of the building. They will provide mutually supporting security to the direct front, security to the building on the far side of the intersection, and intersection security (Figure A-72). The middle (2nd) fire teams provide mutually supporting flank and rear security. On order, the rear (3rd) fire teams move forward, cross the intersection, and assume the lead (1st) fire teams' forward security mission.

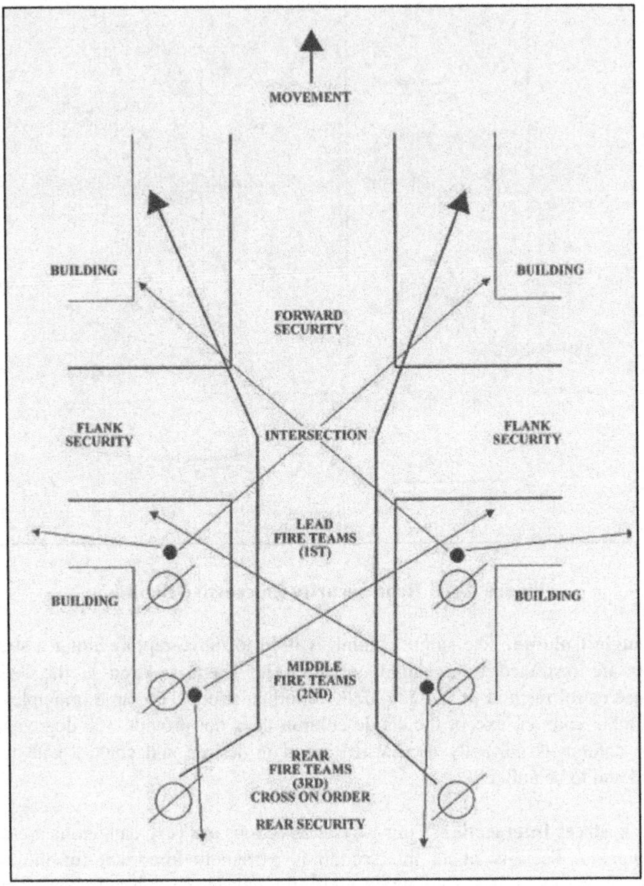

Figure A-72. Crossing an Intersection (Frame 1)

b. Second Fire Team to Cross. The 1st fire teams (now the middle fire teams) continue to provide intersection security and assume rear area security. The 2nd fire teams (now rear fire teams) cross the intersection on order. The second fire teams position themselves to provide rear and intersection security to support the covering of the last fire teams (Figure A-73).

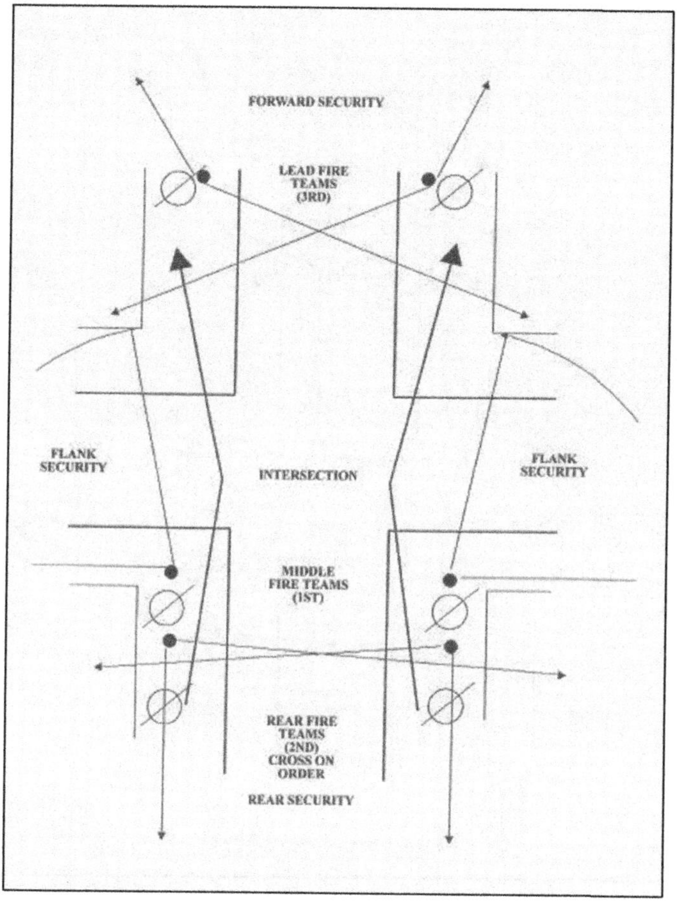

Figure A-73. Crossing an Intersection (Frame 2)

c. Third Fire Team To Cross. The 1st fire teams (now the rear fire teams) provide rear security and cross the intersection on order. The 1st fire teams bypass the 2nd fire teams to become the middle fire teams and provide mutually supporting forward and flank security. The 2nd fire teams become the rear security (Figure A-74).

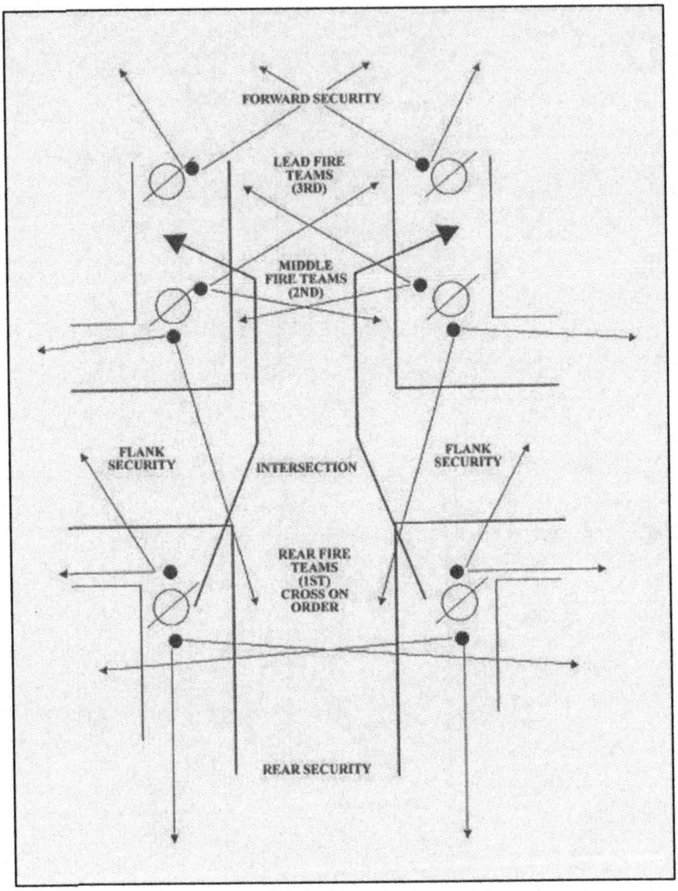

Figure A-74. Crossing an Intersection (Frame 3)

d. The squads and three fire teams have now crossed the intersection, and the fire teams have rotated (Figure A-75). This rotation is convenient because it alternates the lead units, who may quickly become fatigued during MOUT operations. The same procedures are used to cross an intersection with a single column.

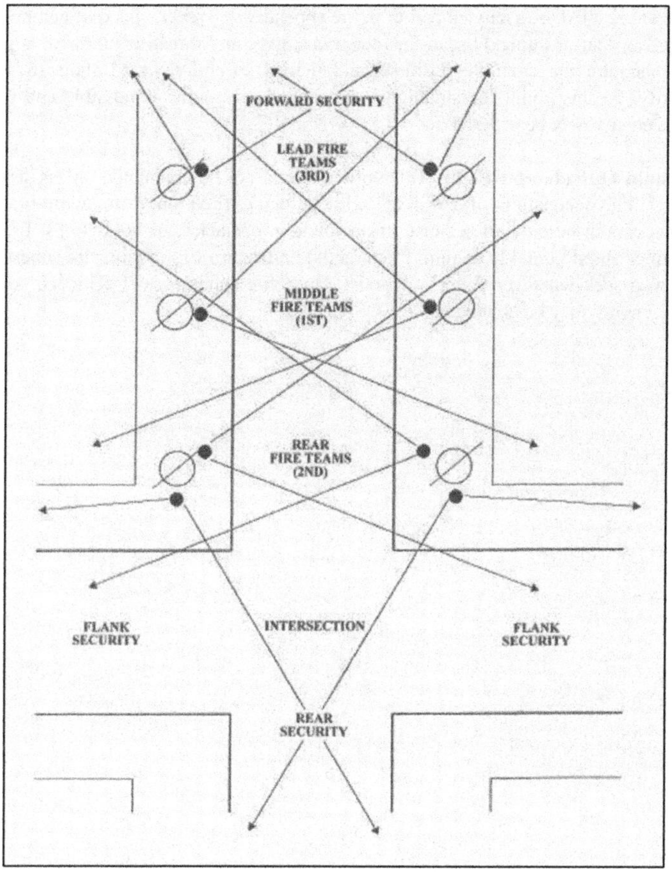

Figure A-75. Crossing an Intersection (Frame 4)

33. Reacting to Enemy Contact. When a platoon or squad is moving along a city street, contact with the enemy could happen at any time, even if the area has been declared secure. If a unit engages or is engaged by the enemy, there are two basic options that Marines may follow. The first and most preferred option is to immediately return fire and conduct a hasty clearing of a structure to seek cover inside of the building to which the unit was adjacent. The second and least preferred method is to remain outside of the building(s) and fight from the street.

a. Hasty Clearing. To conduct a hasty clearing of a structure in order to seek cover, Marines immediately return fire and enter buildings as quickly as possible. The basic principles of entry are applied as discussed earlier in the appendix; however, speed of action is essential. Once inside, Marines quickly scan the area and engage any threat. If no threat is present, the Marines acquire the enemy's positions and deliver carefully aimed shots to achieve fire superiority. At this point, the situation is developed as rapidly as possible and an attack to clear the enemy may be ordered.

b. Remain Outside and Fight. This option should not be conducted unless strict ROE do not allow the occupation of buildings without being fired on from them first or unless buildings cannot be entered because of obstruction, obstacles, or boobytraps. In such cases, the Marines should quickly acquire the target(s) and return fire to gain fire superiority. They should also seek whatever cover may exist. Once fire superiority is achieved, an assault to clear the enemy may be conducted.

Section VIII

Navigation in Built-Up Areas

Navigation in built-up areas present a unique set of challenges. Deep in the city core, the normal terrain features depicted on maps may not apply—buildings become the major terrain features, and units become tied to streets. Fighting in the city destroys buildings, whose rubble then blocks streets. Street and road signs are destroyed during the fighting if they are not switched or removed by the defenders. Operations in subways and sewers present other unique challenges. However, maps and photographs are available to help units overcome these problems. Additionally, the global positioning system (GPS) can help supplement navigational abilities in built-up areas.

34. Military Maps. The military city map is a topographical map of a city that delineates streets and shows street names, important buildings, and other urban elements. The scale of a city map can vary from 1:25,000 to 1:50,000 to 1:100,000 depending on the importance and size of the city, the density of detail, and intelligence information available.

a. Special maps prepared by supporting topographical engineers can assist units in navigating in built-up areas. These maps are designed or modified to provide information that is not covered on a standard military map. This may include maps of road and bridge networks, railroads, built-up areas, and electric power fields. Special maps can be used to supplement military city maps and topographical maps.

b. Once in the built-up area, Marines may use street intersections as reference points as they would use hills and streams in rural terrain. City maps supplement or replace topographical maps as the basis of navigation. These maps enable units moving in the built-up area to know where they are and to move to new locations even though streets have been blocked or a key building has been destroyed.

c. The techniques of compass reading and pace counting can still be used, especially in a blacked-out city where street signs and buildings are not visible. The presence of steel and iron in the MOUT environment may cause inaccurate compass readings. Sewers may be navigated in much the same way. Maps providing the basic layout of the sewer system are maintained by city sewer departments. This information includes directions in which the sewer lines run and distances between manhole covers. Along with basic compass and pace-count techniques, such information enables a unit to move through the city sewers with accuracy.

d. Operations in a built-up area adversely affect the performance of some types of communications-electronic devices such as the GPS, the Position Location Reporting System (PLRS), or other data-distribution systems. These systems function the same as some communications equipment—by LOS. They cannot determine underground locations or positions within a building. These systems should be employed on the tops of buildings, in open areas, and down streets where obstacles will not affect LOS readings.

e. City utility workers are assets to units fighting in built-up areas. They can provide maps of sewers and electrical fields and information about the city. This is especially important with regard to the use of the sewers. Sewers can contain pockets of methane gas that are highly toxic to humans. City sewer workers know the locations of these danger areas and can advise a unit on how to avoid them.

35. Aerial Photographs. Current aerial photographs are also excellent supplements to military city maps and can be substituted for a map. A topographical map or military city map could be obsolete if compiled many years ago. A recent aerial photograph shows changes that have taken place since the map was made. This could include destroyed buildings and streets that have been blocked by rubble as well as enemy defensive preparations. More information can be gained by using aerial photographs and maps together than by using either one alone.

Section IX

Camouflage

To survive and win in combat in built-up areas, a unit should supplement cover and concealment with camouflage. To properly camouflage men, carriers, and equipment, Marines should study the surrounding area and make their fighting positions blend with the local terrain.

36. Application. Only the proper amount of material needed for camouflaging a position should be used. Excess material could reveal the position. Material should be obtained from a wide area around the position. For example, Marines defending a cinder-block building do not strip the front, sides, or rear of the building to camouflage a position.

a. Buildings provide numerous opportunities for establishing concealed positions. Armored vehicles can often find prospective positions under archways or inside small industrial or commercial structures. Thick masonry, stone, or brick walls offer excellent protection from direct fire and may provide concealed routes into or out of the position.

b. After camouflage is completed, the Marine inspects his position from the enemy's viewpoint. He makes routine checks to see whether the camouflage remains natural looking and actually conceals the position. If it does not look natural, the Marine should rearrange or replace it until a natural look is achieved.

c. Positions should be progressively camouflaged as they are prepared. Work should continue until all camouflage is complete. If the enemy has air superiority, work may be possible only at night. Shiny or light-colored objects that attract attention from the air should be hidden.

d. Camouflage face paint is issued in three standard, two-tone sticks. When issue-type face-paint sticks are not available, burnt cork, charcoal, or lampblack can be used to tone down exposed skin. Mud may be used as a last resort because it dries and may peel off, leaving the skin exposed, and may also contain harmful bacteria.

37. Use of Shadows. Buildings throw sharp shadows that can be used to conceal vehicles and equipment (Figure A-76). Marines should avoid movement or positions in areas that are not in shadows. Vehicles may have to be moved periodically as shadows shift during the day. Emplacements inside buildings provide better concealment and cover as well.

38. Color and Texture. Standard camouflage-pattern painting of equipment is not as effective in built-up areas as a solid, dull, dark color hidden in shadows. Because repainting vehicles before entering a built-up area is not always practical, the lighter sand-colored patterns should be subdued with mud or dirt.

Figure A-76. Use of Shadows for Concealment

a. The need to break up the silhouette of helmets and individual equipment exists in built-up areas the same as it does elsewhere. However, burlap or canvas strips are a more effective camouflage than foliage (Figure A-77). Predominant colors are normally browns, tans, and sometimes grays rather than greens, but each camouflage location should be evaluated to determine ideal colors and patterns.

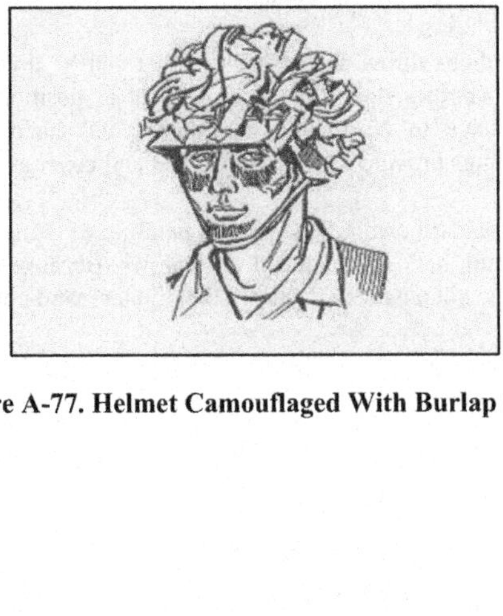

Figure A-77. Helmet Camouflaged With Burlap Strips

b. CPs and logistics emplacements are easier to camouflage and better protected if located underground. Antennas can be remoted to upper stories or to higher buildings based on remote capabilities. Field telephone wire should be laid in conduits, in sewers, or through buildings.

c. Marines should consider the background to ensure that they are not silhouetted or skylined, but rather blend into their surroundings. To defeat enemy urban camouflage, Marines should be alert for common camouflage errors such as:

- Tracks or other evidence of activity
- Shine or shadows
- Unnatural color or texture
- Muzzle flash, smoke, or dust
- Unnatural sounds and smells
- Movement.

d. Dummy positions can be used effectively to distract the enemy and make him reveal his position by firing on them.

e. Built-up areas afford cover, resources for camouflage, and locations for concealment. Following are the basic rules of cover, camouflage, and concealment to which Marines should adhere:

(1) Use the terrain and alter camouflage habits to suit your surroundings.

(2) Employ deceptive camouflage of buildings.

(3) Continue to improve positions. Reinforce fighting positions with sandbags or other fragment- and blast-absorbent material.

(4) Maintain the natural look of the area.

(5) Keep positions hidden by clearing away minimal debris for fields of fire.

(6) Choose firing ports in inconspicuous spots when available.

Note: Remember that a force that *covers* and *conceals* itself has a significant advantage over a force that does not.

Appendix B

Employment and Effects of Weapons

Success against the enemy in urban operations to a large extent depends on the effective employment of weapons. Marines at all levels must understand the effects that their weapons will produce in urban combat.

1. Effectiveness of Weapons and Demolitions. The characteristics and nature of combat in built-up areas affect the results and employment of weapons. Leaders at all levels must consider the following factors in various combinations when choosing weapons to employ.

a. Hard, smooth, flat surfaces are characteristic of urban targets. Rounds rarely impact perpendicular to these flat surfaces, but rather tend to impact at some angle of obliquity. This reduces the effect of a round and increases the threat of ricochets. The tendency of rounds to strike glancing blows against hard surfaces means that up to 25 percent of impact-fuzed explosive rounds may not detonate when fired into rubbled areas.

b. Engagement ranges in MOUT are close. Studies and historical analyses have shown that only 5 percent of all targets are more than 100 meters away. About 90 percent of all targets are located 50 meters or less from the identifying Marine. Few personnel targets will be visible beyond 50 meters, and most occur at 35 meters or less. Minimum arming ranges and unit safety from backblast or fragmentation effects must be considered.

c. Engagement times are short because of the close fight. Enemy personnel present only fleeting targets. Enemy-held buildings or structures are normally covered by mutually supporting fires and often cannot be engaged with deliberate, well-aimed shots without prolonging exposure to their observation and weapons.

d. Depression and elevation limits for some weapons create dead space. Tall buildings form deep canyons that are often safe from indirect fires. Some weapons, such as the M203, can fire rounds to ricochet behind cover and inflict casualties. Target engagement from oblique angles, both horizontal and vertical, demands superior marksmanship skills.

e. Smoke from burning buildings, dust from explosions, shadows from tall buildings, and the lack of light penetrating inner rooms all combine to reduce visibility and to increase a sense of isolation. Added to this is the masking of fires caused by rubble and manmade structures. Targets, even at close range, tend to be indistinct.

f. Urban fighting often leads to confused melees with several small units attacking on converging axes. The risks from friendly fires, ricochets, and fratricide must be considered during the planning phase of operations, and control measures must be continually adjusted to lower these risks. Marines and leaders must maintain a sense of situational awareness and clearly mark their progress in accordance with unit SOPs to avoid fratricide.

g. The firer and target may each be inside or outside the same or separate buildings. The enclosed nature of combat in built-up areas means that a weapon's signature effect, such as a muzzle blast or backblast, must be considered, as well as the round's impact on the target.

h. Usually the manmade structure that is protecting the enemy must be attacked before enemy personnel inside can be engaged. Therefore, weapons and demolitions can be chosen for employment based on their effects against masonry and concrete rather than against enemy personnel.

i. Modern engineering and design improvements mean that most large buildings constructed since World War II are resilient to the blast effects of bomb and artillery attack. Even though modern buildings may burn easily, they often retain their structural integrity and remain standing. Once high-rise buildings burn out, they are still useful to the military and are almost impossible to damage further. A large structure can take 24 to 48 hours to burn out and become cool enough for people to enter.

j. The most common worldwide building type is the 12- to 24-inch brick building. Table B-1 lists the frequency of occurrence of building types worldwide.

Type of Building	Frequency of Occurrence (Percentage)
30-inch stone	1
8- to 10-inch reinforced concrete	6
12- to 24-inch brick	63
6-inch wood	16
14-inch steel and concrete (heavy clad)	2
7-inch steel and concrete (light clad)	12

Table B-1. Types of Buildings and Frequency of Occurrence

2. M16 Rifle and M249 Squad Automatic Weapon. The M16A2 rifle and the M249 squad automatic weapon are the most common weapons used in combat in built-up areas. The M16A2 rifle and the M249 are used to kill enemy personnel, to suppress enemy fire and observation, and to penetrate light cover. Leaders can use 5.56-mm tracer fire to designate targets for other weapons.

a. Employment. Close combat is the predominant characteristic of urban engagements. Marine riflemen must be able to hit small, fleeting targets from bunker apertures, windows, and loopholes. This requires pinpoint accuracy with weapons fired in the semiautomatic mode. Killing an enemy through an 8-inch loophole at a range of 50 meters is a challenge, but one that may be common in combat in built-up areas.

(1) When fighting and engaging targets inside buildings, the weapons handling and firing techniques discussed in Appendix A are used.

(2) Within built-up areas, burning debris, reduced ambient light, strong shadow patterns of varying density, and smoke all limit the effectiveness of night vision and sighting devices. Aiming stakes and noise/light boobytraps coupled with night vision goggles and firing techniques can be used in the defense. During the offense, night vision goggles and illumination munitions can be used. Any Marine using night vision goggles should be teamed with at least one Marine not wearing them.

b. Weapon Penetration. The penetration that can be achieved with a 5.56-mm round depends on the range to the target and the type of material being fired against. Single 5.56-mm rounds are not effective against structural materials (as opposed to partitions) when fired at close range—the closer the range, the less the penetration.

(1) For the 5.56-mm round, maximum penetration occurs at 200 meters. At ranges of less than 25 meters, penetration is greatly reduced. At 10 meters, penetration by the M16 round is poor as a result of the tremendous stress placed on this high-speed round, which causes it to yaw upon striking a target. Stress causes the projectile to break up, and the resulting fragments are often too small to penetrate.

(2) Even with reduced penetration at short ranges, interior walls made of thin wood paneling, sheetrock, or plaster offer no protection against 5.56-mm rounds. Common office furniture such as desks and chairs cannot stop these rounds, but a layer of books 18 to 24 inches thick can.

(3) Wooden-framed buildings and single cinder-block walls offer little protection from 5.56-mm rounds. When clearing such structures, Marines must ensure that friendly casualties do not result from rounds passing through walls, floors, or ceilings.

(4) Armor-piercing rounds are slightly more effective than ball ammunition in penetrating urban targets at all ranges. They are also more likely to ricochet than ball ammunition, especially when the target presents a high degree of obliquity.

c. Protection. The following common barriers in built-up areas stop a 5.56-mm round fired at less than 50 meters:

- One thickness of sandbags
- A 2-inch concrete wall (unreinforced)
- A 55-gallon drum filled with water or sand
- A small ammunition can filled with sand

- A cinder block filled with sand (block will probably shatter)
- A plate-glass windowpane at a 45-degree angle (glass fragments will be thrown behind the glass)
- A brick veneer
- A car body (5.56-mm round will penetrate but normally not exit).

d. Wall Penetration. Although most structural materials repel single 5.56-mm rounds, continued and concentrated firing can breach some typical urban structures (See Table B-2).

Type	Penetration	Rounds Required
8-inch reinforced concrete	Initial Loophole	35 250
14-inch triple brick	Initial Loophole	90 160
12-inch cinder block with single-brick veneer	Loophole Breach hole	60 250
9-inch double brick	Initial Loophole	70 120
16-inch tree trunk or log wall	Initial[1]	1 to 3
12-inch cinder block (filled w/sand)	Loophole	35
24-inch double sandbag wall	Initial[1]	220
3/8-inch mild steel door	Initial[1]	1

[1]Penetration only, no loophole

Table B-2. Structure Penetration Capabilities of the 5.56-mm Round Against Typical Urban Targets (Range 25 to 100 Meters)

(1) The best method for breaching a masonry wall is by firing short bursts (3 - 5 rounds) in a U-shaped pattern. The distance from the gunner to the wall should be minimized for best results—ranges as close as 25 meters are relatively safe from ricochet. Ballistic eye protection, a protective vest, and a helmet should be worn.

(2) Ball ammunition and armor-piercing rounds produce almost the same results, but armor-piercing rounds are more likely to fly back at the firer. The 5.56-mm round can be used to create either a loophole (about 7 inches in diameter) or a breach hole (large enough for a man to enter). When used against reinforced concrete, the M16 rifle and M249 cannot cut the reinforcing bars.

3. Medium/Heavy Machine Guns and Sniper Rifles (7.62 mm and .50 caliber)

a. Machine Guns. In the urban environment, the .50-caliber machine gun and the 7.62-mm M240G machine gun provide high-volume, long-range, automatic fires for the suppression or destruction of targets. They provide final protective fires along fixed lines and can be used to penetrate light structures; the .50-caliber machine gun is most effective in this role. Tracers from both machine guns are likely to start fires, but the .50-caliber tracer is more apt to do so.

(1) Employment. The primary consideration affecting the employment of machine guns within built-up areas is the limited availability of long-range fields of fire. Although machine guns should be emplaced at the lowest level possible, grazing fire at ground level is often obstructed by rubble.

(a) The .50-caliber machine gun is often employed on its vehicular mount during both offensive and defensive operations. If necessary, it can be mounted on the M3 tripod for use in the ground role or in the upper level of buildings. When mounted on a tripod, the .50-caliber machine gun can be used as an accurate, long-range weapon and can supplement sniper fires.

(b) The M240G machine gun is cumbersome, making it difficult to use inside while clearing a building. However, it is useful outside to suppress and isolate enemy defenders. The M240G can be fired from either the shoulder or the hip to provide a high volume of assault and suppressive fires. The use of the long sling to support the weapon and ammunition is preferred.

(c) Because of their reduced penetration power, M240G machine guns are less effective against masonry targets than .50-caliber machine guns. However, their availability and light weight make them well suited to augment heavy machine gun fire, to be used in areas where .50-caliber machine guns cannot be positioned, or as a substitute when heavy machine guns are not available. The M240G machine gun can be employed on its tripod to deliver accurate fire along fixed lines and can then quickly be converted to bipod fire to cover alternate fields of fire.

(2) Penetration. The ability of the 7.62-mm and .50-caliber rounds to penetrate is also affected by the range to the target and type of material against which the rounds are fired. Close ranges affect penetration of the 7.62-mm round less than the 5.56-mm round; the .50-caliber round's penetration is reduced the least.

(a) At 50 meters, the 7.62-mm ball round cannot penetrate a single layer of sandbags. It can penetrate a single layer at 200 meters, but not a double layer. The armor-piercing round does only slightly better against sandbags. It cannot penetrate a double layer but can penetrate up to 10 inches of sand at 600 meters.

(b) The penetration of the 7.62-mm round is best at 600 meters, but most urban targets are closer. The longest engagement range is usually 200 meters or less. Table B-3 lists the penetration capabilities of a single 7.62-mm (ball) round at closer ranges.

Range (meters)	Pine Board (inches)	Dry, Loose Sand (inches)	Cinder Block (inches)	Concrete (inches)
25	13	5	8	2
100	18	4.5	10	2
200	41	7	8	2

Table B-3. Penetration Capabilities of a Single 7.62-mm (Ball) Round

(c) The .50-caliber round is also optimized for penetration at long ranges (about 800 meters). For hard targets, .50-caliber penetration is affected by obliquity and range. Both armor-piercing and ball ammunition penetrate 14 inches of sand or 28 inches of packed earth at 200 meters if the rounds impact perpendicular to the flat face of the target. Table B-4 depicts the effect of a 25-degree obliquity on a .50-caliber penetration.

Thickness (feet)	100 Meters (rounds)	200 Meters (rounds)
2	300	1,200
3	450	1,800
4	600	2,400

Table B-4. Number of Rounds Needed To Penetrate a Reinforced Concrete Wall at a 25-Degree Obliquity

(3) Protection. Barriers that offer protection against 5.56-mm rounds are also effective against 7.62-mm rounds with some exceptions. The 7.62-mm round can penetrate a windowpane at a 45-degree obliquity, a hollow cinder block, or both sides of a car body. It can also easily penetrate wooden-framed buildings. The .50-caliber round can penetrate all of the commonly found urban barriers except a sand-filled 55-gallon drum.

(4) Wall Penetration. Continued and concentrated machine gun fire can breach most typical urban walls. Such fire cannot breach thick reinforced concrete structures or dense natural stone walls. Internal walls, partitions, plaster, floors, ceilings, common office furniture, home appliances, and bedding can be easily penetrated by both 7.62-mm and .50-caliber rounds (Tables B-5 and B-6).

(a) The M240G machine gun can be hard to hold steady enough to repeatedly hit the same point on a wall. The dust created by the bullet strikes also makes precise aiming difficult. Firing with a tripod is usually more effective than without, especially if sandbags are used to steady the weapon. Short bursts of three to five rounds fired in a U-type pattern are best.

(b) Breaching a brick veneer over cinder block presents a special problem for the M240G machine gun. Rounds penetrate the cinder block but leave a net-like structure of unbroken block. Excessive ammunition is required to destroy the net because most rounds only pass through a previously eroded hole. One or two minutes of work with an entrenching tool, crowbar, or ax can remove this web and allow entry through the breach hole.

(c) The .50-caliber machine gun can be fired accurately from the tripod using the single-shot mode. This is the most efficient method for producing a loophole. Automatic fire in three- to five-round bursts in a U-type pattern is more effective in producing a breach.

Type	Thickness (Inches)	Hole Diameter (Inches)	Rounds Required
Reinforced concrete	8	7	100
Triple brick wall	14	7	170
Concrete block with single brick veneer	12	6 and 24	30 and 200
Cinder block (filled)	12	N/A[1]	18
Double brick wall	9	N/A[1]	45
Double sandbag wall	24	N/A[1]	110
Log wall	16	N/A[1]	1
Mild steel door	3/8	N/A[1]	1

[1]Penetration only, no loophole

Table B-5. Structure Penetrating Capabilities of 7.62-mm Round (NATO) Against Typical Urban Targets (Range 25 Meters)

Type	Thickness (Inches)	Hole Diameter (Inches)	Rounds Required
Reinforced concrete	10	12	50
		24	100
	18	7	140
Triple brick wall	12	8 and 26	15 and 50
Concrete block with single brick veneer	12	10 and 33	25 and 45
Armor plate	1	N/A[1]	1
Double sandbag wall	24	N/A[1]	5
Log wall	16	N/A[1]	1

[1]Penetration only, no loophole

Table B-6. Structure Penetrating Capabilities of .50-caliber Ball Against Typical Urban Targets (Range 35 Meters)

b. Sniper Rifles. The 7.62-mm match ball ammunition fired from the M-40A1 sniper rifle will perform similarly to the 7.62-mm ball round. The M-82A1A special application sniper rifle (SASR) can fire the .50-caliber armor-piercing incendiary (API) ammunition, which will have the same performance as when fired from the M-2 HB. The M-82A1A also fires the .50-caliber RAUFOSS ammunition, which contains a tungsten penetrator and a more powerful explosive charge than the API ammunition. When the RAUFOSS penetrator punches through the metal, the explosive detonates inside, acting as an incendiary round. Although the SASR's exact penetration capabilities have yet to be defined, it has penetrated an inch of steel at 2000 yards. Both sniper rifles are organic to the Marine infantry battalion.

4. Grenade Launchers, 40-mm (M203 and MK19). Both the M203 dual-purpose weapon and the MK19 grenade machine gun fire 40-mm HE and HE dual-purpose (HEDP) ammunition. Ammunition for these weapons is not interchangeable, but the grenade and fuze assembly that actually hits the target is identical. Both weapons provide point and area destructive fires as well as suppression. The MK19 has a much higher rate of fire and a longer range. The M203 is much lighter and more maneuverable.

a. Employment. The main consideration affecting the employment of 40-mm grenades within built-up areas is the typically short engagement range. The 40-mm grenade has a *minimum arming range of 14 to 28 meters*. If the round strikes an object before it is armed, it will not detonate. Both the HE and HEDP rounds have 5-meter burst radii against exposed forces, which means that the *minimum safe firing range for combat is 33 meters*. The 40-mm grenades can be used to suppress the enemy in a building or inflict casualties by firing through apertures or windows. The MK19 can use its high rate of fire to concentrate rounds against light structures. This concentrated fire can create extensive damage. The 40-mm HEDP round can penetrate the armor on the flank, rear, and top of Soviet-made BMPs and BTRs. Marines can use the M203 from upper stories to deliver accurate fire against the top decks of armored vehicles. Multiple hits are normally required to achieve a kill.

b. Weapon Penetration. The 40-mm HEDP grenade has a small shaped charge that penetrates better than the HE round. It also has a thin wire wrapping that bursts into a dense fragmentation pattern, creating casualties out to 5 meters. Because they explode on contact, 40-mm rounds achieve the same penetration regardless of range. Table B-7 lists the penetration capabilities of the HEDP round.

Target	Penetration (Inches)
Sandbags	20
Sand-filled cinder block	16
Pine logs	12
Armor plate	2

Table B-7. Penetration Capabilities of the HEDP Round

(1) If projected into an interior room, the 40-mm HEDP can penetrate all interior partition-type walls. It splinters plywood and plaster walls, making a hole large enough to fire a rifle through. It is better to have HEDP rounds pass into a room and explode on a far wall, even though much of the round's energy is wasted penetrating the back wall (Figure B-1). The fragmentation produced in the room causes more casualties than the high-explosive jet formed by the shaped charge.

(2) The fragments from the HEDP round do not reliably penetrate interior walls. They are also stopped by office furniture, sandbags, helmets, and protective vests (flak jackets). The M203 dual-purpose weapon has the inherent accuracy to place grenades into windows at 125 meters and bunker apertures at 50 meters. These ranges are significantly reduced as the angle of obliquity increases. Combat experience shows that M203 gunners cannot consistently hit windows at 50 meters when forced to aim and fire quickly.

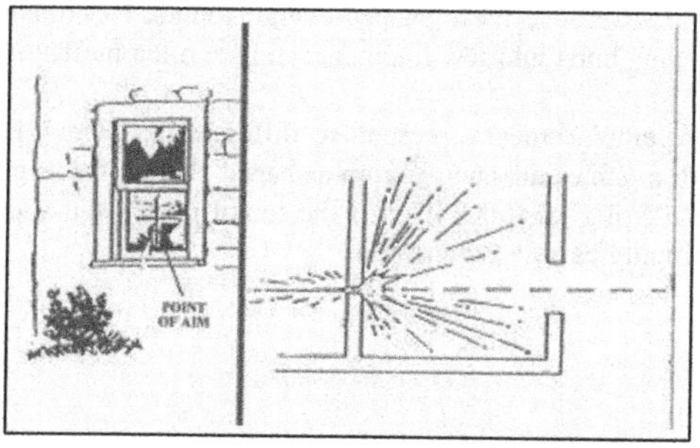

Figure B-1. Aim Point for 40-mm HEDP

c. Wall Penetration. The M203 cannot reasonably deliver the rounds needed to breach a typical exterior wall. The MK19 can concentrate its fire and achieve wall penetration. Firing from a tripod using a locked-down traversing and elevating mechanism is best for this role. Brick, cinder block, and concrete can be breached by using the MK19; individual HEDP rounds can penetrate 6 to 8 inches of brick. The only material that has proven resistant to concentrated 40-mm fire is dense stone such as that used in some European building construction. No precise data exist as to the number of rounds required to produce loopholes or breach holes with the MK19. However, the rounds' explosive effects are dramatic and should exceed the performance of the .50-caliber machine gun.

5. Light and Medium Recoilless Weapons. Light and medium recoilless weapons are used to attack enemy personnel, field fortifications, and LAVs. They have limited capability against main battle tanks, especially those equipped with reactive armor, except when attacking from the top, flanks, or rear. This category of weapons includes the AT4, the M47 Dragon, and the SMAW.

a. Employment. Other than defeating LAVs, the most common task for which light recoilless weapons are used is to neutralize fortified firing positions. Because of the design of the warhead and the narrow blast effect, these weapons are not as effective in this role as heavier weapons such as a tank main gun round. Their light weight allows Marines to carry several AT4s. Light recoilless weapons can be fired from the tops of buildings or from areas with extensive ventilation.

(1) Light and medium recoilless weapons with the exception of the SMAW employ shaped-charge warheads. As a result, the hole they punch in walls is often too small to use as a loophole. The fragmentation and spall that these weapons produce are limited. Normally, shaped-charge warheads do not neutralize enemy forces behind walls unless they are located directly in line with the point of impact.

(2) Against structures, shaped-charge weapons should be aimed about 6 inches below or to the side of a firing aperture (Figure B-2). This enhances the probability of killing the enemy behind the wall. A round that passes through a window wastes much of its energy on the back wall. Because these shaped-charge rounds lack the wire wrapping of the 40-mm HEDP, they burst into few fragments and are often ineffective casualty producers.

(3) Sandbagged emplacements present a different problem (Figure B-3). Because sandbags absorb much of the energy from a shaped charge, the rounds should be aimed at the center of the firing aperture. Even if the round misses the aperture, the bunker wall area near it is usually easiest to penetrate.

Figure B-2. Point of Aim for a Shaped-Charge Weapon Against a Masonry Structure

Figure B-3. Point of Aim for Sandbagged Emplacement

(4) Light and medium recoilless weapons obtain their most effective short-range antiarmor shots by firing from upper stories or from the flanks and rear. When firing at main battle tanks, these weapons should always be employed against these weaker areas in volley or paired firing. They normally require multiple hits to achieve a kill on a tank. Flank, top, and rear shots hit the most vulnerable parts of armored vehicles. Firing from upper stories protects the firer from tank main gun and coaxial machine gun fire because tanks cannot sharply elevate their cannons. The BMP-2 *can* elevate its 30-mm cannon to engage targets in upper stories. The BTR-series armored vehicles *can* also fire into upper stories with their heavy machine gun.

(5) Modern IFVs such as the BMP-2 and the BTR-80 have significantly improved frontal protection against shaped-charge weapons. Many main battle tanks have some form of reactive armor in addition to their thick armor plate. Head-on, ground-level shots against these vehicles have little probability of obtaining a kill. Even without reactive armor, modern main battle tanks are hard to destroy with a light antiarmor weapon.

(6) The easiest technique to improve the probability of hitting and killing an armored vehicle is to increase the firing depression angle. A 45-degree downward firing angle doubles the probability of a first-round hit as compared to a ground-level shot.

b. Backblast. Backblast characteristics must be considered when employing all recoilless weapons. During combat in built-up areas, the backblast area in the open is more hazardous because of the loose rubble and the channeling effect of the narrow streets and alleys. Figure B-4 shows the backblast areas of Marine light recoilless weapons in the open.

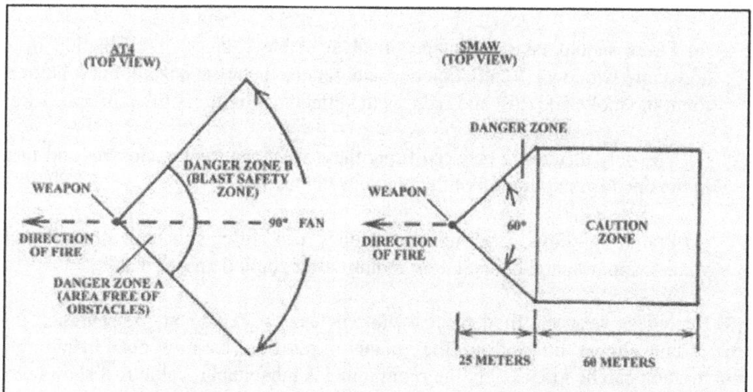

Figure B-4. Backblast Areas of Light Recoilless Weapons in the Open

(1) When firing recoilless weapons in the open, Marines should protect themselves from blast and burn injuries caused by the backblast. All personnel should be out of the danger zone. Anyone not able to vacate the caution zone should be behind cover. Marines in the caution zone should wear helmets, protective vests, and eye protection. The firer and all Marines in the area should wear earplugs.

(2) Extensive testing on the effects of firing recoilless weapons from within enclosures has been conducted since 1948. These tests showed that firing these weapons from enclosures presented no serious hazards, even when the overpressure was enough to produce structural damage to the building. The following were other findings of these tests.

(a) Little hazard exists to the gunnery or crew from any type of flying debris. Loose items were not hurled around the room.

(b) No substantial degradation occurs to the operator's tracking performance as a result of obscuration or blast overpressure.

(c) The most serious hazard that can be expected is hearing loss. This must be evaluated against the advantage gained in combat from firing from cover. To place this hazard in perspective, a gunner wearing earplugs and firing the loudest combination (the Dragon from within a masonry building) is exposed to less noise hazard than if he fired a LAW or AT4 in the open without earplugs.

(d) The safest place for other Marines in the room with the firer is against the wall from which the weapon is fired. Plastic ignition plugs are a hazard to anyone standing directly behind a LAW or TOW when it is fired.

(e) Firers should take advantage of all available sources of ventilation by opening doors and windows. Ventilation does not reduce the noise hazard, but it helps clear the room of smoke and dust and reduces the effective duration of the overpressure.

(f) The only difference between firing these weapons from enclosures and firing them in the open is the duration of the pressure fluctuation.

(g) Frame buildings, especially small ones, can suffer structural damage to the rear walls, windows, and doors. Large rooms suffer slight damage, if any.

(3) Recoilless weapons fired from within enclosures create some obscuration inside the room but almost none from the gunner's position looking out. Inside the room, obscuration can be intense, but the room remains inhabitable. Table B-8 shows the effects of smoke and obscuration.

Building	Weapon	From Gunner's Position Looking Out	Inside the Room	From Outside at a Distance
Masonry	LAW	None	Moderate	Slight smoke
	Dragon	Slight	Moderate	Small flash

Bunker	Dragon	None	Slight	Moderate flash
	TOW	None	Slight	Moderate smoke
Small frame	LAW	None	Moderate	Moderate
	Dragon	None	Severe	Moderate
Medium frame	LAW	None	Slight	Moderate
	Dragon	None	Severe	Slight flash
Large frame	LAW	None	Slight	None
	Dragon	Slight	Severe	Slight flash
	TOW	None	Severe	Slight smoke

Table B-8. Smoke and Obscuration

(4) The Dragon causes the most structural damage, but only in frame buildings. There does not seem to be any threat of injury to the gunner because the damage is usually to the walls away from the gunner. The most damage and debris is from flying plaster chips and pieces of wood trim. Large chunks of plasterboard can be dislodged from ceilings. The backblast from a LAW, AT4, Dragon, or TOW rarely displaces furniture. Table B-9 shows the test results of structural damage and debris.

		Damage		
Building	**Weapon**	**Structure**	**Wall Covering**	**Debris Movement**
Masonry	LAW	None	Slight	Slight
	Dragon	None	Slight	Slight
Bunker	Dragon	None	None	None
	TOW	None	None	Leaves and dust disturbed
Small frame	LAW	None	Slight	None
	Dragon	Severe	Severe	None
Medium frame	LAW	None	None	Slight
	Dragon	Slight	Slight	Lamp and chair overturned
Large frame	LAW	None	Slight	Slight
	Dragon	Slight	Moderate	None
	TOW	Slight	Severe	None

Table B-9. Structural Damage and Debris Movement

(5) To fire a LAW, AT4, or SMAW from inside a room, the following safety precautions must be taken (Figure B-5).

(a) The building should be of a sturdy construction.

(b) The ceiling should be at least 7 feet high with loose plaster or ceiling boards removed.

(c) The floor size should be at least 15 feet by 12 feet. (The larger the room, the better.)

(d) At least 20 square feet of ventilation (room openings) should exist to the rear or side of the weapon. An open 7- by 3-foot door would provide minimum ventilation.

(e) All glass should be removed from windows, and small, loose objects should be removed from the room.

(f) Floors should be wet to prevent dust and dirt from blowing around and obscuring the gunner's vision.

(g) All personnel in the room should be forward of the rear of the weapon.

(h) All personnel in the room should wear helmets, protective vests, ballistic eye protection, and earplugs.

(i) If the gunner is firing from the prone position, his lower body must be perpendicular to the bore of the weapon, or the blast could cause injury to his legs.

Figure B-5. Firing an AT4 From Inside a Building

c. Weapon Penetration. The most important tasks to be performed against structures are the neutralization of fortified firing positions, personnel, and weapons behind barriers. Recoilless weapons can be used in this role; none, however, are as effective as heavy, direct-fire weapons or standard demolitions. Each recoilless weapon has different penetrating ability against various targets. Penetration does not always mean the destruction of the integrity of a position. Usually only those enemy soldiers directly in the path of the spall from a high-explosive antitank (HEAT) round become casualties. Other soldiers inside a fortification could be deafened, dazed, or shocked but could eventually return to action.

(1) AT4. The AT4 is heavier than the LAW and has a diameter of 84 millimeters, which gives the warhead much greater penetration. The AT4 can penetrate more than 17.5 inches (450 mm) of armor plate. Its warhead produces highly destructive results behind the armor. Tests against typical urban targets are still ongoing, but the AT4 should penetrate at least as well as the 90-mm recoilless rifle, if not better. The AT4 has a *minimum arming distance of 10 meters*, which allows it to be fired successfully against close targets. Firers should be well covered and protected when firing at close targets.

(2) MK153 83-mm SMAW. The SMAW is a lightweight assault weapon that is easily carried and placed into action by one man. It is used against fortified positions, but it is also effective against LAVs. The SMAW has a 9-mm spotting rifle and a 3.8-power telescope that ensure accuracy over ranges common to combat in built-up areas. The SMAW has excellent incapacitating effects for enemies behind walls and inside bunkers and can arm within 10 meters. Its fuze has the ability to distinguish between armor and soft earth, maximizing its capabilities against buildings, bunkers, or light armor. The warhead detonates immediately against hard targets but delays detonation against soft targets and burrows in to explode inside. The SMAW can destroy most bunkers with a single hit. The 83-mm HE warhead can create an 8-inch mousehole in reinforced concrete but will not cut the steel reinforcing bar. The SMAW is an outstanding urban support weapon that can be decisive during an urban assault. The SMAW allows for quick reduction of obstacles and bunkers as well as creation of entry points. SMAWs fired in a volley can be devastating to a building.

d. Wall Breaching. Wall breaching is a common combat task in built-up areas for which light recoilless weapons can be used. Breaching operations improve mobility by providing access to building interiors without using existing doors or windows. Breaching techniques can also be used to create loopholes for weapons positions or to allow hand grenades to be thrown into defended structures. Breaching holes for unit mobility should be about 24 inches (60 centimeters) in diameter. Loopholes should be about 8 inches (20 centimeters) in diameter (Figure B-6). Only the SMAW provides a one-shot wall breaching ability. However, in breaching walls a number of shots should be planned.

Figure B-6. Tactical Use of Holes in Masonry Walls

(1) Of all the common building materials, heavy stone is the most difficult to penetrate. The AT4 normally will not penetrate a heavy European-style stone wall. Surface cratering is usually the only effect it will have.

(2) Layered brick walls are also difficult to breach with light recoilless weapons. Some brick walls can be penetrated by multiple firings, especially if they are less than three bricks thick. Multiple AT4 rounds fired at the same spot on an 8-inch (double-brick) wall normally produce a loophole. The SMAW produces a hole in brick walls that is often large enough to be a breach hole.

(3) Wooden structural walls offer little resistance to light recoilless weapons. Even heavy timbered walls are penetrated and splintered. Multiple AT4 rounds fired at the same area of a wood-framed wall usually produce a man-sized hole. The SMAW has a devastating effect against wood-framed walls. A single round produces a breach hole as well as significant spall.

(4) Because of its high velocity, the AT4 may penetrate a soft target such as a car body or frame building before exploding.

(5) None of the light recoilless weapons are as effective against structural walls as demolitions or heavier weapons such as tank main guns or field artillery. Of all the light recoilless weapons, the SMAW is the most effective.

6. Antitank Guided Missiles. ATGMs are used mainly to defeat main battle tanks and other armored combat vehicles. They have a moderate capability against bunkers, buildings, and other fortified targets commonly found during combat in built-up areas. This category of weapons includes the TOW and Dragon missiles.

a. Employment. TOWs and Dragons provide overwatch antitank fires during the attack of a built-up area and an extended range capability for the engagement of armor during the defense. Within built-up areas, they are best employed along major thoroughfares and from the upper stories of buildings to attain long-range fields of fire. Their *minimum firing range of 65 meters* could limit firing opportunities in the confines of densely built-up areas.

(1) Obstacles. Rubble or other obstacles could interfere with the flight of missiles when fired from street level. At least 3.5 feet (1 meter) of vertical clearance over such obstacles must be maintained. Figure B-7 shows the most common obstacles to ATGM flight that are found in built-up areas. *Power lines are a special obstacle and present a unique threat to ATGM gunners. If the power in the lines has not been interrupted, the ATGM guidance wires could create a short circuit. This would allow extremely high voltage to pass to the gunner in the brief period before the guidance wires melted. This voltage could either damage the sight and guidance system or injure the gunner. Before any ATGM is fired over a power line, an attempt must be made to determine whether or not the power to that line has been interrupted.*

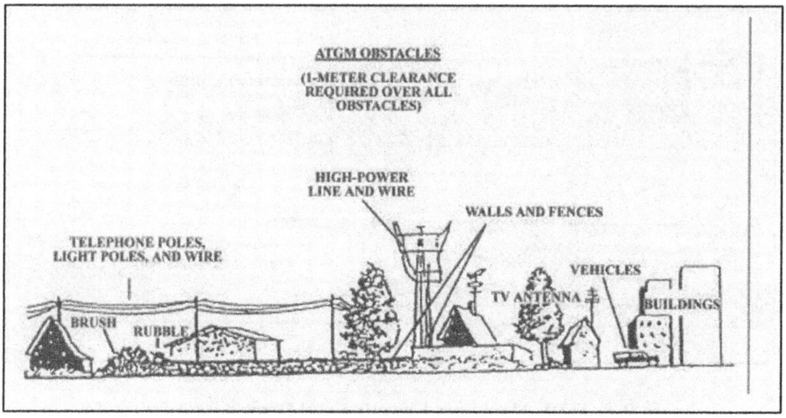

Figure B-7. Common Obstacles to ATGM Flights

(2) Dead Space. Three aspects of dead space that affect ATGM fires are arming distance, maximum depression, and maximum elevation.

(a) Both the Dragon and TOW missiles have a minimum arming distance of 65 meters, which severely limits their use in built-up areas. Few areas in the inner city permit fires much beyond the minimum arming distance. However, ground-level, long-range fires down streets or rail lines and across parks or plazas are possible. ATGMs may be used effectively from upper stories or roofs of buildings to fire into other buildings.

(b) The TOW is limited much more than the Dragon by its maximum depression and elevation. The maximum depression and elevation limits of the TOW mount could result in dead space and preclude the engagement of close targets (Figure B-8). A target located at the minimum arming range (65 meters) cannot be engaged by a TOW crew located any higher than the sixth floor of a building due to maximum depression limits. At 100 meters, the TOW crew can be located as high as the ninth floor and still engage the target.

(3) Backblast. Backblast for ATGMs is of more concern during combat in built-up areas than in open country. Any loose rubble in the caution zone could be picked up and thrown by the backblast, and the channeling effect of walls and narrow streets is even more pronounced due to the greater backblast. If the ATGM backblast strikes a wall at an angle, it can pick up debris or be deflected and cause injury to unprotected personnel (Figure B-9). Both types of ATGM can be fired from inside some buildings. In addition to the helmet and protective vest, eye protection and earplugs should be worn by all personnel in the room.

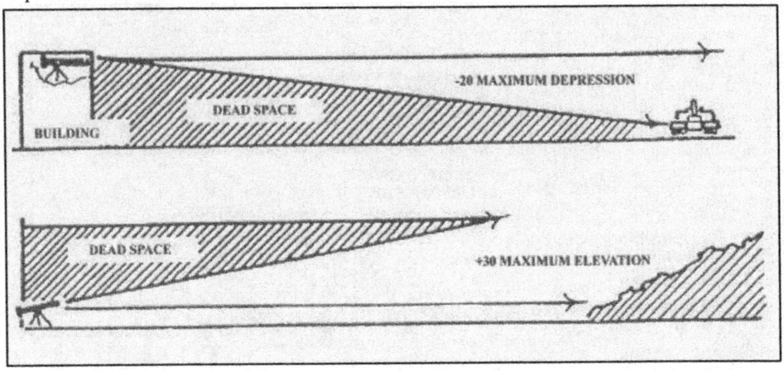

Figure B-8. TOW Maximum Elevation and Depression Limitations

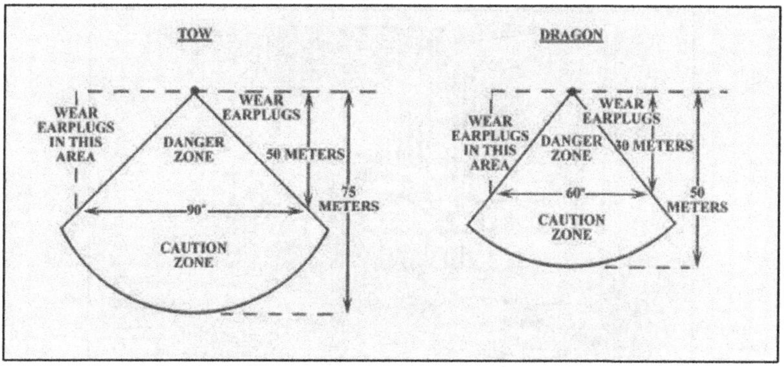

Figure B-9. ATGM Backblast in an Open Street

(a) To fire a TOW from inside a room, the following safety precautions must be taken (Figure B-10):

- The building must be of sturdy construction.
- The ceiling should be at least 7 feet high.
- The floor size of the room should be at least 15 by 15 feet, or larger if possible.
- At least 20 square feet of room ventilation should exist, preferably to the rear of the weapon. An open 7- by 3-foot door is sufficient. Additional ventilation can be created by removing sections of interior partitions.
- All glass must be removed from the windows and all small, loose objects removed from the room.
- All personnel in the room should be forward of the rear of the TOW.
- All personnel in the room should wear ballistic eye protection and earplugs.
- A clearance of 9 inches (23 centimeters) must exist between the launch tube and the aperture from which it is fired. (See FM 23-34, *TOW Weapon System*, for more detailed safety information.)

Figure B-10. TOW Fired From Inside a Room

(b) To fire a Dragon from inside a room, the following safety precautions must be taken.

- The building must be of sturdy construction.
- The ceiling should be at least 7 feet high.
- The floor size should be at least 15 by 15 feet, or larger if possible.
- At least 20 square feet of ventilation should exist (room openings), preferably to the rear of the weapon. An open 7- by 3-foot door would provide minimum ventilation.
- All glass should be removed from windows, and small, loose objects should be removed from the room.
- The room should be clean or the floors must be wet to prevent dust and dirt (kicked up by the backblast) from obscuring the vision of other soldiers in the room.
- All personnel in the room must be forward of the rear of the weapon.
- All personnel in the room must wear ballistic eye protection and earplugs.
- At least a 6-inch clearance must exist between the launch tube and the aperture from which it is fired.

b. Weapon Penetration. ATGMs can penetrate and destroy heavily armored tanks. They have large warheads that contain shaped charges. Because of their size, these warheads can achieve significant penetration against typical urban targets. Penetration, however, does not mean a concurrent destruction of the structural integrity of a position. The shaped-charge warhead produces relatively little spall. Enemy personnel not standing directly behind or near the point of impact of an ATGM may escape injury.

(1) Standard TOW Missiles. The basic TOW missile can penetrate 8 feet of packed earth, 4 feet of reinforced concrete, or 16 inches of steel plate. The improved TOW (ITOW), the TOW 2, and the TOW 2A have all been modified to improve their penetration. They all penetrate better than the basic TOW. All TOW missiles can defeat triple sandbag walls, double layers of earth-filled 55-gallon drums, and 18-inch log walls.

(2) TOW 2B. The TOW 2B uses a different method of defeating enemy armor. It flies over the target and fires an explosively formed penetrator down onto the top armor, which is thinner. *Because of this design feature, the TOW 2B missile cannot be used to attack nonmetallic structural targets. When using the TOW 2B missile against enemy armor, gunners must avoid firing directly over other friendly vehicles, disabled vehicles, or large metal objects such as water or oil tanks.*

(3) Dragon Missile. The Dragon missile can penetrate 8 feet of packed earth, 4 feet of concrete, or 13 inches of steel plate. It can attain effective short-range fire from upper stories or from the rear or flanks of a vehicle. These engagements are targeted against the most vulnerable parts of tanks and can entrap tanks in situations where they are unable to counterfire. Elevated firing positions increase the first-round hit probability. Firing down at an angle of 20 degrees increases the chance of a hit by 67 percent at 200 meters. A 45-degree downward angle doubles the first-round hit probability, compared to a ground-level shot.

c. Breaching Structural Walls. Firing ATGMs is the least efficient means to defeat structures. Because of their small basic load and high cost, ATGMs are better used against tanks or enemy-fortified firing positions. They can be effective against bunkers or other identified enemy firing positions.

7. Flame Weapons. Flame weapons are characterized by both physical and psychological casualty-producing abilities. Flame does not need to be applied with pinpoint accuracy, but it also must not spread to structures needed by friendly forces. Large fires in built-up areas are catastrophic. If they burn out of control, fires can create an impenetrable barrier for hours. The Marine Corps does not possess flame-generating equipment. However, the M34 WP grenade can be used to ignite and destroy flammable objects, especially wooden structures. The Army's M2A1-7 portable flamethrower is stored in war reserve status as a standard "C" item. Its availability is limited; however, Marines could be issued flame-support weapons for urban operations depending on the status of international agreements on the use of flame weapons that are in effect at the time.

a. Employment. Flame weapons used against fortified positions should be aimed directly at the aperture. Even if the round or burst misses, enough flaming material enters the position to cause casualties and disrupt the enemy occupants. The M34 WP grenade is difficult to throw far or into a small opening such as a bunker aperture. However, its effects are dramatic when thrown into a room or building.

b. Effects. The M34 is used to ignite and destroy flammable objects, especially wooden structures. It is also used to create an immediate smoke cloud to conceal movement across a narrow open space such as a street. Its smoke is not toxic but can cause choking in heavy concentrations.

(1) The grenade's explosion, bright flash, smoke, and burning WP particles all combine to make the M34 one of the most effective psychological weapons available. The M34 hand grenade throws WP fragments up to 35 meters from the point of detonation. These fragments can attach to clothing or skin and continue burning. Because of its weight, most infantrymen can throw this grenade only 30 to 40 meters.

(2) The Marine must avoid injury from friendly use of the M34. The M34 can ignite if the WP inside is exposed to the air. Bullets and shell fragments have been known to strike and rupture M34 grenades; therefore, grenades must be protected from enemy fire.

(3) The M34 WP grenade is an effective weapon against enemy armored vehicles when used in the close confines of combat in built-up areas. It can be thrown or dropped from upper stories onto enemy vehicles. The M34 can be combined with flammable liquids, detonating cords, blasting caps, and fuze igniters to create the eagle fireball, a field-expedient antiarmor device. (See FM 21-75, *Combat Skills of the Soldier*, Appendix H.)

(4) The M34 is also excellent as a screening device. A grenade can be thrown from behind cover into an open street or plaza. When it explodes, the enemy's observation is temporarily obscured. Thus, friendly forces can quickly cross the open area—if the enemy fires, it is unaimed and presents less of a danger. If screening smoke is used to cover a squad's movement across short open areas, it will reduce expected casualties from small-arms fire by about 90 percent.

8. Hand Grenades. Hand grenades are used extensively during combat in built-up areas. Smoke grenades are used for screening and signaling. Riot control grenades are used to drive the enemy out of deep fortifications. Fragmentation and concussion grenades are used to clear the enemy out of rooms and basements. Hand grenades are the most-used explosive munition during intense combat in built-up areas. In World War II, it was common for a battalion fighting in a city to use more than 500 hand grenades each day.

a. Employment. Smoke and riot-control grenades have similar employment techniques. Fragmentation grenades are used to produce enemy casualties.

(1) The AN-M8 HC grenade produces a dense white or grey smoke. It burns intensely and cannot be easily extinguished once it ignites. The smoke can be dangerous in heavy concentrations because it makes breathing difficult and causes choking. The M8 grenade is normally used for screening. It produces a slowly building screen of longer duration than the M34 WP grenade without the problem of collateral damage caused by scattered burning particles.

(2) The M18-series smoke grenades produce several different colors of smoke that are used for signaling. Yellow smoke is sometimes difficult to see in built-up areas.

(3) The M7A3 CS riot control grenade can be used to drive enemy troops out of fortifications when civilian casualties or collateral damage constraints are considerations. Built-up areas often create variable and shifting wind patterns. When using CS grenades, Marines must prevent the irritating smoke from affecting friendly forces. The CS grenade burns intensely and can ignite flammable structures. Enemy troops wearing even rudimentary chemical protective masks can withstand intense concentrations of CS gas.

(4) The MK3A2 offensive hand grenade, commonly referred to as the concussion grenade, produces casualties during close combat while minimizing the danger to friendly personnel. For this reason, it is the preferred hand grenade for use during offensive operations in a MOUT environment. The grenade produces severe concussion effects in enclosed areas. It can be used for light blasting and demolitions and for creating breach holes in interior walls. The concussion produced by the MK3A2 is much greater than that of the fragmentation grenade. It is very effective against enemy soldiers within bunkers, buildings, and underground passages.

(5) The M67 fragmentation grenade is the most commonly available grenade during combat in built-up areas. It provides suppression during room-to-room or house-to-house fighting, and it is used while clearing rooms of enemy personnel. The M67 grenade has a 3- to 5-second delay fuze. When used at close ranges, it can be cooked-off for two seconds to deny the enemy the time to throw it back. The fragmentation grenade can be rolled, bounced, or ricocheted into areas that cannot be reached by 40-mm grenade launchers.

b. Effects. Each type of hand grenade has its own specific effect during combat in built-up areas.

(1) Both HC smoke and CS gas *displace* oxygen from enclosed areas. Gas masks will not prevent this from occurring because they only filter air, not provide oxygen. Smoke grenades produce dense clouds of colored or white smoke that remain stationary in the surrounding area. They can also cause fires if used indiscriminately. If trapped and concentrated within a small space, smoke grenades and CS gas *can suffocate Marines*.

(2) The fragmentation grenade has more varied effects in combat in built-up areas. It produces a large amount of small, high-velocity fragments that can penetrate sheetrock partitions and are lethal at short ranges (15 to 20 meters). Fragments lose their velocity quickly and are less effective beyond 25 meters.

The fragments from a fragmentation grenade cannot penetrate a single layer of sandbags, a cinder block, or a brick building, but they can perforate wood-framed and tin buildings if exploded close to their walls.

(3) Inside rooms, fragmentation barriers consisting of common office furniture, mattresses, doors, or books can be effective against the fragmentation grenade. For this reason, a room should never be considered safe just because one or two grenades have been detonated inside. Fragmentation grenades detonated on the floor not only throw fragments laterally, but also send fragments and spall downward to lower floors. Predicting how much spall will occur is difficult because flooring material varies, but wooden floors are usually affected the most.

(4) Some foreign grenades throw fragments that are much larger than those of the U.S.-made M67. Light barriers and interior walls would probably be less effective against these grenades than against the M67. A major problem with the U.S.-made fragmentation grenade is its tendency to bounce back off of hard targets. Grenades are often directed at window openings on the ground floor or second floor. At ranges as close as 20 meters, a thrower's chances of missing a standard 1-meter by 1-meter window are high. The fragmentation grenade normally breaks through standard window glass and enters a room. If the grenade strikes at a sharp angle or the glass is thick plate, the grenade could be deflected without penetrating.

(5) Hand grenades are difficult weapons to use. They involve a high risk of fratricide. Commanders should conduct precombat training with hand grenades as part of normal preparations. Marines must be very careful when throwing hand grenades up stairs to avoid having the grenade roll back down on them.

(6) Once pulled, the pins of both fragmentation and concussion grenades can be replaced if the thrower decides not to use the weapon. This pin replacement must be done carefully (see FM 23-30, *Grenades and Pyrotechnic Signals*) and prior to the spoon being released or detaching.

(7) An estimate of the situation (METT-T) and existing ROE will dictate what type of grenade will be used to clear each room. Because of the high rate of expenditure of grenades, units should use assault packs to carry additional grenades of all types. Additional grenades can also be carried in empty ammunition or canteen pouches.

9. 25-mm Automatic Gun. The 25-mm automatic gun mounted on the LAV-25 is an effective weapon during combat in built-up areas. The primary roles of LAV-25s during combat in built-up areas are to provide suppressive fire and to breach exterior walls and fortifications.

a. Obliquity. The 25-mm gun produces its best urban target results when fired perpendicular to the hard surface (zero obliquity). In combat in built-up areas, however, finding a covered firing position that permits low-obliquity firing is unlikely unless the streets and gaps between buildings are wide. Most shots impact the target at an angle, which normally reduces penetration. With the armor-piercing, discarding, sabot with tracer (APDS-T) round, an angle of obliquity of up to 20 degrees can actually improve breaching. The rounds tend to dislodge more wall material for each shot but do not penetrate as deeply into the structure.

b. Target Types. The 25-mm gun has different effects when fired against different urban targets.

(1) Reinforced Concrete. Reinforced concrete walls, which are 12 to 20 inches thick, present problems for the 25-mm gun when trying to create breach holes. It is relatively easy to penetrate, fracture, and clear away the concrete, but the reinforcing rods remain in place. These create a "jail window" effect by preventing entry but allowing grenades or rifle fire to be placed behind the wall. Steel reinforcing rods are normally 3/4 of an inch thick and 6 to 8 inches apart—there is no quick way of cutting these rods. They can be cut with demolition charges, cutting torches, or special power saws. Firing with either APDS-T or high-explosive, incendiary with tracer (HEI-T) rounds from the 25-mm gun will not always cut these rods.

(2) Brick Walls. Brick walls are more easily defeated by the 25-mm gun regardless of their thickness, and they produce the most spall.

(3) Bunker Walls. The 25-mm gun is devastating when fired against sandbag bunker walls. Obliquity has the least effect on the penetration of bunker walls. Bunkers with earth walls up to 36 inches thick are easily penetrated. At short ranges typical of combat in built-up areas, defeating a bunker should be easy, especially if the 25-mm gun can fire at an aperture.

c. Burst Fire. The 25-mm gun's impact on typical urban targets seems to be magnified if the firing is in short bursts. At close ranges, the gunner might need to shift his point of aim in a spiral pattern to ensure that the second and third bursts enlarge the hole. Even without burst fire, sustained 25-mm gunfire can defeat almost all urban targets.

d. Weapon Penetration. The penetration achieved by the two combat rounds (HEI-T and APDS-T) differs slightly, but both are effective.

(1) APDS-T. The APDS-T round penetrates urban targets by retaining its kinetic energy and blasting a small hole deep into the target. The APDS-T round gives the best effects behind the wall, and the armor-piercing core often breaks into two or three fragments that can create multiple enemy casualties. The APDS-T needs as few as four rounds to achieve lethal results behind walls. Table B-10 lists the number of APDS-T rounds needed to create different-sized holes in common urban walls.

Target	Loophole	Breach Hole
3-inch brick wall at 0-degree obliquity	22 rounds	75 rounds
3-inch brick wall at 45-degree obliquity	22 rounds	35[1] rounds
5-inch brick wall at 0-degree obliquity	32 rounds	50[1] rounds
8-inch reinforced concrete at 0-degree obliquity	22 rounds	75 rounds (Note: Reinforcing rods still in place)
8-inch reinforced concrete at 45-degree obliquity	22 rounds	40[1] rounds (Note: Reinforcing rods still in place)

[1] Obliquity and depth tend to increase the amount of wall material removed.

Table B-10. Breaching Effects of APDS-T Rounds

(a) When firing single rounds, the APDS-T round provides the greatest capability for behind-the-wall incapacitation. The APDS-T round can penetrate more than 16 inches of reinforced concrete with enough energy left to cause enemy casualties. It penetrates through both sides of a wood-framed or brick veneer building. Field fortifications are easily penetrated by APDS-T rounds. Table B-11 lists the number of APDS-T rounds needed to create different-sized holes in commonly found bunkers.

(b) The APDS-T round creates a hazardous situation for exposed personnel because of the pieces of sabot that are thrown off of the round.

Type Bunker	Obliquity	Penetration	Loophole	Small Breach Hole
36-inch sand/timber	0 degree	1 round	25 rounds	40 rounds
36-inch sand/6-inch concrete	0 degree	6 rounds	6 rounds	20 rounds

Table B-11. Number of APDS-T Rounds Needed To Create Different-Sized Holes in Bunkers

WARNING

Personnel not under cover forward of the 25-mm gun's muzzle and within the danger zone could be injured or killed by these sabots, even if the penetrator passes overhead to hit the target.

The danger zone extends at an angle of about 10 degrees below the muzzle level, out to at least 100 meters, and about 17 degrees left and right of the muzzle. Figure B-11 shows the hazard area of the APDS-T round.

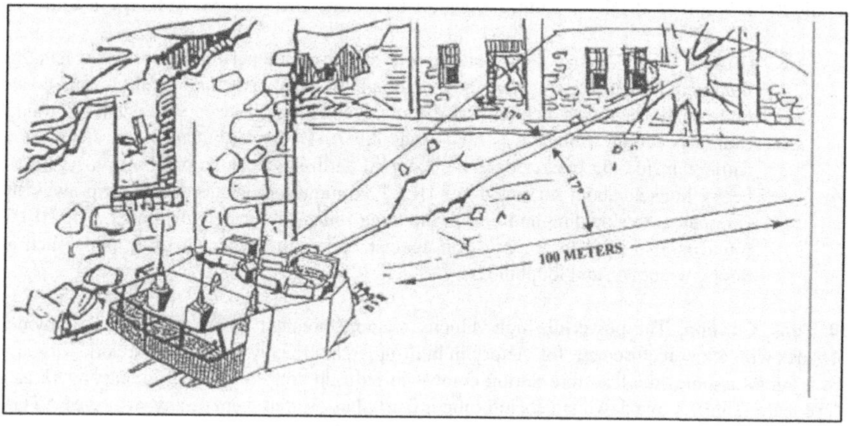

Figure B-11. APDS-T Danger Zone

(2) HEI-T. The HEI-T round penetrates urban targets by blasting away chunks of material.

(a) The HEI-T round does not penetrate an urban target as well as the APDS-T, but it does strip away a greater amount of material for each round. The HEI-T does more damage to an urban target when fired in multiple short bursts because the cumulative impact of multiple rounds is greater than the sum of individual rounds. Table B-12 lists the number of HEI-T rounds needed to create different-sized holes.

Target	Loophole	Breach Hole
3-inch brick wall at 0-degree obliquity	10 rounds	20 rounds
3-inch brick wall at 45-degree obliquity	20 rounds	25 rounds
5-inch brick wall at 0-degree obliquity	30 rounds	60 rounds

8-inch reinforced concrete at 0-degree obliquity	15 rounds	25 rounds
8-inch reinforced concrete at 45-degree obliquity	15 rounds	30 rounds

Table B-12. Number of HEI-T Rounds Needed To Create Different-Sized Holes in Bunkers

(b) The HEI-T round does not provide single-round perforation or incapacitating fragments when used against any external masonry structural wall. It can create first-round fragments behind wood-framed and brick veneer walls. HEI-T rounds cannot penetrate a bunker as quickly as APDS-T rounds, but they can create more damage inside the bunker once the external earth has been stripped away. Against a heavy bunker, about 40 rounds of HEI-T ammunition are needed to strip away the external earth shielding and breach the inner lining of concrete or timber. The HEI-T round is also used for suppression against known or suspected firing ports such as doors, windows, and loopholes.

10. Tank Cannon. The powerful, high-velocity cannon mounted on the M1A1 tank provides Marines with a key requirement for victory in built-up areas—heavy direct-fire support. Although the infantry assumes the lead role during combat in built-up areas, tanks and infantry work as a close team. Tanks move down streets after the infantry has cleared them of any suspected ATGM positions and, in turn, support the infantry with fire. The tank is one of the most effective weapons for heavy fire against structures. The primary role of the tank cannon during combat in built-up areas is to provide heavy direct fire against buildings and strongpoints that are identified as targets by the infantry. The wall and fortification breaching effects of the 120-mm tank cannon are major assets to Marines fighting in built-up areas.

a. Obliquity. The tank cannon produces its best urban target effects when fired perpendicular to the hard surface (zero obliquity). During combat in built-up areas, however, finding a covered firing position that permits low-obliquity firing is unlikely. Most shots strike the target at an angle that would normally reduce penetration. With tank cannon APDS rounds, obliquity angles of up to 25 degrees have little effect, but angles greater than 45 degrees greatly reduce penetration.

b. Ammunition. Armor-piercing, fin-stabilized, discarding sabot (APFSDS) rounds are the most commonly carried tank ammunition. These rounds work best against armored vehicles. The 120-mm cannon also carries an effective high-explosive, antitank multipurpose (HEAT-MP) round.

c. Characteristics. The 120-mm tank cannon has two specific characteristics that affect its employment in built-up areas: limited elevation and depression and short arming ranges. In addition, the M1A1 tank has another characteristic that is not involved with its cannon but that affects Marines working with or around the tank—*extremely hot turbine exhaust.*

(1) The M1A1 tank can elevate its cannon 20 degrees and depress it 10 degrees. The lower depression limit creates a 35-foot (10.8-meter) dead space around a tank. On a street 16 meters wide, this dead space extends to the buildings on each side (Figure B-12). Similarly, there is a zone overhead in which the tank cannot fire (Figure B-13). This dead space offers ideal locations for short-range antiarmor weapons and allows hidden enemy gunners to fire at the tank when the tank cannot fire back. It also exposes the tank's most vulnerable areas: the flanks, rear, and top. Infantry must move ahead, alongside, and to the rear of tanks to provide close protection. The extreme heat produced immediately to the rear of the M1A1 tank prevents dismounted infantry from following closely, but protection from small-arms fire and fragments is still provided by the tank's bulk and armor. The M1A1 tank also has a blind spot caused by the zero-degree depression available over part of the back deck. To engage any target in this area, the tank must pivot to convert the rear target to a flank target.

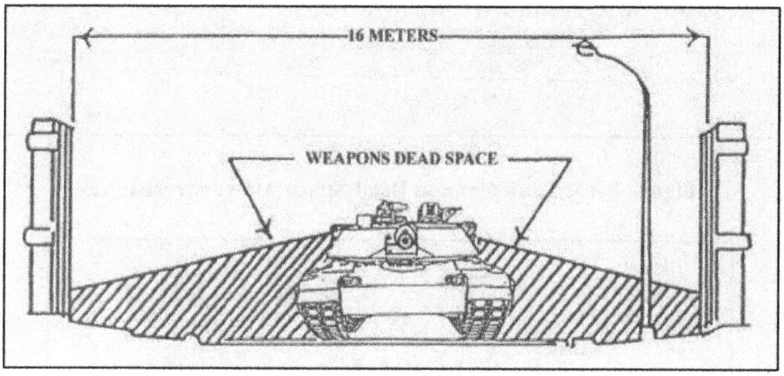

Figure B-12. Tank Cannon Dead Space at Street Level

(2) The 120-mm HEAT-MP round arms at about 36 feet. This arming distance allows the tank to engage targets from short ranges. The armor of the tank protects the crew from both the blowback effects of the round and enemy return fire. The APFSDS round does not need to arm and can, therefore, be fired at almost any range. The discarding portions of the round can be lethal to exposed infantry forward of the tank.

d. Target Effects. HEAT rounds are most effective against masonry walls. The APFSDS round can penetrate deeply into a structure but does not create as large a hole or displace as much spall behind the target. In contrast to lighter HEAT rounds, tank HEAT rounds are large enough to displace enough spall to inflict casualties inside a building. One HEAT round normally creates a breach hole in all but the thickest masonry construction—brick veneer and wood-framed constructions are demolished by a single round. Even the 120-mm HEAT round cannot cut all of the reinforcing rods, which are usually left in place, often hindering entry through the breach hole (Figure B-14). Both HEAT and APFSDS rounds are effective against all field fortifications. Only large earth berms and heavy mass-construction buildings can provide protection against tank fire.

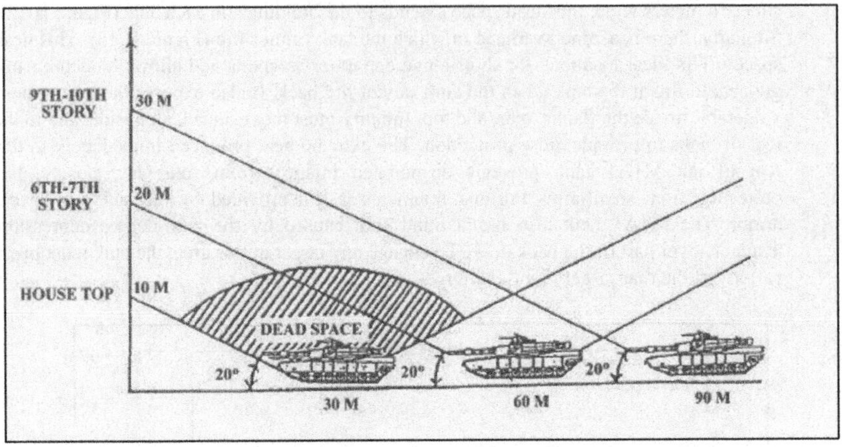

Figure B-13. Tank Cannon Dead Space Above Street Level

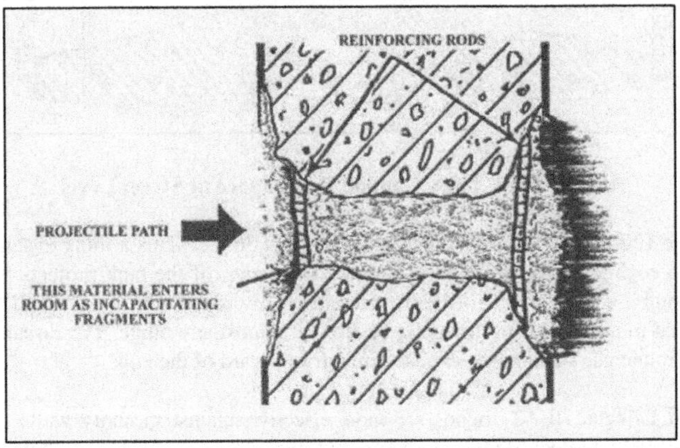

Figure B-14. Tank HEAT Round Effects on Reinforced Concrete Walls

e. Employment. Tank-heavy forces could be at a severe disadvantage during combat in built-up areas, but a few tanks working with the infantry can be most effective, especially if they work well together at the small-unit level. Tank, infantry, and engineer task forces are normally formed to attack a fortified area. The tank platoon or section can work together with rifle squads or platoons.

(1) Tanks need infantry on the ground to provide security in built-up areas and to designate targets. Against targets protected by structures, tanks should be escorted forward to the most covered location that provides a clear shot. On-the-spot instructions by infantry leaders ensure that the tank's fire is accurate and its exposure is limited. The tank commander may have to halt in a covered position, dismount, and reconnoiter his route forward into a firing position.

(2) When the tank main gun fires, it creates a large fireball and smoke cloud. In the confines of a built-up area, dirt and masonry dust are also picked up and added to this cloud. The target is further obscured by the smoke and dust of the explosion. Depending on the local conditions, this obscuration could last as long as two or three minutes. Marines can use this period to reposition or advance unseen by the enemy. Caution must be exercised, however, because the enemy might also move.

(3) A tank cannon creates an overpressure and noise hazard to exposed Marines. All dismounted Marines working near tanks should wear Kevlar helmets and protective vests, as well as ballistic eye protection. If possible, they should also wear earplugs and avoid the tank's frontal 60-degree arc during firing (Figure B-15).

WARNING

- **The overpressure from the tank 120-mm cannon can kill a marine found within a 90-degree arc extending from the muzzle of the gun tube out to 200 meters.**
- **From 200 to 1,000 meters along the line of fire, on a frontage of about 400 meters, dismounted Marines must be aware of the danger from discarding sabot petals, which can kill or seriously injure personnel.**
- **Personnel outside the tank should remain at least 50 meters from the tank in all directions as they may receive damaging effects from firing noise and overpressure.**
- **Personnel must also wear hearing protection when operating within 704 meters of a tank that is firing its main gun.**

TM 9-2350-264-10-2, *Operator's Manual for Operation Under Usual and Unusual Conditions for Tank, Combat, Full-Tracked: 120mm Gun, M1A1 General Abrams*, with Change 4

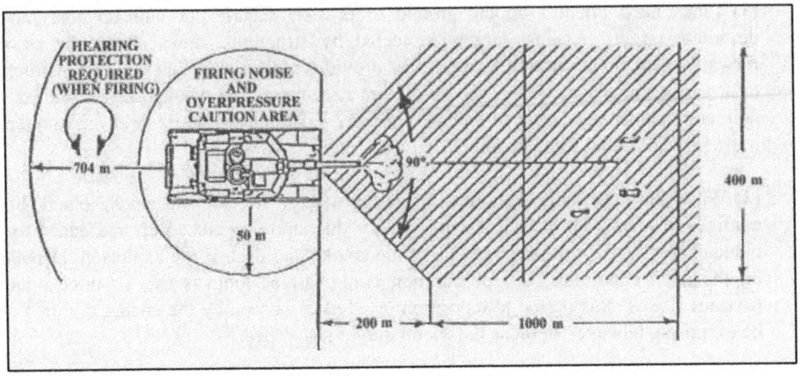

Figure B-15. Danger Areas Around a Tank Firing a 120-mm Gun

(4) Tanks are equipped with powerful thermal sights that can be used to detect enemy personnel and weapons that are hidden in shadows and behind openings. Dust, fires, and thick smoke significantly degrade these sights.

(5) Tanks have turret-mounted grenade launchers that project screening smoke grenades. The grenades use a bursting charge and burning RP particles to create this screen. *Burning particles can easily start uncontrolled fires and are hazardous to dismounted Marines near the tank.* The tank commander and the infantry small-unit leader must coordinate when and under what conditions these launchers can be used. Grenade launchers are a useful feature for protecting the tank but can cause significant problems if unwisely used.

(6) The tank's size and armor can provide dismounted Marines with cover from direct-fire weapons and fragments. With coordination, tanks can provide moving cover for Marines as they advance across small open areas. However, enemy fire that strikes but does not penetrate a tank is a major threat to nearby Marines. Fragmentation generated by antitank rounds and ricochets off of tank armor have historically been a prime cause of casualties while infantry was working with tanks in built-up areas.

11. Artillery. A major source of fire support for infantry forces fighting in built-up areas is field artillery weapons. If the built-up area is near the coast, NGF can be used. Field artillery employment can be in either the indirect- or direct-fire mode.

a. Indirect Fire. Low-angle indirect artillery fire is not effective in attacking targets within walls and masonry structures. It tends to impact on roofs or upper stories rather than on structurally critical wall areas or pillars. High-angle fire can achieve the trajectories necessary to strike targets in urban areas.

Weapons of at least 155 mm are required against thick reinforced concrete, stone, or brick walls. Even with heavy artillery, large expenditures of ammunition are required to knock down buildings of any size. Tall buildings also create areas of indirect-fire dead space, which are areas that cannot be engaged by indirect fire because of a combination of building height and angle of fall of the projectile (Figure B-16). Usually the dead space for low-angle indirect fire is about five times the height of the highest building over which the rounds must pass.

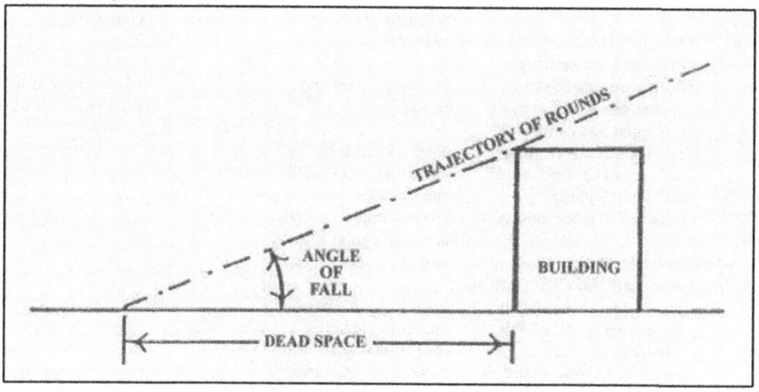

Figure B-16. Indirect-Fire Dead Space (Low Angle)

b. Direct Fire. A likely use of Marine artillery in an urban direct-fire role is to reinforce infantry against tough or important urban targets. Artillery should be used in this role only after an analysis of the need for heavy direct fire and the tradeoff involved in the extreme decentralization of artillery firepower. Artillery has the same need for close security and target designation as tanks.

c. Target Effects. Medium-caliber (155-mm) direct fire has a devastating effect against masonry construction and field fortifications. Smaller artillery (105-mm) pieces in the direct-fire mode are much less destructive than the larger caliber weapons. However, all Marine artillery is towed and, therefore, more difficult to employ in the direct-fire mode.

The 155-mm (M198) towed howitzer is effective because of its rate of fire and penetration. HE rounds can penetrate up to 38 inches of brick and unreinforced concrete. Projectiles can penetrate up to 28 inches of reinforced concrete with considerable damage beyond the wall. HE rounds with concrete-piercing fuzes provide an excellent means of penetrating strong reinforced concrete structures. One round can penetrate up to 46 inches. Five rounds are needed to reliably create a 1.5-meter breach in a wall that is 1 meter thick. About 10 rounds are needed to create the same size breach in a wall that is 1.5 meters thick. Superquick fuzing causes the rubble to be blown into the building, whereas delay fuzing tends to blow the rubble outward into the street. (See Table B-13 for employment considerations for artillery.)

Organization for Combat	Movement/Positioning	Delivery of Fire	Security	Command and Control
Centralized control is required during initial phases; decentralized control is required during later phases to support semi-independent actions of small units.	Movement should occur during night or periods of reduced visibility when possible. There are few displacements, often by platoon or section. Positions should be selected that minimize masking, provide several routes of escape, and afford as much cover and concealment as possible. Use of existing structures (garages, office buildings, highway overpasses) is recommended. Special techniques for emplacing howitzers, such as spades against a curb when the ground is not suitable for emplacement, may be required. Explosives may be required to soften emplacement of howitzers. Reconnaissance, selection, and occupation of position (RSOP) elements should be well armed because they may have to clear areas to be occupied. Extensive route reconnaissance is required. Target acquisition devices are somewhat degraded. Radars should be emplaced to cover likely areas of enemy indirect-fire weapon employment. Radars should not be placed in the midst of an urban area because of masking.	Both direct and indirect fires are delivered for supported units. Destruction of fortifications may require assault fire techniques. High-angle fires may be required. Need for accurate meteorological (MET) and survey data increases because most targets are point targets. Improved conventional munition and variable time (fuze) effects are reduced by structures, although they are effective against personnel on rooftops. HE delay is used for penetration effects. Illumination, chemical incendiary ammunition, and smoke are effective. Ammunition expenditures will be heavy. Lasers and PGMs permit destruction of targets with minimal rubbling of adjacent buildings. Tall buildings may hamper laser use. Batteries must be prepared for hasty survey techniques. Magnetic instruments are impaired.	Positions must be fortified.	Radio communications are impaired by buildings. Wire can usually be run overhead. Make use of civilian communications. A greater use of messengers and prearranged audio and visual signals is required.

Table B-13. Artillery Employment Considerations in Built-Up Areas

12. Mortars. The urban environment greatly restricts low-angle indirect fires because of overhead masking. While all indirect-fire weapons are subject to overhead masking, mortars are less affected than field artillery weapons because of the mortar's higher trajectory. For low-angle artillery fire, dead space is about five times the height of the building behind which the target sits. For mortar fire, dead space is only about one-half the height of the building. Because of these advantages, mortars are even more important than field artillery to the infantry during combat in built-up areas.

a. Employment. Not only can mortars fire into the deep defilade created by tall buildings, but they can also fire out of it. Mortars emplaced behind buildings are difficult for the enemy to accurately locate and even harder to hit with counterfire. Because of their light weight, even heavy mortars can be hand-carried to firing positions that may not be accessible to vehicles.

(1) The 60-mm and 81-mm mortars have limited effects on structural targets. Even with delay fuzes, they seldom penetrate more than the upper stories of light buildings. However, their wide area coverage and multioption fuzes make them useful against an enemy force advancing through streets, through other open areas, or over rubble.

(2) Mortar platoons often operate as separate firing sections during combat in built-up areas. The lack of large open areas can preclude establishing a platoon firing position. Figure B-17 shows how two mortar sections that are separated by only one street can be effective in massing fires and be protected from countermortar fire by employing defilade and dispersion.

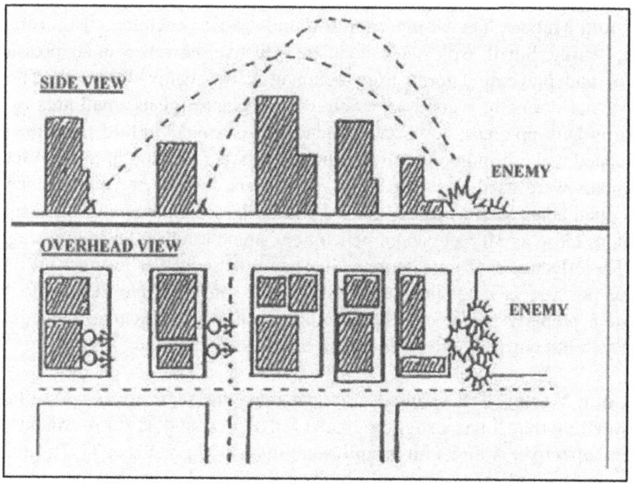

Figure B-17. Split-Section Mortar Operations on Adjacent Streets

(3) All three of the standard mortar projectiles are useful during combat in built-up areas. HE fragmentation rounds are most commonly used. WP is effective for starting fires in buildings and forcing the enemy out of cellars and light-framed buildings, and it is also the most effective mortar round against dug-in enemy tanks. Even near-misses blind and suppress the tank crew, forcing them to button up. Hits are difficult to achieve but are effective.

(4) Because the artificial roughness of urbanized terrain reduces wind speed and increases atmosphere mixing, mortar smoke tends to persist longer and give greater coverage in built-up areas than in open terrain.

(5) Urban masking has an impact on the use of illumination. In built-up areas, it is often necessary to plan illumination behind friendly positions, which places friendly troops in shadows and enemy troops in the light. Illumination rounds are difficult to adjust and are often of limited use because of the deep canyon nature of the urbanized area. Rapidly shifting wind currents in built-up areas also affect mortar illumination, making it less effective.

b. Effects of Mortar Fire. The multioption fuze on newer U.S. mortar rounds makes them effective weapons on urbanized terrain. Delay settings can increase penetration slightly, and proximity bursts can increase the lethal area covered by fragments. Tall buildings can cause proximity-fuzed mortar rounds to detonate prematurely if they pass too closely to the buildings.

(1) 60-mm Mortar. The 60-mm mortar round cannot penetrate most rooftops, even with a delay setting. Small explosive rounds are effective, however, in suppressing snipers on rooftops and preventing roofs from being used by enemy observers. The 60-mm WP round is not normally a good screening round because of its small area of coverage. In combat in built-up areas, however, the tendency of smoke to linger and the small areas to be screened make it more effective. During the battle for Hue in South Vietnam, 60-mm WP rounds were used to create small, short-term smoke screens to conceal movement across open areas such as parks, plazas, and bridges. Fragments from 60-mm HE rounds landing as close as 10 feet cannot penetrate a single sandbag layer or a single-layer brick wall. The effect of a 60-mm mortar HE round that achieves a direct hit on a bunker or fighting position is equivalent to 1 - 2 pounds of TNT. Normally, the blast will not collapse a properly constructed bunker, but it can cause structural damage. The 60-mm mortar will not normally create craters in a hard-surfaced road.

(2) 81-mm Mortar. The 81-mm mortar has much the same effect against urban targets as the 60-mm mortar. It has a slightly greater lethal area, and its smoke rounds (WP and RP) are more effective. A direct hit is equivalent to about 2 pounds of TNT. The 81-mm round cannot cause significant craters in a hard-surfaced road. With a delay setting, the 81-mm round can penetrate the roofs of light buildings.

13. Naval Gunfire. The primary NGF weapon, the 5-inch/54, has a high rate of fire and is roughly equivalent to the 155-mm howitzer in target effect. Target engagement considerations are the same as for artillery; however, the 5-inch/54's flat trajectory is even more affected by terrain masking and a constantly changing gun-target line. NGF can achieve greater target engagement by firing reduced charges at high angles; however, range will be reduced.

14. Aerial Weapons. Both rotary- and fixed-wing aircraft can quickly deliver large volumes of firepower over large built-up areas. Specific targets are hard to distinguish from the air. Good ground-to-air communications are vital in successfully employing aerial firepower. Planners have historically tended to overestimate the effects of HE on defenders. Modern, large buildings are remarkably resistant to damage from bombs and rocket fire.

a. Rotary-Wing Aircraft. Attack helicopters can be used to engage targets in built-up areas. Enemy armored vehicles in small parks, boulevards, or other open areas are good targets for attack helicopters.

(1) The HELLFIRE missile has a larger warhead and greater range than the TOW, but it too is a shaped-charge warhead and is not specifically designed for use against masonry targets. Laser target designation for the HELLFIRE may be difficult as a result of laser reflections off of glass and shiny metal surfaces. The use of attack helicopters to deliver ATGMs against targets in the upper stories of high buildings is sometimes desirable.

(2) The 2.75-inch folding-fin aerial rocket and the 20-mm cannon on Marine attack helicopters are good area weapons to use against enemy forces in the open or under light cover. They are usually ineffective against a large masonry target. The 20-mm cannon produces many ricochets, especially if armor-piercing ammunition is fired into built-up areas.

b. Fixed-Wing Aircraft. Close air support to ground forces fighting in built-up areas is a difficult mission for fixed-wing aircraft. Targets are hard to locate and identify, enemy and friendly forces could be intermingled, and enemy short-range air defense weapons are hard to suppress.

(1) Because enemy and friendly forces may be separated by only one building, accurate delivery of ordnance is critical. Marking panels, lights, electronic beacons, smoke, or some other positive identification of friendly forces is needed.

(2) General-purpose bombs that weigh from 500 to 2,000 pounds are moderately effective in creating casualties among enemy troops located in large buildings. High-dive-angle bomb runs increase accuracy and penetration but also increase the aircraft's exposure to antiaircraft weapons. Low-dive-angle bomb runs using high-drag (retarded) bombs can be used to get bombs into upper stories. Penetration is not good with high-drag bombs. Sometimes aerial bombs pass completely through light-clad buildings and explode on the outside.

(3) Aerial rockets and 20-mm cannons are only moderately effective against enemy soldiers in built-up areas because rockets lack the accuracy to concentrate their effects. The 20-mm cannon rounds penetrate only slightly better than the .50-caliber round, 20-mm armor-piercing rounds can ricochet badly, and tracers can start fires.

(4) The AC-130 aircraft has weapons that can be most effective during combat in built-up areas. This aircraft can deliver accurate fire from a 20-mm Vulcan cannon, 40-mm rapid-fire cannon, and 105-mm howitzer. The 105-mm howitzer round is effective against the roofs and upper floors of buildings. The AC-130 is accurate enough to concentrate its 40-mm cannon and 105-mm howitzer fire onto a single spot to create a rooftop breach that allows fires to be directed deep into the building.

(5) Laser and optically guided munitions can be effective against high-value targets. Specially developed, heavy, laser-guided bombs can penetrate hardened targets. However, problems associated with dense smoke and dust clouds hanging over the built-up area and laser scatter can restrict their use. If the launching aircraft can achieve a successful laser designation and lock-on, these weapons have devastating effects, penetrating deep into reinforced concrete before exploding with great force. If launched without a lock-on, or if the laser spot is lost, these weapons are unpredictable and can travel long distances before they impact.

15. Demolitions. Combat in built-up areas requires the extensive use of demolitions. All Marines, not just engineers, should be trained to employ demolitions. See FM 5-250, *Explosives and Demolitions*, for specific information on the safe use of demolitions.

a. Demolitions. Demolitions come in two types: TNT and C4. Exposed Marines must take cover or move at least 300 meters away from bulk explosives that are being used to breach walls.

(1) TNT comes in 1/4-, 1/2-, and 1-pound blocks. About 5 pounds of TNT are needed to breach a nonreinforced concrete wall that is 12 inches thick if the explosives are laid next to the wall and are not tamped. If the explosives are tamped, about 2 pounds are sufficient.

(2) C4 comes in many different-sized blocks. Normally, it is found in 1 1/4- and 2 1/2-pound blocks within prepackaged satchel charges. About 10 pounds of C4 placed between waist and chest high will blow a hole in the average masonry wall large enough for a man to walk through.

b. Shaped Charges. There are two sizes of shaped charges: a 15-pound M2A3 and a 40-pound M3A3. The M3A3 is the shaped charge that is most likely to be used in built-up areas. It can penetrate 5 feet of reinforced concrete. The hole tapers from 5 inches down to 2 inches. The amount of spall thrown behind the target wall is considerable. There is also a large safety hazard area for friendly forces.

c. Satchel Charges. There are two standard satchel charges: the M183 and the M37. Both come in their own carrying satchel with detonators and blasting cords. Each weighs 20 pounds. The M183 has 16 individual 1 1/4-pound blocks that can be used separately. When used untamped, a satchel breaches a 3-foot-thick concrete wall. Satchel charges are very powerful. Debris is thrown great distances. Friendly forces must move away and take cover before detonation.

d. Cratering Charges. The standard cratering charge is a 43-pound cylinder of ammonium nitrate. This explosive does not have the shattering effect of bulk TNT or C4. It is more useful in deliberate demolitions than in hasty ones.

Appendix C

Attacking and Clearing Buildings

Offensive operations on urbanized terrain will require Marines to attack and clear buildings.

1. **Introduction.** At the rifle platoon and squad level, the major offensive tasks for combat in built-up areas are attacking and clearing buildings. This involves:

- Suppressing defensive fires
- Advancing Marine assault forces
- Assaulting buildings
- Clearing buildings
- Reorganizing assault forces.

To accomplish these tasks, the assaulting unit task organizes into assault, security, and support forces or elements. (See Appendix A)

2. **Requirements.** There are five interrelated requirements for attacking a defended building: *fire support, movement, assault, clearing,* and *reorganization.* Proper application and integration of these requirements reduce casualties and hasten accomplishment of the mission. The proper integration of these requirements are determined by METT-T. For example, larger cities may have high-rise apartment buildings and industrial complexes that are separated by parking areas or parks. Increased fire support may be required to suppress and obscure enemy gunners covering the open terrain and spaces between buildings. Conversely, a built-up area with few open areas can provide numerous covered routes that may decrease fire support requirements.

3. **Fire Support.** Fire support necessary to advance the assault force may include:

- Suppressing and obscuring enemy gunners within the objective building(s) and adjacent structures
- Isolating the objective building(s) with fires to prevent enemy withdrawal, reinforcement, or counterattack
- Breaching walls en route to and in the objective structure
- Destroying enemy positions with direct-fire weapons

- Securing cleared portions of the objective
- Providing replacements for the assault force
- Providing resupply of ammunition and explosives
- Evacuating casualties and prisoners.

a. The size of the support force is determined by the type and size of the objective building(s), whether the adjacent terrain provides open or covered approaches, and the organization and strength of enemy defenses.

b. The support force for a Marine rifle squad may consist of one fire team that could be reinforced with heavy machine guns, antitank weapons, and/or armored vehicles. In situations involving a larger assault force, a rifle platoon or company reinforced with armored vehicles, engineers, and/or artillery may be required to support the movement and assault of an adjacent platoon or company.

c. After seizing objective buildings, the assault force reorganizes and may be required to provide supporting fires for a subsequent assault. Each weapon is assigned a target or area to cover. Individual small-arms weapons place fires on likely enemy weapon positions—loopholes, windows, and roof areas. Snipers are best employed in placing accurate fire through loopholes or engaging long-range targets. M203 fires are directed through windows or loopholes.

d. AT4s and demolitions are employed to breach walls, doors, barricades, and window barriers on the ground level of structures. The tank cannon or LAV-mounted 25-mm gun engages first-floor targets and breaches walls for attacking Marines. Armored vehicle- or wheeled vehicle-mounted machine guns can engage suspected enemy positions on upper floors and in adjacent structures. In addition to destroying or weakening structures, the tank's main gun rounds cause casualties by their explosive effects and by hurling debris throughout the interior of structures.

e. NGF, artillery, and mortars should initially use time fuzes to initially clear exposed enemy personnel, weapons, OPs, and radio sites from rooftops. They should then use delayed fuze action to cause casualties among the defenders inside the structure as a result of shrapnel and falling debris. They may also provide smoke to cover the movement and assault phases of the attack. Artillery can also be used in the direct-fire mode.

f. In support of the movement and assault phases of the attack, buildings can be attacked by fixed-wing aircraft or by attack helicopters.

4. Movement. The assault force minimizes the effects of enemy defensive fires during movement by:

- Using covered routes
- Moving after defensive fires have been suppressed or obscured
- Moving at night or during other periods of reduced visibility
- Selecting routes that will not mask friendly suppressive fires
- Crossing open areas (e.g., streets and spaces between buildings) quickly under the concealment of smoke and suppressive fires
- Moving on rooftops that are not covered by enemy direct fires.

a. The requirement for speed may dictate movement through streets and alleys without clearing all buildings. When moving in this manner, the maneuver force should employ infantry to lead the column, and supported by AAVs, LAVs, wheeled vehicle-mounted machine guns, or tanks following.

b. Marines should move along each side of a street with the leading squad staying abreast of armored vehicles. When not accompanied by armored vehicles, Marines move in a single file along one side of the street under cover of fires from supporting weapons. Individual Marines should be detailed to observe and cover a specific area, such as second-floor windows on the opposite side of the street.

5. Assault. Assault forces quickly and violently execute their assault and subsequent clearing operations. Momentum is maintained to prevent the enemy from organizing a more determined resistance on other floors or in other rooms. Small-unit leaders should keep the assault force moving without allowing the operation to become disorganized.

a. An assault in a built-up area involves the elementary skills of close combat. Leaders must:

- Be trained in the required techniques to defeat the enemy in a face-to-face encounter
- Keep themselves in excellent physical condition
- Have confidence in their abilities and those of their unit.

b. The composition of the assault force will vary depending on METT-T; however, its basic organization at the squad level consists of two assault elements carrying a fighting load of equipment and as much ammunition as possible, especially grenades.

6. Clearing. Entering at the top and fighting downward is the preferred method of clearing a building. Clearing a building is easier from an upper story because gravity and building construction become assets to the assault elements when throwing hand grenades and moving from floor to floor. However, this method is only feasible when access to an upper floor or rooftop can be gained from the windows or roofs of adjoining buildings and assault elements can be transported to the rooftops by helicopter (Figure C-1). Helicopters should land only on those buildings that have special heliports or parking garages on the roofs. When these are unavailable, Marines can rappel or fastrope onto the roof or dismount as the helicopter hovers a few feet above the roof. Assault elements then breach the roof or common walls with explosives and use ropes to enter the lower floors. Stairs are guarded by a covering team or security force when not in use to protect the assault force from a flank attack.

Figure C-1. Helicopters Used To Clear Buildings

a. Although the top-to-bottom method is preferred for clearing a building, assaulting the bottom floor and clearing upward is a common method. When using this technique, the assault force closes on the flank(s) or rear of the building. It then clears each room on the ground floor and then, moving upward, begin a systematic clearance of the remaining floors.

b. Entry can be gained through walls breached by explosives or gunfire. Assault elements should avoid windows and doors because they are usually covered by fire or boobytrapped. If tanks are attached to the company, they can breach the wall by firing their main gun (Figure C-2).

Figure C-2. Main Gun Used To Breach Exterior

c. Shortly before the assault, suppressive fires on the objective should be increased until masked by the advancing assault force. Once masked, fires are shifted to upper windows and continued until the assault force has entered the building. At that time, fires are shifted to adjacent buildings to prevent enemy withdrawal or reinforcement.

d. Hand grenades should be used to clear the space before assault element members enter through a breached wall. Immediately after the explosion, the assault element clearing team enters and uses three round bursts to fire into the interior of the room, concentrating on areas that are possible enemy positions.

e. Once inside the building, the assault force must cover staircases and seize rooms that overlook approaches to the building. These actions are required to isolate enemy forces within the building and to prevent reinforcement. Assault elements should clear each ground floor room and then the basement. Sequence for the assault is that:

(1) The assault force leader determines which room(s) to clear first.

(2) The support force provides suppressive fire while the assault force systematically clears the building. It also provides suppressive fire on adjacent buildings to prevent enemy reinforcements or withdrawal. The support force destroys any enemy trying to exit the building.

(3) The security force isolates the immediate area.

(4) After assault elements establish a foothold in the building, the assault force leader proceeds with the systematic clearing of the building.

f. When clearing a building from the top floor down, assault forces should gain entrance through a common wall or the roof of an adjoining building (Figure C-3). Demolition charges are used to breach the wall and gain entrance to the top floor. Access to lower floors and rooms may be gained by breaching holes in the floor and having the Marines jump or slide down ropes to the lower floors. Stairs can be used if they are first cleared.

g. When using the top-to-bottom method of clearing, requirements remain the same as for other methods. After the floor is breached to gain access to a lower floor, a grenade is allowed to cook-off and is dropped to the lower room. A Marine then sprays the lower room with gunfire using three-round bursts and drops through the mousehole.

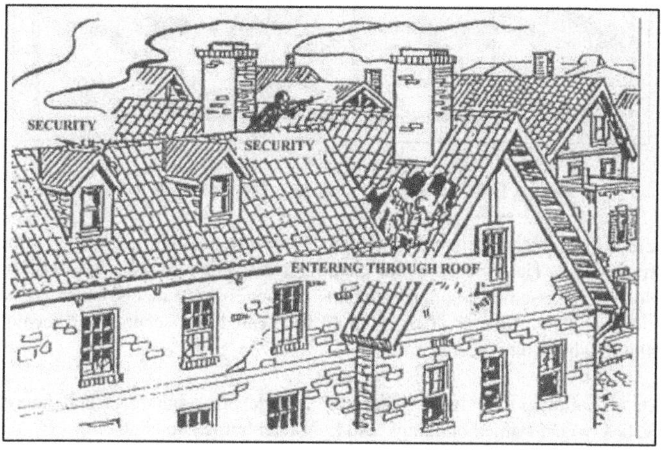

Figure C-3. Upper Floors Secured

h. Marines should avoid predictability when entering and clearing rooms. They should vary entry through doors and mouseholes to keep the enemy off guard (Figure C-4). As rooms are cleared, doors should be left open, and a predetermined mark (cloth, tape, spray paint) should be placed on the doorjamb or over the door, as established by unit SOPs.

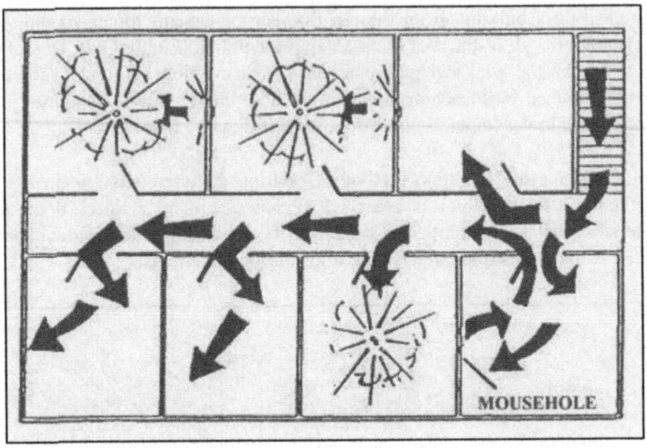

Figure C-4. Varying Techniques for Clearing Rooms

i. If there is a basement, it should be cleared as soon as possible, preferably at the same time as the ground floor. The procedures for clearing a basement are the same as for any room or floor. However, basements often contain entrances to tunnels such as sewers and communications cable tunnels. These should be cleared and secured to prevent the enemy from infiltrating back into cleared areas.

j. Types of buildings that may be cleared include brick buildings and houses, box-wall buildings, heavy-clad framed buildings, and light-clad framed buildings.

(1) Brick Buildings and Houses. The best way to enter a brick building is to blow a breaching hole in the side by using a tank firing HEAT ammunition (Figure C-5). If tanks are not available, a door or window in the rear of the building usually provides better cover and concealment for entry than one in the front. If there is enough cover and concealment, the assault force should enter the rear of the building at an upper level by using a fire escape or grappling hook.

Brick houses usually have similar floor plans on each floor (Figure C-6); therefore, ground floors are cleared the same way as upper floors.

Converely, the floor plans in larger brick buildings typically are different on ground-floor levels than on upper levels (Figure C-7).

Figure C-5. Entering a Brick Building

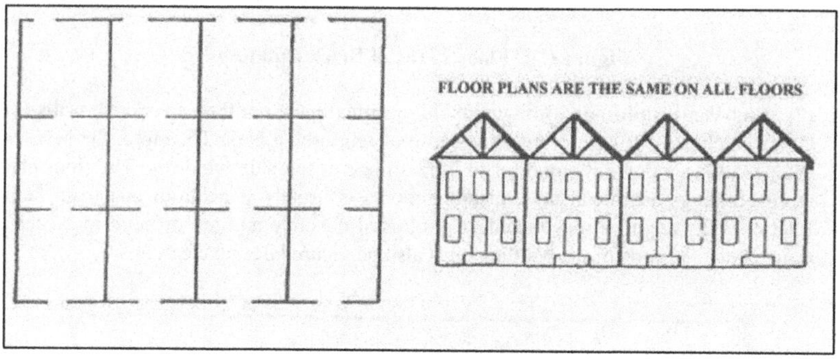

Figure C-6. Brick Houses Have Similar Floor Plans

To clear from building to building, the best method is to move from rooftop to rooftop because the roofs of brick buildings are usually easy to breach. The walls between buildings are at least three bricks thick (total of six bricks between buildings) and require large quantities of demolitions to breach. Walls are normally easier to breach on an upper floor than on a lower floor because the walls are thinner on upper floors. If rooftops are covered by fire and there are not enough demolitions to breach walls between buildings, clearing from rear to rear of buildings is safer than clearing from front to front.

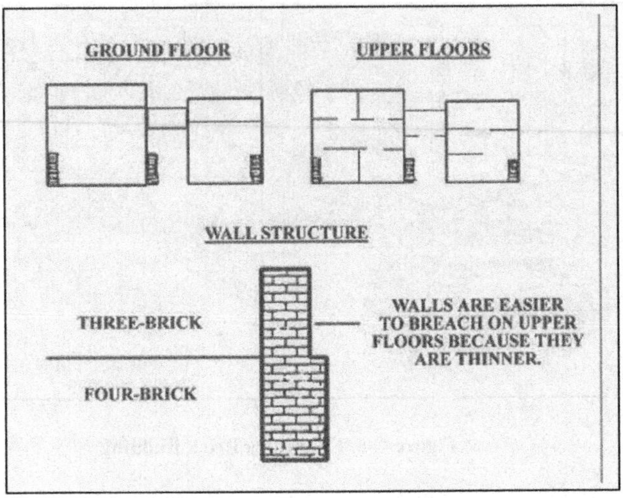

Figure C-7. Floor Plans of Brick Buildings

(2) Box-Wall Buildings. Box-wall buildings often have reinforced concrete walls (Figure C-8), which are difficult to breach because of reinforcing bars. Therefore, the best way to enter is to blow down the door or to blow in one of the side windows. The floor plans of these buildings are predictable; clearing rooms is usually done from one main hallway. Interior walls are also constructed of reinforced concrete and are difficult to breach. The stairways at the ends of the building must also be secured during clearing.

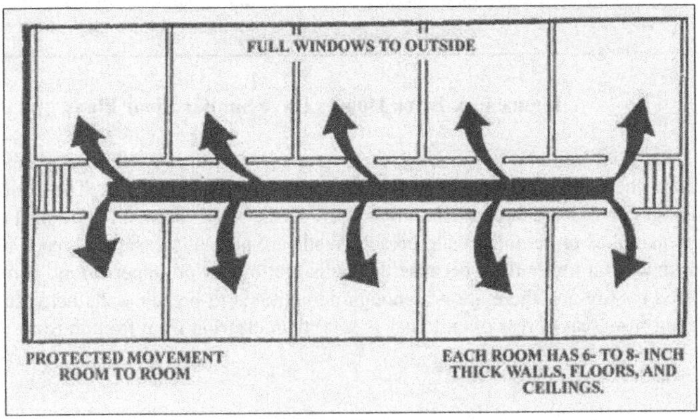

Figure C-8. Box-Wall Buildings

(3) **Heavy-Clad Framed Buildings.** Tanks can be used to breach a hole in heavy-clad framed buildings. Their floor plans are normally oriented around a stairway or elevator. this must be secured during clearing (Figure C-9). The interior walls of these buildings can be breached, although they may require use of demolitions.

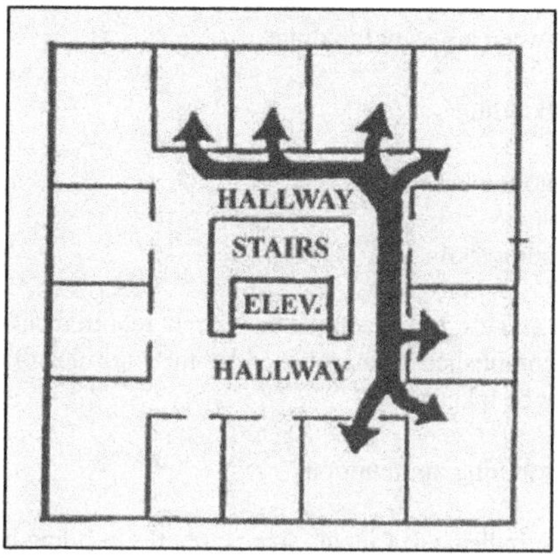

Figure C-9. Heavy-Clad Framed Buildings

(4) **Light-Clad Framed Buildings.** Light-clad framed buildings (Figure C-10), can be cleared in the same manner as heavy-clad buildings. Secure the central stairway and clear in a circular pattern. Walls are usually thin enough to be breached with an ax.

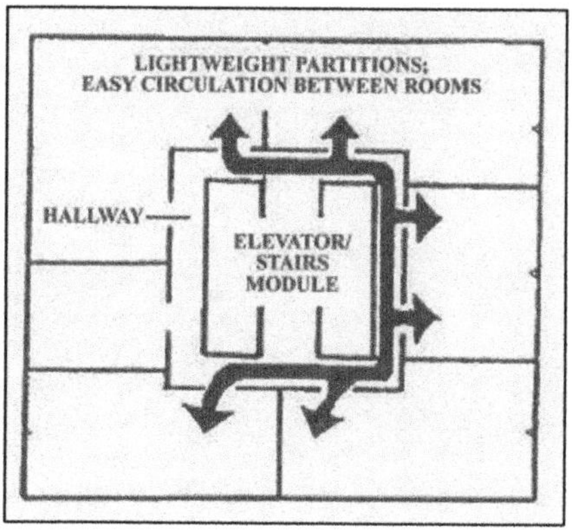

Figure C-10. Light-Clad Framed Buildings

7. **Consolidation and Reorganization.** Consolidation and reorganization of the assault force in a cleared building must be accomplished quickly to repel enemy counterattacks and to prevent the

enemy from infiltrating back into the cleared building. After securing a floor (bottom, middle, or top), selected members of the assault force are assigned to cover potential enemy counterattack routes to the building. Those sentinels alert the assault force and place a heavy volume of fire on enemy forces approaching the building. They should cover:

- Enemy mouseholes between adjacent buildings
- Covered routes to the building
- Underground routes into the basement
- Approaches over adjoining roofs.

As the remainder of the assault force completes search requirements, defensive positions are assigned and tasks are accomplished as established by the priorities of work in the defense. The following actions may then be taken:

- Resupplying and redistributing ammunition
- Marking the building to indicate to friendly forces that the building has been cleared
- Assuming an overwatch mission and supporting an assault on another building
- Treating and evacuating wounded personnel
- Further developing a defensive position if the building is to be occupied for any period.

Appendix D

Fighting Positions

A critical defensive task in combat in built-up areas is the selection and preparation of fighting positions.

1. Considerations. Leaders should consider the following factors when establishing fighting positions:

a. Protection. Leaders should select buildings that provide protection from direct and indirect fires. Reinforced concrete buildings with three or more floors provide suitable protection, while buildings constructed of wood, paneling, or other light materials require reinforcement to gain sufficient protection. One- to two-story buildings without strongly constructed cellars are vulnerable to indirect fires and require construction of overhead protection for each firing position.

b. Dispersion. Normally, a position should not be established in a single building when it is possible to occupy two or more buildings that permit mutually supporting fires. A position in one building without mutual support is vulnerable to bypass, isolation, and subsequent destruction from any direction.

c. Concealment. The fighting position should be concealed. Buildings that arc obvious defensive positions should not be selected. The occupation of exposed buildings may be necessary because of requirements for security and fields of fire. Therefore, reinforcement of the structure must be accomplished to provide suitable protection within the building.

d. Fields of Fire. Positions should be mutually supporting and have interlocking fields of fire in all directions. Clearing fields of fire may require the destruction of adjacent buildings by using explosives, engineer equipment, and field expedients. Care should be taken to avoid highlighting the building as a defensive position.

e. Covered Routes. Defensive positions should have at least one covered route that permits resupply, MEDEVAC, reinforcement, or withdrawal from the building. The route can be established by one of the following means:

- Through walls to adjacent buildings
- Through underground systems
- Through communications trenches
- Behind protective buildings.

f. Observation. The building should permit observation of enemy avenues of approach and adjacent defensive sectors.

g. Fire Hazard. Avoid selecting positions in buildings that are a fire hazard. If flammable structures must be occupied, the danger of fire can be reduced by wetting the structure, laying an inch of sand on the floors, and providing fire extinguishers and firefighting equipment. Routes of escape should also be prepared in case of fire.

h. Time. The time available to prepare the defense could be the most critical factor. If enough time is insufficient, buildings that require extensive preparation should not be used. Conversely, buildings located in less desirable areas that require little improvement may be a better choice for a defensive position.

2. Preparation. Preparation of fighting positions depends on proper selection and construction.

a. Selecting Positions. Each weapon should be assigned a primary sector of fire to cover enemy approaches. Alternate positions that overwatch the primary sector should also be selected. These positions are usually located in an adjacent room on the same floor (Figures D-1A, D-1B, and D-1C). Each weapon must be assigned a supplementary position (to engage attacks from other directions) and an FPL.

b. Building Positions. There are many ways to establish a fighting position in a building.

(1) Window Position. Marines should kneel or stand on either side of a window. To fire downward from upper floors, tables or similar objects can be placed against the wall to provide additional elevation, but they must be positioned to prevent the weapon from protruding through the window (Figure D-2). Leaders should inspect positions to determine the width of sector that each position can engage.

(2) Loopholes. To avoid establishing a pattern of always firing from windows, loopholes should be prepared in walls. Marines should avoid firing directly through loopholes to enhance individual protection.

(a) Several loopholes are usually required for each weapon (primary, alternate, and supplementary positions). The number of loopholes should be carefully considered because they can weaken walls and reduce protection. Engineers should be consulted before numerous loopholes are made. Loopholes should be made by punching or drilling holes in walls and should be placed where they are concealed. Blasting loopholes can result in a large hole that can easily be seen by the enemy.

(b) Loopholes should be cone-shaped to obtain a wide arc of fire, to facilitate engagement of high and low targets, and to reduce the size of the exterior aperture (Figure D-3). The edges of a loophole splinter when hit by bullets; therefore, protective linings such as an empty sandbag held in place by wire mesh will reduce spalling effects. When not in use, loopholes should be covered with sandbags to prevent the enemy from firing into or observing through them.

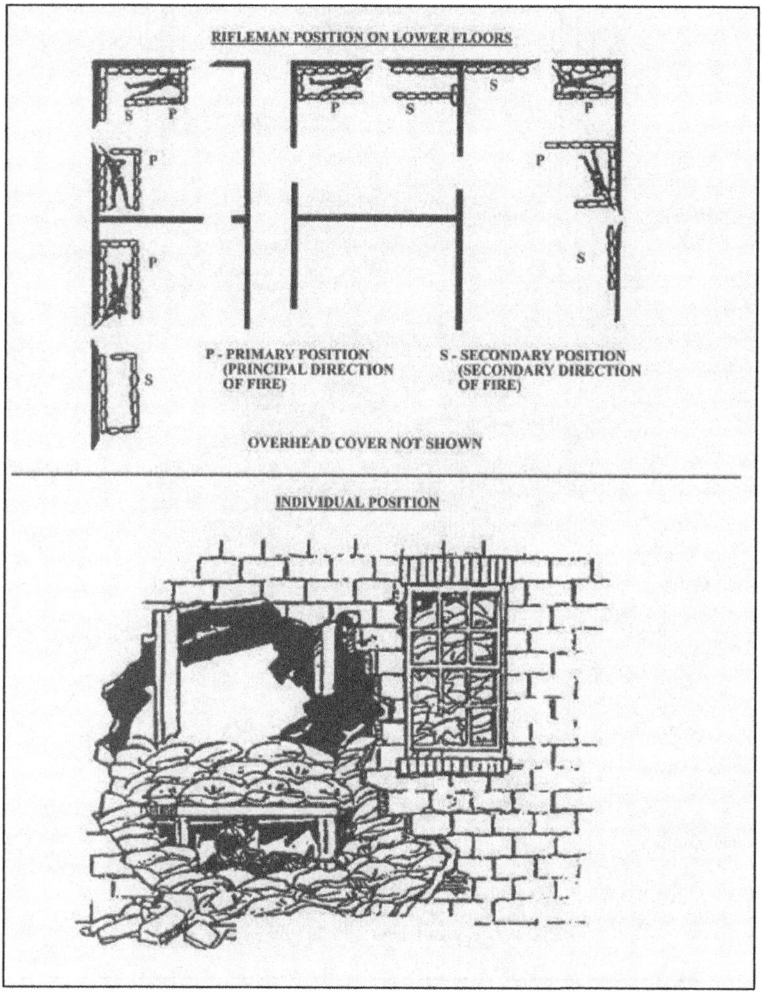

Figure D-1A. Weapon Positions

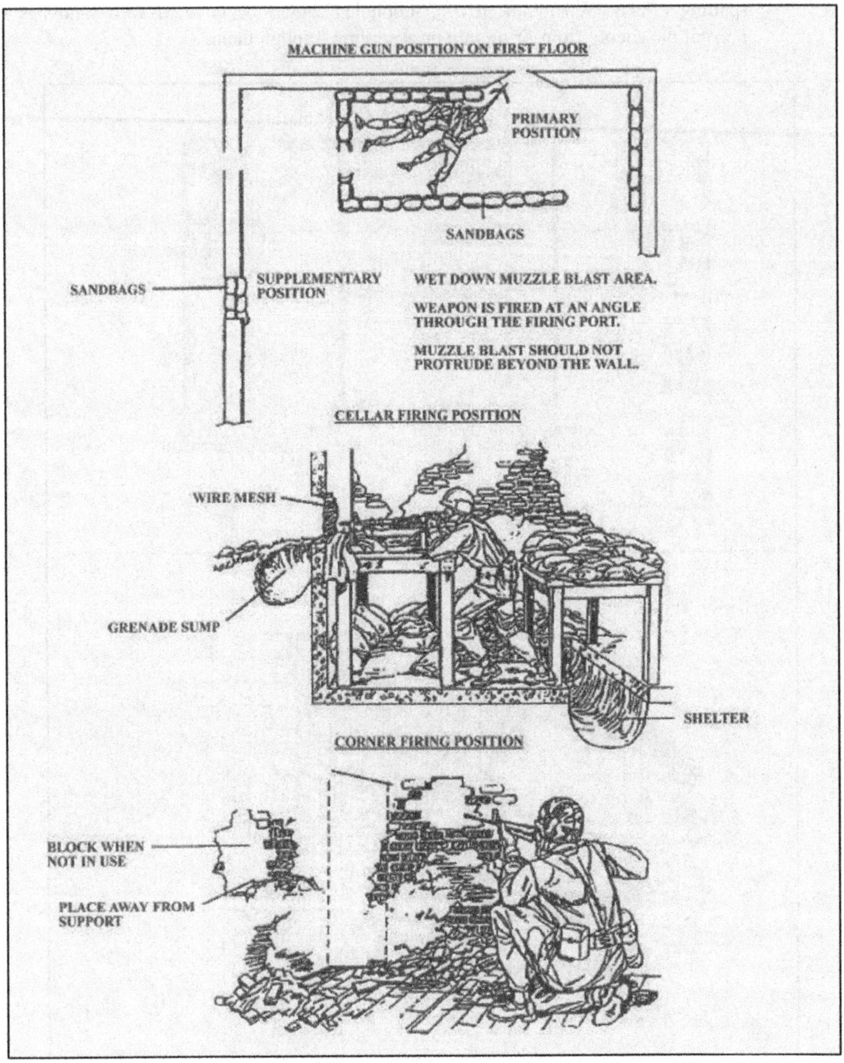

Figure D-1B. Weapon Positions (Continued)

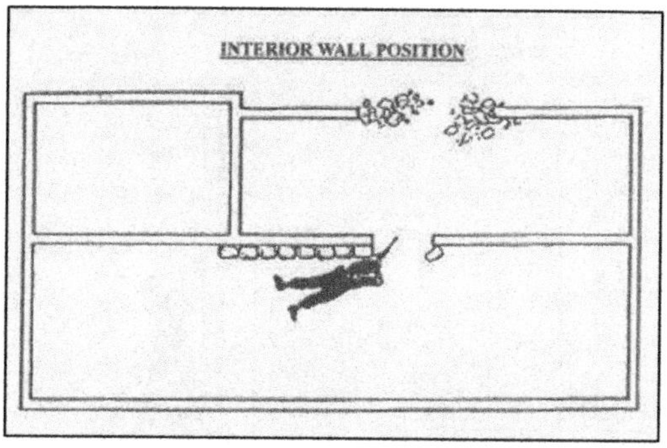

Figure D-1C. Interior Wall Position

Figure D-2. Window Position

(c) Loopholes should also be prepared in interior walls and ceilings of buildings to permit fighting within the position. Interior loopholes should enable overwatch of stairs, halls, and unoccupied rooms and be concealed by pictures, drapes, or furniture. Loopholes in floors permit the defender to engage enemy personnel on lower floors with small-arms fire and grenades.

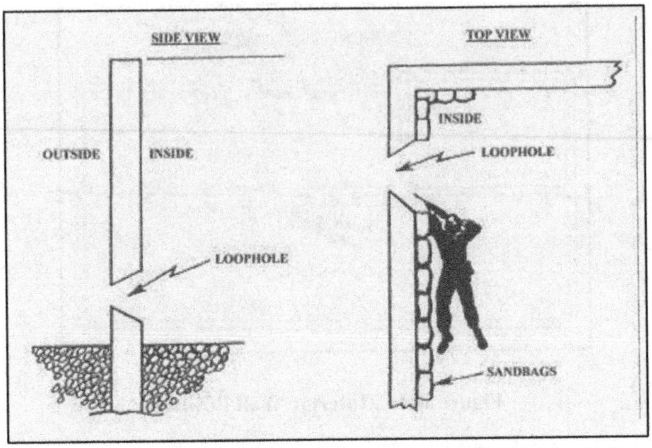

Figure D-3. Cone-Shaped Loopholes

(d) Although walls provide some frontal protection, they should be reinforced with sandbags, furniture filled with dirt, or other expedients. Each position should have overhead and all-around protection (Figure D-4).

Figure D-4. Position With Overhead and All-Around Protection

c. Other Construction Tasks. Other construction tasks may need to be performed. These include work on:

(1) Basements and Ground Floors. Basements require preparation similar to that of the ground floor. Any underground system that is not used by the defender but that could provide enemy access to the position must be blocked.

(a) Doors. Unused doors should be locked, nailed shut, blocked, and reinforced with furniture, sandbags, or other field expedients. Outside doors can be boobytrapped by engineers or other personnel.

(b) Hallways. If not required for the defender's movement, hallways should be blocked with furniture and tactical wire (Figure D-5). If authorized, boobytraps should be employed to impede enemy movement.

(c) Stairs. Defenders should block stairs not used by the defense with furniture and tactical wire (Figure D-5) or remove them. If possible, all stairs should be blocked, and ladders should be used to move from floor to floor and then removed when not being used. Boobytraps should also be employed on stairs.

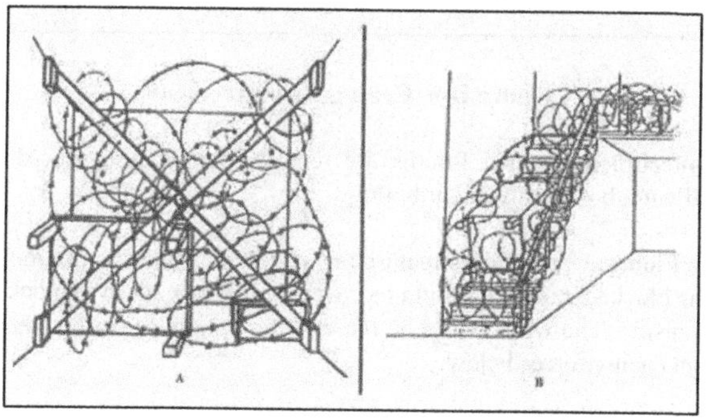

Figure D-5. Blocking Hallways and Stairs

(d) Windows. All glass should be removed. Windows not being used should be blocked with boards or sandbags.

(e) Floors. Fighting positions should be constructed on the ground or basement floors. If there is no basement, fighting positions on the first floor must be constructed to provide additional protection from heavy direct-fire weapons.

(f) Ceilings. Supports that can withstand the weight of rubble from upper floors should be placed under ceilings to provide overhead cover (Figure D-6).

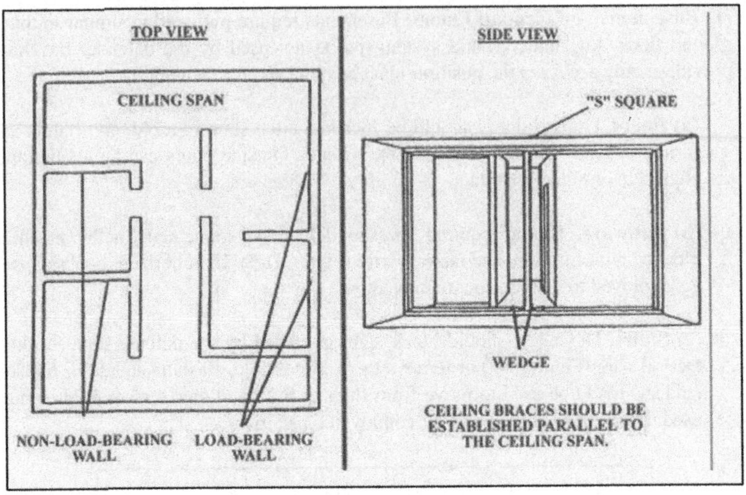

Figure D-6. Ceiling Reinforcement

(g) Unoccupied Rooms. Rooms not required for defense should be blocked with tactical wire, boobytrapped, or both.

(2) Upper Floors. Upper floors require the same preparation as ground floors. Windows need not be blocked, but they should be covered with wire mesh to block grenades thrown from the outside. The wire should be loose at the bottom to permit the defender to drop grenades on enemy forces below.

(3) Interior Routes. Routes are required that permit defending forces to move within the building to engage enemy forces from any direction or reinforce other forces (Figure D-7). Escape routes should also be planned and constructed to permit rapid evacuation of a room or the building. Small holes (called mouseholes) should be made through interior walls to permit movement between rooms. Once the defender has withdrawn to another level, such holes should be clearly marked for both day and night identification. All personnel must be briefed as to where the various routes are located. Rehearsals should be conducted so that everyone becomes familiar with the routes.

(4) Fire Prevention. Buildings that have wooden floors and raftered ceilings require extensive fire prevention measures. The attic and other wooden floors should be covered with about 1 inch of sand or dirt, and buckets of water should be positioned for immediate use. Firefighting materials (dirt, sand, fire extinguishers, and blankets) should be placed on each floor for immediate use. Water basins and bathtubs should be filled as a reserve for firefighting. All electricity and gas should be turned off. Fire breaks can be created by destroying buildings adjacent to the defensive position.

Figure D-7. Movement Between Floors

(5) Communications. Telephone lines should be laid through adjacent buildings or underground systems or buried in shallow trenches. Radio antennas can be concealed by placing them among civilian television antennas, along the sides of chimneys and steeples, or out windows that direct FM communications away from enemy electronic warfare sources and ground observation. Telephone lines within the building should be laid through walls and floors.

(6) Rubbling. Rubbling parts of the building may provide useful cover and concealment for weapons emplacements. Rubbling should be performed only by trained engineers.

(7) Rooftops. Positions in flat-roofed buildings require obstacles that restrict helicopter landings (See Figure F-2). Rooftops that are accessible from adjacent structures should be covered with tactical wire or other expedients and must be guarded. Entrances to buildings from rooftops can be blocked if compatible with the overall defensive plan. Any structure on the outside of a building that could assist in scaling the buildings to gain access to upper floors or to the rooftop should be removed or blocked.

(8) Obstacles. Obstacles should be positioned adjacent to buildings to stop or delay tanks and infantry.

(9) Fields of Fire. Fields of fire should be improved around the defensive position. Selected buildings may be destroyed to enlarge fields of fire. Obstacles to ATGMs, such as telephone or electrical wires, should be cleared. Dead space should be covered with mines and obstacles.

3. Tank and Other Armored Vehicle Positions. Fighting positions for tanks and other armored vehicles are essential to a complete and effective defensive plan in built-up areas.

a. Armored Vehicle Positions. Armored vehicle positions are selected and developed to obtain the best cover, concealment, observation, and fields of fire while retaining the vehicle's ability to move.

(1) If fields of fire are restricted to streets, hull-down positions should be used to gain cover and to fire directly down streets (Figure D-8). From those positions, armored vehicles are protected and can rapidly move to alternate positions. Buildings collapsing from enemy fires are a minimal hazard to armored crews.

Figure D-8. Hull-Down Position

(2) The hide position (Figure D-9) covers and conceals the vehicle until time to move into position for engagement of targets. Because the crew will not be able to see advancing enemy forces, an observer from the vehicle or a nearby infantry unit must be concealed in an adjacent building to alert the crew. The observer acquires the target and signals the armored vehicle to move to the firing position and fire. After firing, the vehicle moves to an alternate position to avoid compromising one location.

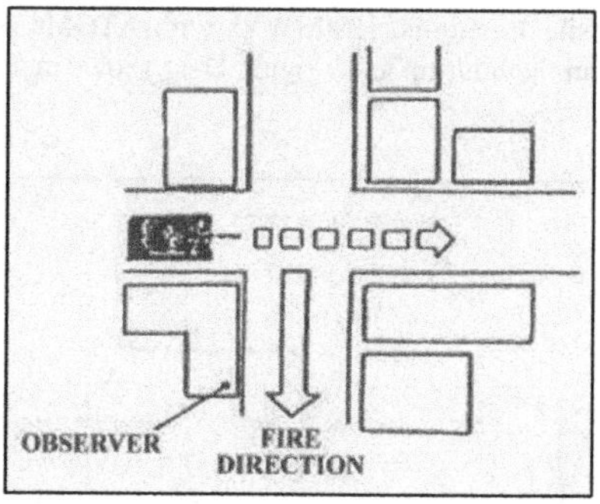

Figure D-9. Hide Position

(3) The building hide position (Figure D-10) conceals the vehicle inside a building. If basement hide positions are inaccessible, engineers can evaluate the building's floor strength and prepare for the vehicle. Once the position is detected, it should be evacuated to avoid enemy fires.

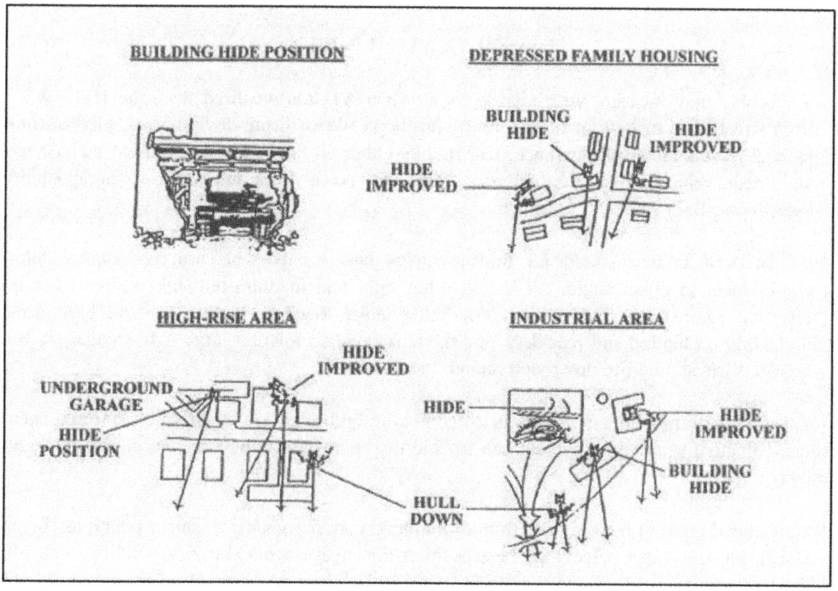

Figure D-10. Building Hide Position

.3

4. Antitank Guided Missile Positions. HMMWVs with ATGMs are employed in areas that maximize their capabilities in the built-up area. Figure D-11 shows mounted and dismounted firing positions.

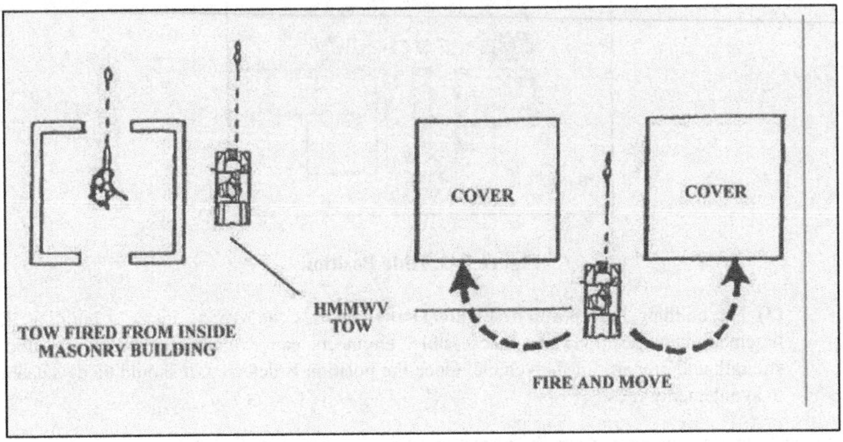

Figure D-11. ATGM Positions

a. Rubble may interfere with missile flight when ATGMs are fired from the HMMWV or from street-level or bottom-floor fighting positions. When firing down streets, missiles must have at least 1 meter of clearance over rubble. Other obstacles to missile flight include trees and brush, vehicles, television antennas, buildings, power lines and wires, walls, and fences (See Figure B-7).

b. An AT4 is best suited for built-up areas because its short arming distance allows employment at close range. AT4s and other light and medium antitank weapons are not effective against the front of modern battle tanks. Because tanks have the least armor protection on the top and rear deck and the tank presents a larger target when engaged from above, AT4s should fire down onto tanks.

5. Sniper Positions. Snipers contribute to combat in built-up areas by firing on selected enemy targets. Effective sniper employment can trouble the enemy far more than its relative personnel numbers reflect.

a. General areas (a building or group of buildings) are designated as sniper positions (Figure D-12), but the sniper selects the best position for engagement. Masonry buildings that offer the best protection, long-range fields of fire, and all-around observation are preferred. The sniper also selects several secondary and supplementary positions to cover his areas of responsibility.

b. Engagement priorities for snipers are determined by the relative importance of the targets to the effective operations of the enemy. Sniper targets may include tank commanders, crewmen of direct fire support weapons, crewmen of crew-served weapons, officers, FOs, and radiotelephone operators.

c. The urban environment often limits snipers to firing down or across streets. Snipers can be employed to cover rooftops, obstacles, dead space, and gaps in final protective fires.

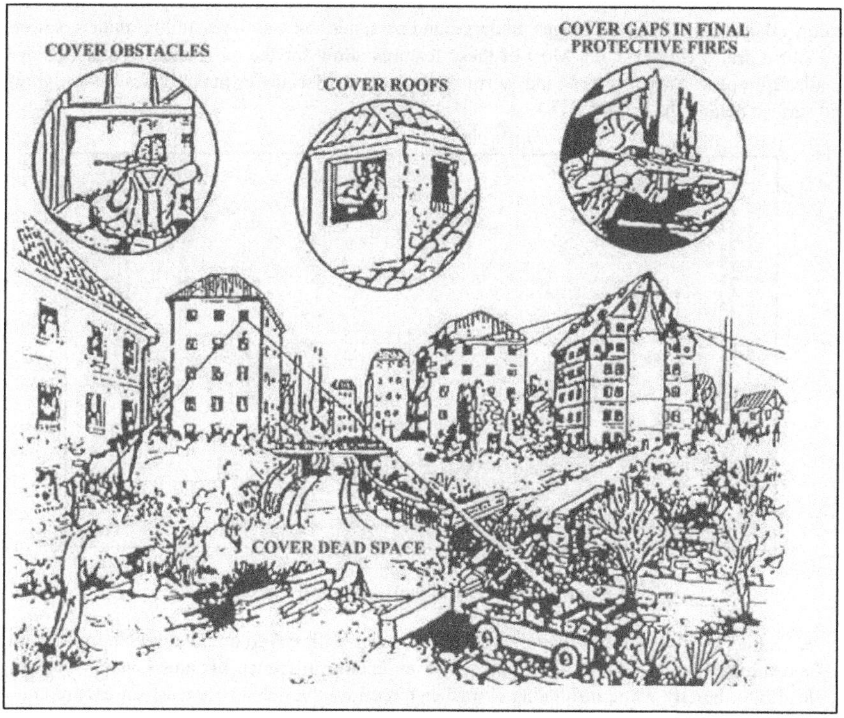

Figure D-12. Sniper Positions

Appendix E

Subterranean Operations

Knowledge of the nature and location of underground facilities is valuable to both the urban attacker and defender.

1. Tactical Value. Fighting in MOUT is multidimensional. Larger cities may have subterranean features that include sunken garages, underground passages, subway lines, utility tunnels, sewers, and storm drains (Figure E-1). Most of these features allow for the movement of units. Even in smaller European towns, sewers and storm drains permit Marines to move beneath the fighting and surface behind the enemy.

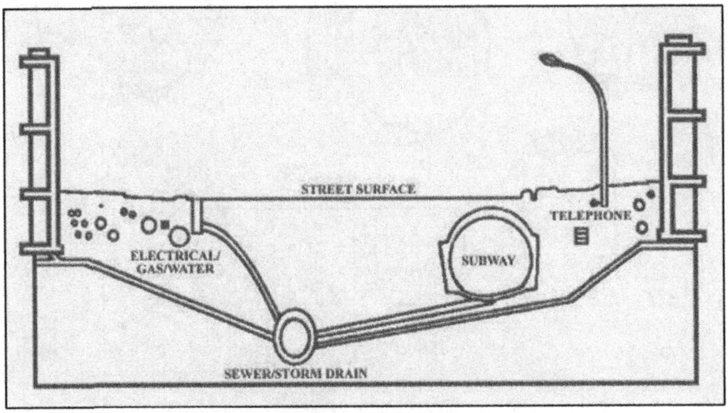

Figure E-1. Potential Underground Passages

a. Subterranean passages may provide the attacker with covered and concealed routes into and through built-up areas. This may allow an enemy to launch his attack along roads that lead into the city while infiltrating a smaller force into the defender's rear. An enemy unit in the defender's rear area may disrupt his defense and obstruct the avenues of withdrawal for his forward defense.

b. An attack along a subterranean avenue of approach could become the main effort. Even if the subterranean effort is not immediately successful, it forces the defender to fight on atleast two levels and to extend resources to more than just street-level fighting.

c. The existence of subterranean passages forces the defender to defend the built-up area above and below ground. While these passages are usually more of a disadvantage to the defender than to the attacker they may offer some advantages in the defense. When thoroughly reconnoitered and controlled by the defender, subterranean passages provide excellent covered and concealed routes to move reinforcements or to launch counterattacks. They also provide ready-made covered lines of communication for the movement of supplies and evacuation of casualties and provide places to cache supplies for forward companies. Subterranean passages also offer the defender a ready-made conduit for communications wire, with protection from armored vehicles and indirect fires.

2. Denial to the Enemy. Subterranean passages are useful to the defender only to the extent that the attacker can be denied their use. The defender has an advantage in that, given the confining, dark environment of these passages, a small group of determined Marines in a prepared position can defeat a numerically superior force.

a. Tunnels afford the attacker little cover and concealment except for the darkness and any manmade barriers. The passageways provide tight fields of fire and amplify the effects of grenades. Obstacles at intersections in tunnels support excellent ambush sites and may turn the subterranean passages into a maze. These obstacles can be created quickly by using chunks of rubble, furniture, and parts of abandoned vehicles interspersed with M18 claymore mines.

b. A thorough reconnaissance of the subterranean or sewer system is required. To be effective, obstacles must be located at critical intersections in the passage network so that they trap attackers in a kill zone but allow defenders freedom of movement (Figure E-2).

3. Subterranean Reconnaissance Techniques. The mission of local reconnaissance for a platoon- or company-sized area of responsibility should be given to a squad-sized element.

a. The squad leader tasks one fire team with providing forward security. One fire team follows in support, and one fire team provides rear security (Figure E-3). The squad leader moves directly behind the pointman and records data collected by the patrol. If available, an engineer should be with the lead fire team. The rear security fire team protects the ground- and subterranean-level point of entry. This fire team is responsible for detecting enemy movement on the patrol's rear and serves as the communications link between the squad leader and higher headquarters. Organization of the patrol may vary based on METT-T.

b. The squad leader should carry a map, compass, street plan, and notebook in which he has detailed the information that he must gather. One Marine is assigned to carry the tools needed to open manhole covers. If the patrol is to move more than 200 to 300 meters, or if the platoon leader directs, the squad may carry a sound-powered telephone (TA-1) and wire dispenser (XM-306A) for communications. Radios are often unreliable in this environment. Marines (especially the pointman, squad leader, engineer, and fire team leaders) should be equipped with night vision devices to maintain surveillance within the sewer. Thermal night devices are also useful as there may not be enough ambient light for accurate visibility with night vision devices.

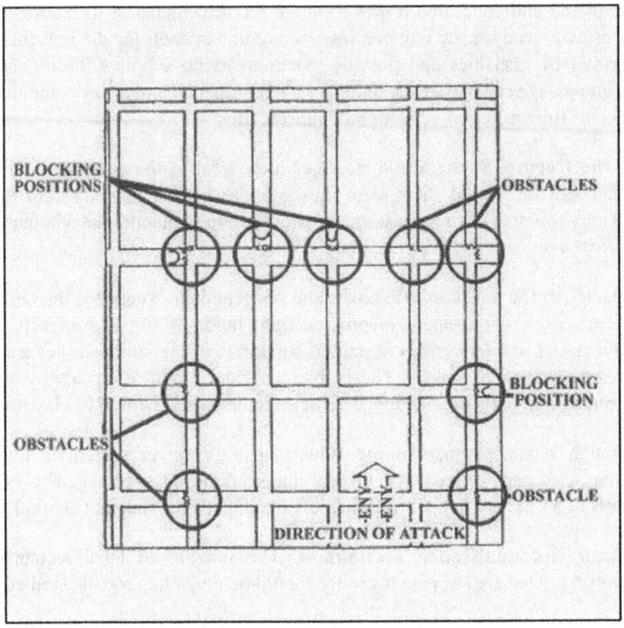

Figure E-2. Defense of a Sewer System

Figure E-3. Organization of the Patrol Unit

c. Marines entering the sewer should carry a sketch of the sewer system that includes magnetic north, azimuths, distances, and manholes. They should also carry protective masks, flashlights, gloves, and chalk (or chemiluminescent lights) for marking features along the route. The patrol should also be equipped with a safety rope to which each man is tied. To improve their footing in slippery sewers and storm drains, the squad members may wrap chicken wire or screen wire around their boots.

d. A constant concern to units conducting a subterranean patrol is chemical defense. Enemy chemical agents used in tunnels may be suddenly encountered in dense concentrations with little dispersion. The pointman should carry an automatic chemical agent alarm system to provide instant warning of the presence of chemical agents. M8 detector paper can also be used to detect chemical agents. At the first indication that harmful gases are present, the squad should mask.

e. In addition to enemy chemical agents, noxious gases from decomposing sewage can also pose a threat. These gases are not detected by chemical agent alarm systems, nor are they completely filtered by the protective mask. Physical signs that indicate their presence in harmful quantities are nausea and dizziness. The squad leader should be constantly alert to these signs and know the shortest route to the surface for fresh air. Some sewer gases deplete oxygen, a situation against which the field protective mask provides no protection.

f. Once the squad is organized and equipped, it moves to the entrance of the tunnel. This may be a manhole. With the manhole cover removed, the squad waits 15 minutes before entry to allow any gases that may be present to dissipate. The pointman then descends into the tunnel to determine whether the air is safe to breathe and whether movement is restricted. The pointman should remain in the tunnel for 10 minutes before the rest of the squad follows. If he becomes ill or is exposed to danger, he can be pulled out by the safety rope.

g. When the squad is moving through the tunnel, the pointman moves about 10 meters in front of the squad leader. Other squad members maintain 5-meter intervals. If water in the tunnel is flowing faster than 2.5 meters per second or the sewer contains slippery obstacles, those intervals should be increased to prevent all patrol members from falling if one man slips. All squad members should stay tied into the safety rope so that they can easily be retrieved from danger. The last Marine in the patrol marks the route with chalk or chemiluminescent lights so other units can find the squad.

h. The squad leader should record the azimuth and pace count of each turn he takes in the tunnel. When he encounters a manhole to the surface, the pointman should open it and determine the location, which the patrol leader then records. The use of recognition signals (Figure E-4) prevents friendly forces from accidentally shooting the pointman as he appears at a manhole.

Figure E-4. Recognition Signals

i. Once the patrol has returned and submitted its report, the platoon leader must decide how, or whether, to use the tunnel. In the offense, a tunnel might provide a covered route to move behind the enemy's defenses. In the defense, a tunnel might provide a covered passage between positions. In any case, the squad members that provided the reconnaissance should act as guides along the route.

j. If the tunnel is to be blocked, the platoon may emplace concertina wire, early warning devices, and antipersonnel mines. A security force should be established at the entrance of the sewer (Figure E-5) to provide warning against any enemy trying to approach the platoon's defense. It should be equipped with command-detonated illumination. The overpressure from grenades, mines, and boobytraps exploding in a sewer or tunnel can have adverse effects on friendly forces, including ruptured eardrums and wounds from flying debris. Also, gases found in sewers can be ignited by the blast effects of these munitions. For these reasons, small-arms weapons should be employed in tunnels and sewers. Friendly personnel should be outside of tunnels or out of range of the effects when mines or demolitions are detonated.

4. Psychological Considerations. Combat operations in subterranean passages are similar to night combat operations. The psychological factors of night operations reduce confidence, cause fear, and increase a feeling of isolation. This feeling of isolation is further magnified by the tight confines of the tunnels. The layout of tunnels could require greater dispersion between positions.

Figure E-5. Security Position Established at the Entrance to a Sewer

a. Leaders must enforce measures to help dispel the feelings of fear and isolation experienced when operating in tunnels. These measures include leadership training, physical and mental fitness, sleep discipline, and stress management.

b. Leaders maintain communication with Marines manning positions in tunnels either through personal visits, messengers, or field telephone. Communication increases situational awareness and provides insight into the mental state of Marines. Marines manning positions below ground should be given as much information as possible on the organization of the tunnels and the importance of the mission. They should be briefed on contingency plans and alternate positions should their primary positions become untenable.

c. Physical and mental fitness can be maintained by periodically rotating Marines out of tunnels so they can stand and walk in fresh air and sunlight. Stress management is also a factor of operations in tunnels. Historically, combat in built-up areas has been one of the most stressful forms of combat. Fighting in continuous darkness and restricted maneuver space causes even more stress to Marines than street fighting.

Appendix F

Obstacles, Mines, and Demolitions

Obstacles are used extensively in combat in built-up areas to allow the defender to canalize the enemy, impede his movement, and disrupt his attack.

Section I

Obstacles

1. Introduction. Obstacles are designed to prevent movement by personnel, to separate infantry from tanks, and to slow or stop vehicles. (See MCRP 3-17A, *MAGTF Breaching Operations.*)

2. Types. Antipersonnel mines, barbed wire, boobytraps, and exploding flame devices are used to construct antipersonnel obstacles (Figure F-1). (See FMFRP 13-5/FM 5-34, *Engineer Field Data*; FM 5-250, *Explosives and Demolitions*; and FM 3-11, *Flame, Riot Control Agent, and Herbicide Operations,* for more detailed information.) These obstacles are used to block the following infantry approaches:

- Streets
- Buildings
- Roofs
- Open spaces
- Dead space
- Underground systems.

a. The approval authority to boobytrap buildings is normally the GCE commander. This authority may be delegated to subordinate commanders.

b. The three types of wire obstacles used in defensive operations are protective, tactical, and supplementary. Wire obstacles should be integrated with other obstacles and covered by fire and observation.

(1) Protective wire is usually located beyond hand-grenade range (40 to 100 meters) from the defensive position.

(2) Tactical wire is positioned to increase the effectiveness of friendly weapons fire. Tactical wire is usually positioned on the friendly side of machine gun FPLs.

(3) Supplementary wire is used to break up the pattern of tactical and protective wire to prevent the enemy from easily locating friendly weapons.

c. Obstacles may be placed in dead spaces to restrict infantry movement in areas that cannot be observed and that are protected from direct fires.

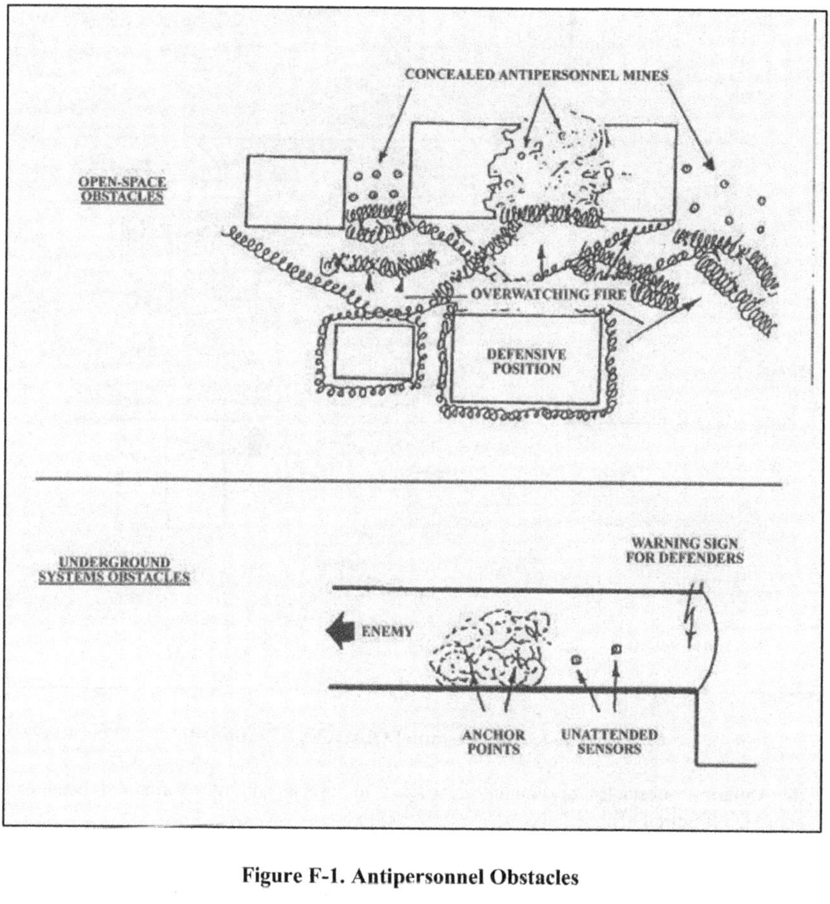

Figure F-1. Antipersonnel Obstacles

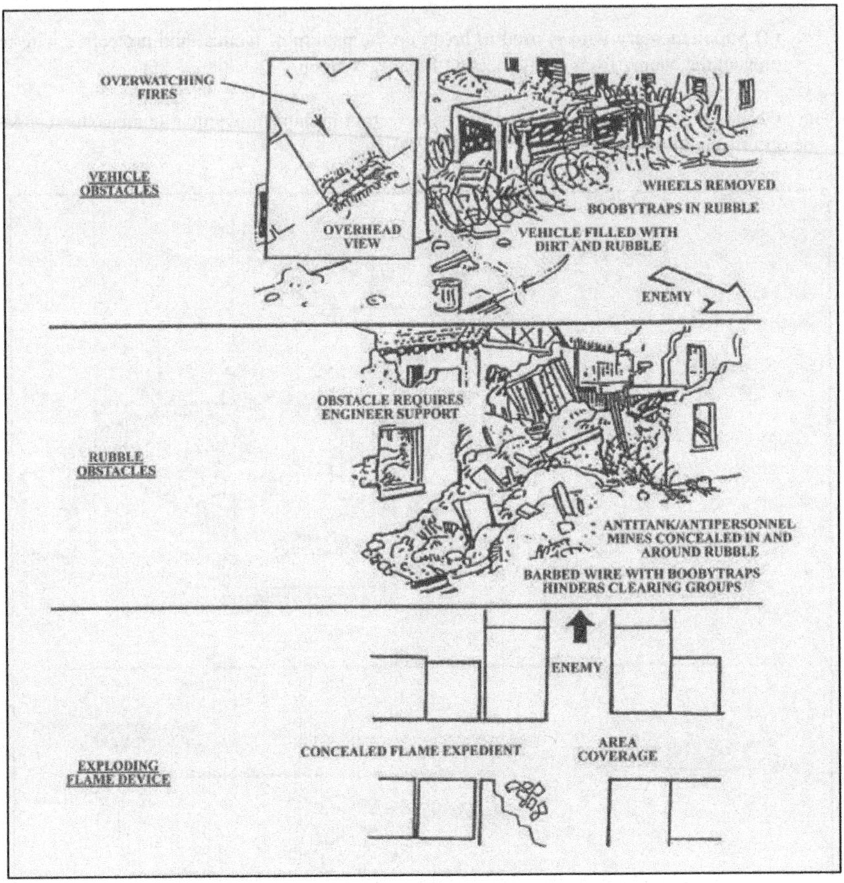

Figure F-1. Antipersonnel Obstacles (Continued)

d. Antiarmor obstacles are mainly restricted to streets and other armored avenues of approach (Figure F-2).

e. Some antitank mines may be sited and employed as antihelicopter mines to cover probable HLZs (Figure F-2).

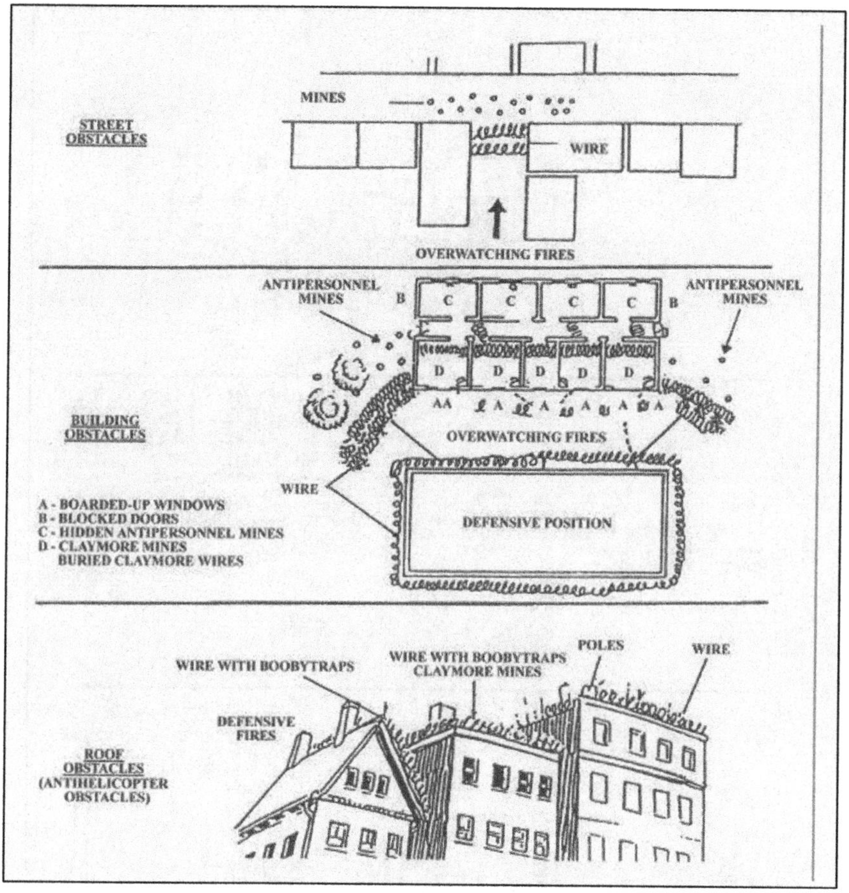

Figure F-2. Antiarmor Obstacles

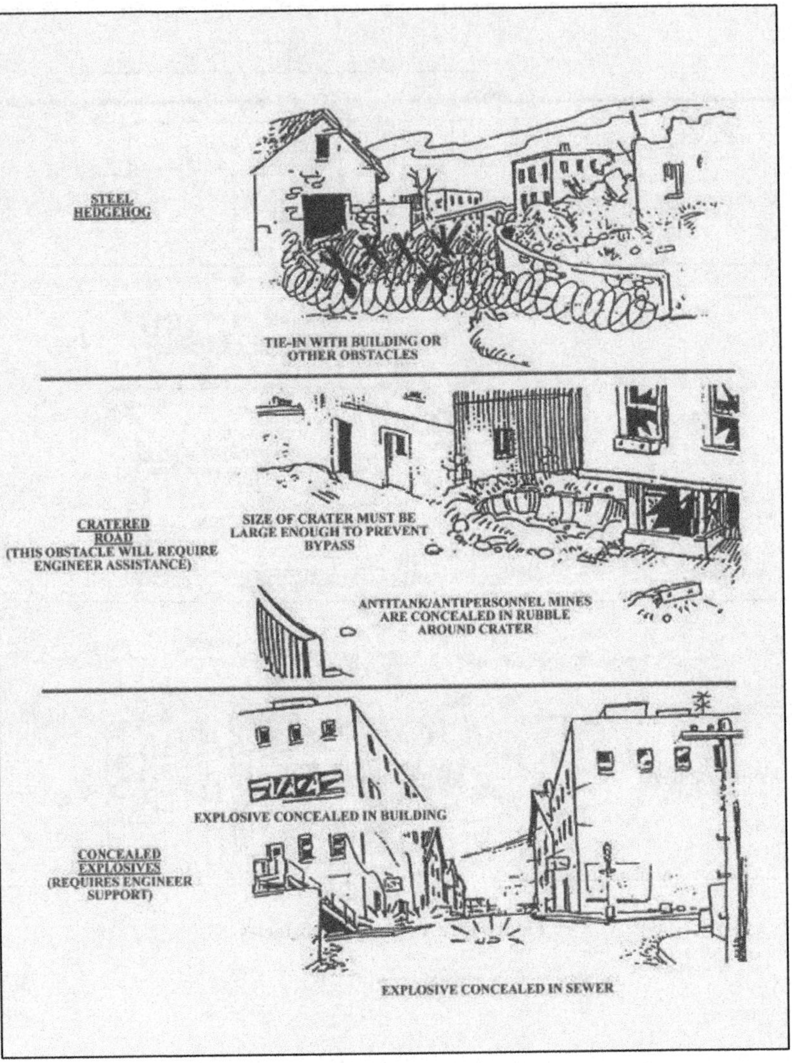

Figure F-2. Antiarmor Obstacles (Continued)

3. Construction of Obstacles. Obstacles are constructed in buildings to deny enemy infantry covered routes and weapons positions that are close to friendly defensive positions. They can be constructed of explosives, by using wire, or by using boobytraps within buildings. The building may be prepared as an explosive or flame trap for command detonation after enemy occupation.

Section II

Mines

4. **Introduction.** The employment or discovery of mines in built-up areas should be recorded on a building sketch (Figure F-3). (See FM 20-32.) The sketch should include the number of the building (taken from a city map, if available) and all floor plans. It should also include the type of mine and firing device. When possible, mined buildings should be marked on the friendly side (Figure F-4). Clearing areas or buildings that have been mined is extremely difficult and dangerous. Therefore, such areas should be considered "no go" areas for infantry. This factor must be carefully considered when planning and authorizing the placement of mines. (See Table F-1 for notional approving authority for minefields, but with the understanding that information in that table is subject to change with international agreements and ROE.)

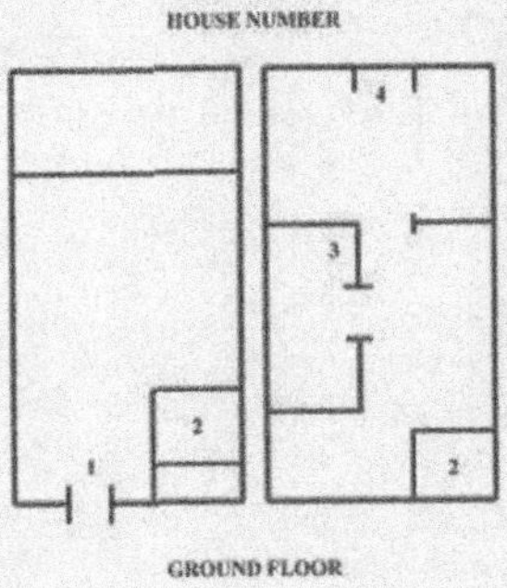

MINE	LOCATION	TYPE	FIRING DEVICE
1	FRONT DOOR	CLAYMORE	COMMAND
2	STAIRS	CLAYMORE	COMMAND
3	CLOSET	CLAYMORE	COMMAND
4	DRESSER	CLAYMORE	COMMAND

Figure F-3. Building Sketch Showing Mines

Figure F-4. Marking Mined Buildings

	Minefields (Type)		FASCAM (Duration)		Other Obstacles	
Level	Tactical	Protective	Long	Short	Tactical	Protective
MEF	X	X	X	X	X	X
Division	X	X	X	X	X	X
Regiment	X	X	X	X	X	X
Battalion	X	X		X	X	X
Company		X				X

Table F-1. Notional Minefield Employment Authority Matrix

5. **Types.** Several types of mines may be encountered or employed in built-up areas.

a. The M14 mine (a plastic type of antipersonnel landmine) cannot currently be used by American forces due to the Leahy Amendment of 1995 and the Certain Conventional Weapons Convention, of which the United States is a signatory nation. It may, however, be encountered in MOUT or be employed by coalition or enemy forces. For this reason, the following information is given as a guide for what U.S. forces may encounter. This type of antipersonnel landmine is normally used in conjunction with metallic antipersonnel, antitank, or chemical mines to confuse and hinder enemy breaching attempts. When encountered, it must be carefully handled because its light weight makes it easy to displace (Figure F-5). However, its size makes it ideal for covering obscure places such as stairs and cellars.

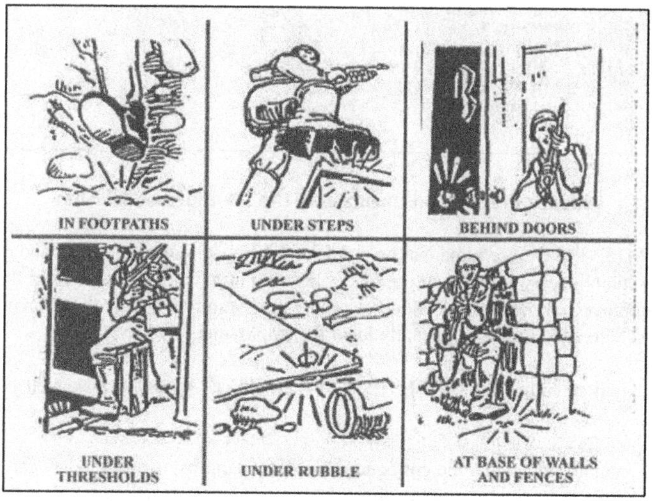

Figure F-5. Typical Employment of the M14 Antipersonnel Mine

b. The M16A2 antipersonnel landmine is subject to the same legal limitations as the M14. It is a boundary type of antipersonnel landmine, which makes it ideal for covering large areas such as rooftops, backyards, parks, and cellars. It can be expediently rigged for command detonation by attaching a rope or piece of communications wire to the pull ring (Figure F-6).

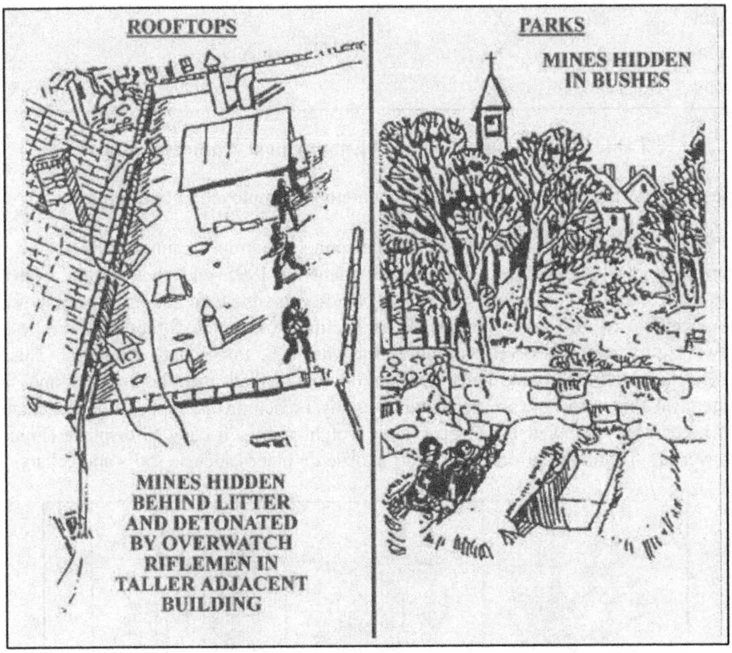

Figure F-6. Emplacements of M16A2 Antipersonnel Mine

c. The M18A1 claymore mine can be employed during the reorganization and consolidation phase on likely enemy avenues of approach. It does not have to be installed in the street, but can be employed on the sides of buildings or any other sturdy structure. The claymore mine, in command-detonated mode, is currently legal for employment by U.S. forces.

(1) Claymore mines can also be used for demolition against thin-skinned buildings and walls.

(2) Claymore mines may be configured for detonation by tripwire. They can help fill the dead space in the final protective fires of automatic weapons (Figure F-7).

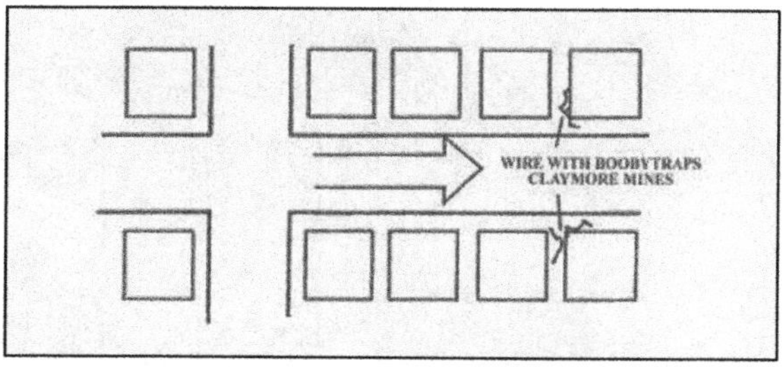

Figure F-7. Claymore Mine Used To Cover Dead Space of Automatic Weapons

(3) Claymore mines can be used in several ways in the offense. For example, if friendly forces are advancing on a city, claymore mines can be used in conjunction with blocking positions to cut off enemy avenues of escape (Figure F-8).

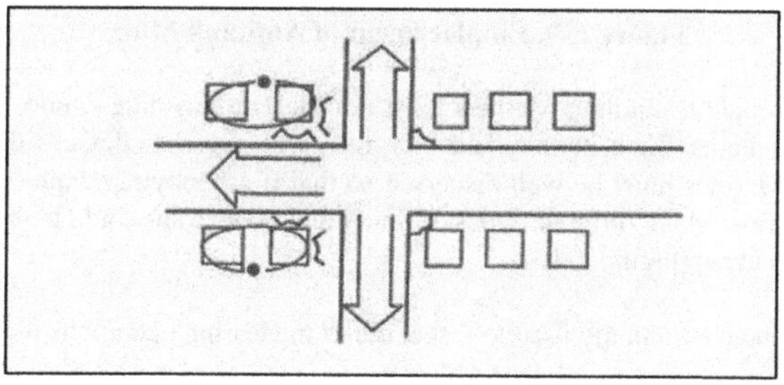

Figure F-8. Claymore Mines Used To Block Enemy Avenues of Escape

d. The M15, M19, and M21 antitank mines are used to stop or turn enemy armor and should be employed (Figure F-9):

- In conjunction with other manmade obstacles and covered by observation and fire
- In streets or alleys to block routes of advance in narrow defiles
- As command-detonated mines with other demolitions.

6. Enemy Mines and Boobytraps. Buildings contain many areas and items that are potential hiding places for boobytraps, for example, doors, windows, telephones, stairs, books, canteens, and so on.

Figure F-9. Emplacement of Antitank Mines

When moving through a building, Marines must not pick up anything—food, valuables, enemy weapons, or other items. Such items could be rigged with pressure devices that explode when they are moved. Marines must be well-dispersed so that if a boobytrap explodes, the number of casualties will be few. Many different types of mines and boobytraps could be encountered during combat in built-up areas (Figure F-10).

a. The equipment commonly used to detect mines in clearing operations includes:

- Mine detectors
- Probes
- Grappling hooks
- Ropes
- Explosives
- Flak vests
- Eye protection
- Engineer tape.

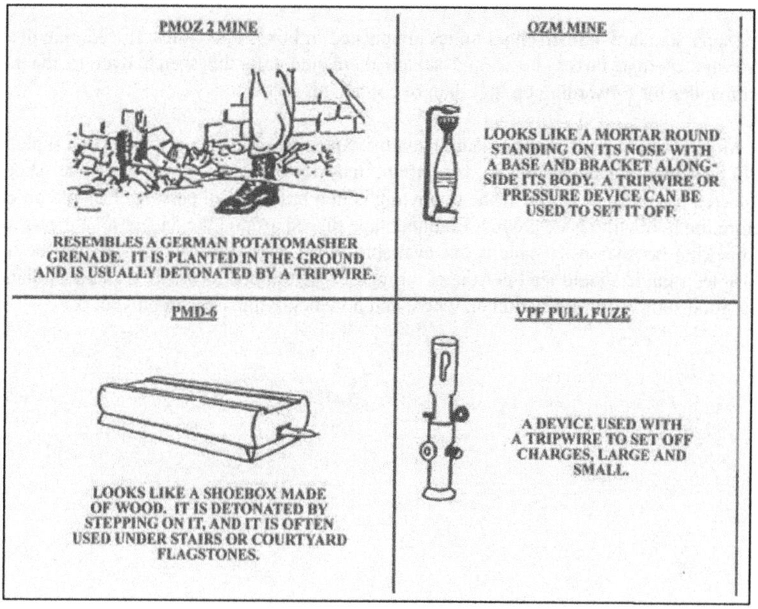

Figure F-10. Common Threat Antipersonnel Mines and Boobytraps

b. To detect tripwires, a Marine can use a 10-foot pole with 5 feet of string tied on one end. A small weight attached to the loose end of the string will snag on the tripwire. This allows the lead Marine to easily detect a tripwire (Figure F-11). Marines can also use a flashlight to cause reflections and shadows to find tripwires. A wand with a dangling thread can be swept over the ground ahead with the thread being watched for any disturbance by a tripwire.

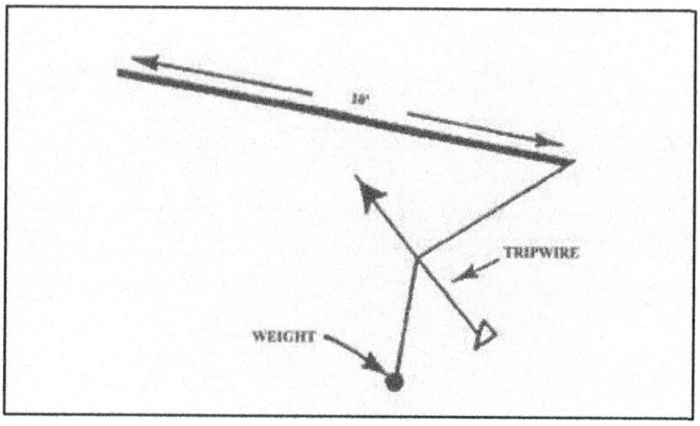

Figure F-11. Tripwire Detection

c. Many standard antipersonnel mines are packed in boxes and crates. If a Marine discovers explosive storage boxes, he should sketch them and turn the sketch over to the platoon commander for forwarding up the chain of command.

d. Most boobytraps should be neutralized by explosive ordnance disposal (EOD) personnel. If EOD teams are not available, boobytraps may be blown in place. Personnel should be protected by adequate cover. If the boobytrap is in a building, all personnel should go outside before the boobytrap is destroyed. Engineer tape placed around the danger area is one method of marking boobytraps. If tape is not available, strips ripped from bedsheets can be used. If possible, a guide should lead personnel through boobytrapped areas. EPWs and civilians may be a good source of information on where and how boobytraps are employed.

Section III

Demolitions

7. Introduction. Demolitions are used more often during combat in built-up areas than during operations in open natural terrain. Demolition operations should be directed by the engineers that support the GCE. However, if engineers are involved in the preparation and execution of the barrier plan, other trained Marines can prepare mouseholes, breach walls, and rubble buildings themselves if assisted and advised by attached engineers.

8. Offensive Use. When assaulting or clearing a built-up area, demolitions may enable the GCE commander to create an avenue of approach through buildings.

a. Every other man in an assault force should carry demolitions, and other selected personnel should carry blasting caps. The same man should *not* carry both the explosives and the blasting caps because of the danger of sympathetic detonation. As the demolitions are expended by the assault force, they should be replaced by explosives carried by the support force.

b. One of the most difficult breaching operations faced by the assault force is the breaching of masonry and reinforced concrete walls. When demolitions must be used, composition C4 is the ideal explosive to use. Normally, building walls are 15 inches thick or less. Assuming that all outer walls are constructed of reinforced concrete, a rule of thumb for breaching is to place 10 pounds of C4 against the target between waist and chest height. When detonated, this normally blows a hole large enough for a man to go through. The amounts of TNT and C4 required to breach concrete are shown in Table F-2.

c. Mouseholes provide the safest method of moving between rooms and floors. They can be created with C4. Because C4 comes packaged with an adhesive backing or can be emplaced using pressure-sensitive tape, it is ideal for this purpose. When using C4 to blow a mousehole in a lath and plaster wall, one block or a strip of blocks should be placed on the wall from neck to knee height. Charges should be primed with detonating cord or electrical blasting caps to obtain simultaneous detonation, which will blow a hole large enough for a man to fit through.

9. Defensive Use. The use of demolitions in defensive operations is similar to their use in offensive operations. When defending a built-up area, demolitions are used to create covered and concealed routes through walls and buildings that can be used for withdrawals, reinforcements, or counterattacks. Demolitions are also used to create obstacles and clear fields of fire.

a. Marines can use demolitions for creating mouseholes and constructing command-detonated mines. Expedient C4 satchel charges can be concealed in firing positions and along movement routes for detonation in the vicinity of the enemy.

Thickness of Material	TNT	C4	Size of Opening
Reinforced Concrete			
Up to 10 cm (4 inches)	.792 lbs	.591 lbs	10 to 15 cm (4 to 6 inches)
10 to 20 cm (4 to 8 inches)	3.4 lbs	2.54 lbs	15 to 30 cm (6 to 12 inches)
Nonreinforced Concrete Masonry			
Thickness of Material	**TNT**	**C4**	**Size of Opening**
Up to 35 cm (14 inches)	3.66 lbs	2.73 lbs	35 cm (14 inches)
35 to 45 cm (14 to 18 inches)	7.5 lbs	5.6 lbs	45 cm (18 inches)
45 to 50 cm (18 to 20 inches)	10.2 lbs	7.61 lbs	50 cm (20 inches)

Table F-2. Amounts of TNT and C4 Required To Breach Concrete

b. The engineers should furnish technical assistance for selective rubbling. Normally, buildings can be rubbled by using shaped charges or C4 on the supports and major beams.

c. Charges should be placed directly against the surface to be breached unless a shaped charge is used. Whenever possible, demolitions should be tamped to increase their effectiveness. Tamping materials could be sandbags, rubble, or desks and chairs (Figure F-12).

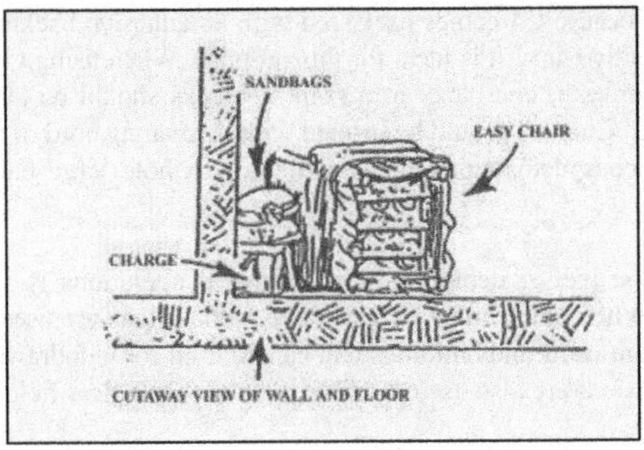

Figure F-12. Chair Used To Tamp Breaching Charge

d. For exterior walls, tamping of breaching charges may be impossible because of enemy fire. Thus, the untamped charge normally requires twice the explosive charge to produce the same effect as a tamped charge (Figure F-13).

METHODS OF PLACEMENT				
THICKNESS OF CONCRETE				
FEET	POUNDS OF TNT	POUNDS OF C4	POUNDS OF TNT	POUNDS OF C4
2	14	11	28	21
2 1/2	27	21	54	41
3	39	30	78	59
3 1/2	62	47	124	93
4	93	70	185	138
4 1/2	132	99	263	196
5	147	106	284	211
5 1/2	189	141	376	282
6	245	186	490	366

Figure F-13. Breaching Reinforced Concrete

e. When enemy fire prevents an approach to the wall, the breaching charge can be attached to a pole and slid into position for untamped detonation at the base of the wall (Figure F-14).

Figure F-14. Charge Placement When Small-Arms Fire Cannot Be Suppressed

f. The internal walls of most buildings function as partitions rather than as load-bearing walls. Therefore, smaller explosive charges can be used to breach them. In the absence of C4 or other military explosives, internal walls can be breached by using a claymore mine (Figure F-15). The claymore should be tamped to increase its effectiveness and to reduce the amount of explosive force directed to the rear.

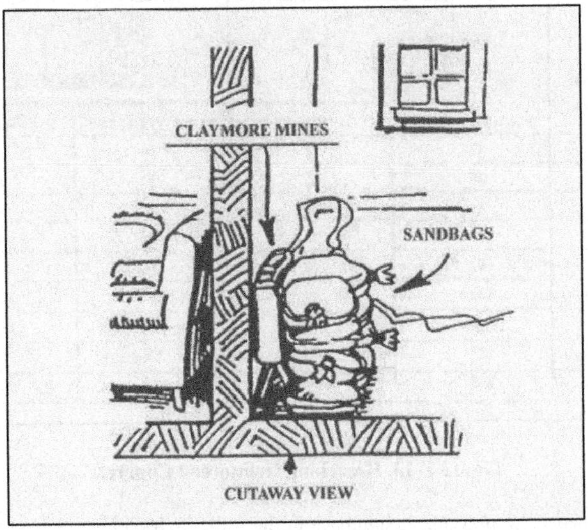

Figure F-15. Tamping a Claymore Mine To Breach Internal Walls

g. The Molotov cocktail (Figure F-16) is an expedient device for disabling both wheeled and tracked vehicles. It is easy to make because the materials are readily available. The results are most effective because of the close engagement in built-up areas. The objective is to ignite a flammable portion of the vehicle such as the fuel or ammunition that it is transporting. The following materials are needed to make a Molotov cocktail:

- Bottle or other glass container
- Gas (60 percent)
- Oil (40 percent)
- Rag for use as a wick.

The gas and oil are mixed thoroughly (60 percent gas to 40 percent oil). The rag is soaked with the mixture, then the mixture is placed into the bottle. The rag is then inserted in the opening of the bottle as a wick. When a target is sighted, the wick is lit and the bottle is thrown hard enough to break.

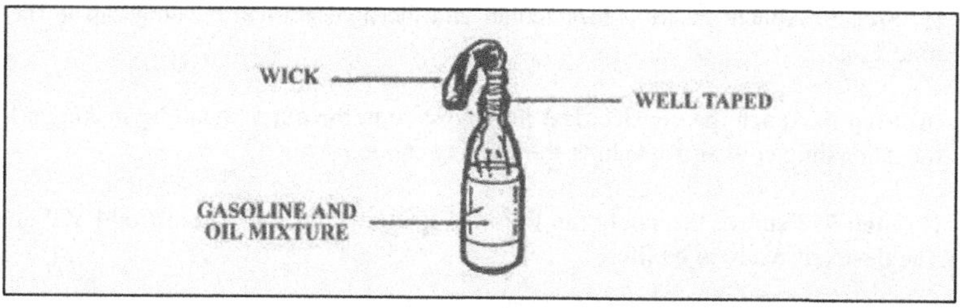

Figure F-16. Molotov Cocktail

WARNING

Ensure that a safe distance is maintained when throwing a Molotov cocktail.

h. The bunker bomb (Figure F-17) is an expedient, explosive flame weapon that is best used against fortified positions or rooms. This expedient munition should be used with a nonelectrical firing system. The following materials are required to make a bunker bomb:

- 1 metal small-arms ammunition container
- 1 gallon of gasoline
- 50 feet of detonating cord
- 1 nonelectrical blasting cap
- 1 M60 fuze igniter
- 7 1/2 feet of M700 time fuze
- 1 M49 trip flare or M34 WP grenade.

(1) Step 1. Fill the ammunition can 3/4 full with fuel and secure the lid.

(2) Step 2. "Hasty whip" the device with 15 turns around the center of the container using 44 feet of detonating cord. Leave 2-foot "pigtails" for attaching the igniter and fuze igniter.

(3) Step 3. Tape the igniter (M49 trip flare or M34 WP grenade) to the container handle.

(4) **Step 4.** Place one detonating cord pigtail end under the igniter spoon handle. Tape it in place.

(5) **Step 5.** Attach the M60 fuze igniter and the nonelectrical blasting cap to the M700 time fuze.

(6) **Step 6.** Attach the nonelectrical firing system to the other pigtail by making a loop in the detonating cord and attaching the blasting cap to it.

(7) **Step 7.** Remove the safety pin from the igniter (M49 trip flare or M34 WP grenade). The device is ready to be fired.

WARNING

Never carry the device by the handle or igniter. Remove the igniter safety pin only when it is time to use the device. Use extreme care when handling or carrying nonelectrical firing systems. Protect blasting caps from shock and extreme heat. Do not allow the time fuze to kink or become knotted. Doing so may disrupt the powder train and may cause a misfire. Prime detonating cord and remove the time fuze igniter safety pin only when it is time to use the device.

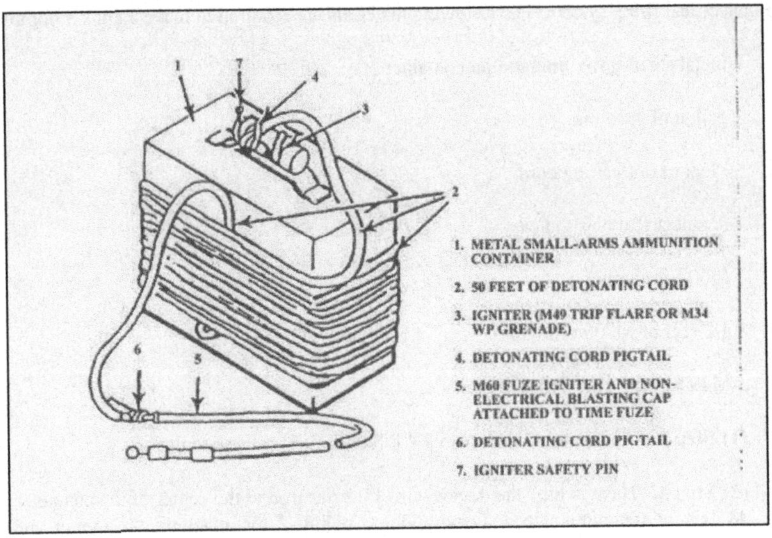

Figure F-17. Bunker Bomb Made From an Ammunition Can

10. Safety. The greatest danger to friendly personnel from demolitions is the debris thrown by the explosion. Leaders must ensure that protective measures are enforced. The safe distance listed in Table F-3 indicates the danger of demolition effects.

Pounds of Explosive	Safe Distance in Meters[1]	Pounds of Explosive	Safe Distance in Meters[1]
1 to 29	300	150	534
30	311	175	560
35	327	200	585
40	342	225	609
45	356	250	630
50	369	275	651
60	392	300	670
70	413	325	688
80	431	350	705
90	449	375	722
100	465	400	737
125	500	425	750

[1]These distances will be modified in combat when troops are in other buildings, around corners, or behind intervening walls.

Table F-3. Minimum Safe Distances for Personnel in the Open

a. The following are the guidelines for using demolitions:

- Keep the blasting machine under the control of a noncommissioned officer.
- Wear helmets and flack jackets at all times while firing explosives.
- Handle misfires with extreme care.
- Clear the room and protect personnel when blowing interior walls.

b. Some charges should be prepared beforehand, without detonators, to save time. Examples include 10- or 20-pound breaching charges of C4 and expedient-shaped charges in No. 10 cans.

- Use C4 to breach hard targets (masonry construction).

- Do not take chances.
- Do not divide responsibility for demolition work.
- Do not mix demolitions and detonators until ready to use.
- Do not carry demolitions and caps together.

Appendix G

Nuclear, Biological, and Chemical Considerations

Current U.S. policy regarding lethal or incapacitating agents is that their use against an armed enemy requires approval at the NCA level. Potential enemies may not operate under the same restrictions. Field commanders must be prepared to assume an adequate NBC defensive posture when engaged in urban fighting.

1. Protection From Nuclear, Biological, and Chemical Weapons. Generally, the lowest floor or basement of a reinforced concrete or steel-formed building offers good protection from nuclear hazards and initial liquid chemical contamination although some chemical agents tend to collect in lower areas. Tanks and LAVs/AAVs also provide protection. (See FMFM 11/MCWP 3-37 series manuals for greater detail.)

a. Biological attacks are difficult to detect and recognize. Biological agents can be disseminated by using aerosols, vectors, and covert methods. Because biological agents can be sprayed or dropped in bomblets, personnel who observe such indicators should promptly report them. Prompt reporting and treatment of the sick speed the employment of medical countermeasures. Although buildings and shelters provide some protection against spraying, they provide little protection against biological agents.

b. Chemical agents cause casualties by being inhaled or absorbed through the skin. They may afford Marines a few seconds to mask. Buildings have a channeling effect and tend to contain the effects of an agent, causing great variations in chemical concentration from room to room or from building to building. Chemical agents usually settle in low places, making sewers and subways hazardous hiding places. A prepared defender should include some collective protective measures in the defensive network. Personnel using fans may be able to put enough overpressure into tunnels to keep some chemical agents from entering. The protective mask and battle dress overgarment provide the best protection against chemical agents.

c. Personal hygiene is a critical defensive measure against infection and disease. Unfortunately, built-up areas are characterized by sophisticated sanitation systems. When those systems are destroyed, the resulting sanitary conditions become much worse than those in areas where sanitation facilities do not exist. Commanders must ensure that personnel employ sanitation measures and that their immunizations are current.

d. Commanders should plan their mission-oriented protective posture (MOPP) with the realization that increased logistics demands in built-up areas also apply to NBC equipment. Protective clothing, detection and decontamination equipment, and sealed containers of food and water must be stockpiled the same as other critical supplies. When operating in protective clothing, commanders must make allowances for the strenuous activities normally associated with combat in built-up areas.

(1) **Detection.** After an NBC attack, battalions should dispatch their detection and survey teams. Detection in built-up areas is complicated by the numerous rooms within buildings. Teams should conduct tests and surveys of major streets, intersections, and buildings in their area for inclusion in initial NBC reports. A systematic survey of all buildings, rooms, and underground facilities must be accomplished before occupation by unmasked personnel. All data should be forwarded using the appropriate NBC report.

(2) **Decontamination.** Personnel begin decontamination operations as soon after an NBC attack as the mission allows. Personnel should conduct individual decontamination of themselves and their personal equipment. Unit commanders determine the need for MOPP gear exchange and the requirements for a hasty or deliberate decontamination operation.

(a) **Nuclear.** Personnel should wear wet-weather gear for certain decontamination operations (hosing down buildings) to prevent radioactive material from touching the skin.

(b) **Chemical and Biological.** Roads, sidewalks, and other hard surfaces are best decontaminated by weathering, if time permits. Agents can also be covered with several inches of dirt or sand to provide protection. Fragment testing should be conducted periodically to ensure that the agent has not seeped through the covering. For critical sections of roads, a truck-mounted M12A1 (power-driven decontamination apparatus (PDDA)) can be used to spray supertropical bleach (STB) slurry; this aids rapid decontamination. Buildings are difficult to decontaminate, especially wooden ones. Some techniques for their decontamination are scrubbing with STB slurry; washing with hot, soapy water; washing or spraying with a soda solution; and airing.

2. Smoke Operations. The use of smoke is an integral part of either offensive or defensive operations in built-up areas. In the offense, smoke can support the maneuver of combat units and deception operations. Smoke employed in the defense obscures enemy air and ground observation, limiting the accuracy of enemy fires and intelligence collection.

a. Smoke should not be used when it degrades the effectiveness of friendly forces. Likewise, an extremely dense concentration of smoke in a closed area displaces the oxygen, suffocating Marines even when they are wearing protective masks.

b. Smoke pots, generators, or artillery smoke munitions should be used to cover the withdrawal of defending forces or the movement of attacking forces. Artillery-delivered WP can be effective against enemy forces by screening, in addition to causing casualties and fires. Consideration should be given to the incendiary effects of WP, smoke (HC and M825), and illumination munitions on the litter and debris of built-up areas.

c. Smoke grenades can be massed to provide a hasty screen for concealing personnel moving across streets and alleys. Smoke grenades can also be used for signaling; those launched by an M203 can be used to mark targets for fixed and rotary-wing air support.

d. The use of smoke in built-up areas is affected by the complex wind patterns caused by buildings. When covering a built-up area with a smoke haze or blanket, personnel must include all buildings. Failure to obscure tall buildings, towers, and steeples provides enemy observers with reference points for fire placement within the built-up area.

3. Riot Control Agents. When authorized, riot control agents such as CS and CN can be used to drive enemy forces from positions or to deny them areas for occupation. Riot control agents are incapacitating but generally have no lasting effects. They may be appropriate when preventing civilian casualties is a planning consideration. However, riot control agents may not be effective against an enemy that is well trained in chemical defense.

Appendix H

MOUT Under Limited-Visibility Conditions

To be successful, leaders must use limited-visibility conditions to their advantage.

1. Advantages. When fighting in built-up areas during night or periods of limited visibility, attacking or defending forces have several advantages.

a. In many cases, Marines may have a technological advantage in thermal imagery and light intensification over their opponents. This enables Marines to identify, engage, and destroy enemy targets before being detected by the enemy.

b. Possessing the capability to conduct operations as well at night or during periods of reduced visibility as during daylight gives Marines a tremendous advantage over an enemy with reduced capability. It also aids the generation of tempo and allows Marines to gain or retain the initiative.

c. Generally, ranges of direct-fire engagements are greatly reduced in the MOUT environment. During periods of limited visibility, effective unaided target acquisition ranges are further reduced. This enables attacking forces to close to shorter ranges, thus increasing the lethality and accuracy of weapons. Attacking forces can also take advantage of the enemy's reduced ability to see and can more effectively engage the enemy before being detected with thermal imagery or light intensification devices.

d. Helicopterborne assaults are best conducted during periods of limited visibility because the enemy's air defenses are degraded.

e. Attacking during periods of limited visibility gives the attacker a greater chance of achieving surprise.

2. Disadvantages. When fighting in built-up areas during periods of limited visibility, attacking and defending forces also face some disadvantages.

a. Command and control is difficult in any operation in a built-up area; periods of limited visibility increase this difficulty.

b. Marines have an instinctive tendency to move close together during periods of limited visibility. Constant attention must be given to prevent Marines from "bunching up."

c. Marines may become disoriented easily because of the combined effects of low visibility and the characteristics of built-up areas.

d. Target identification becomes more difficult in limited-visibility conditions. Depending on the training of the individual, the Marine may fire at anything seen or may hesitate too long before firing. Improper target identification is one of the leading causes of fratricide, so leaders must pay close attention to individual target engagement.

3. Fratricide Avoidance. The risk of fratricide is much greater during periods of limited visibility. The key to avoiding fratricide is increased situational awareness by leaders and individuals, coupled with realistic training in target identification. (See *Fratricide: Reducing Self-Inflicted Losses*, Newsletter, No. 92-4, April 92, Center for Army Lessons Learned, U.S. Army Combined Arms Command, Fort Leavenworth, Kansas.) Other considerations include:

a. Graphic control measures should be clearly defined and obvious. Examples include distinctive buildings, large boulevards, rivers, and so forth.

b. Leaders must exercise firm control when engaging targets. Movements should also be coordinated and controlled.

c. Cleared rooms and buildings should be distinctly marked to identify cleared areas and friendly forces to base-of-fire elements supporting the maneuver.

d. Visible markers (for example, glint tape or thermal strips) should be attached to individual Marines for rapid identification.

e. Far and near recognition symbols should be coordinated and used properly.

f. Units employing close air support must exercise firm control. Failure to do so may lead to the pilot becoming disoriented and engaging friend and foe alike.

4. Urban Environmental Effects on Night Vision Devices. Built-up areas affect standard night vision devices and sights differently than do open areas. This may cause some confusion because the images Marines receive through their night vision devices are unusual compared to those they may be used to.

a. Most built-up areas have electric power; therefore, street lights and/or building lights may "white out" some light intensification devices unless the power is disrupted or lights are turned off.

b. Fires may be burning in the urban area. Open flames cause problems not only for light intensification devices, but also for thermal devices.

c. Subterranean areas and the interiors of buildings will not have ambient light if the power is off. Passive night vision devices must then use an artificial light source, such as infrared radiation, to provide enough ambient light for the devices to work effectively.

d. The many reflective surfaces found in built-up areas may cause false images, particularly for laser range finders and laser target designators.

e. Thermal imaging devices may not be able to see through large amounts of dust particles suspended in the air.

f. Smoke and dust degrade the effectiveness of night vision devices.

g. Fog degrades long-range target acquisition from thermal sights.

h. Weapons flashes within enclosed areas appear to be much brighter. This causes Marines to lose their night vision and washes out light intensification devices.

5. Considerations. The environment of built-up areas presents special challenges and considerations during periods of limited visibility.

a. The use of glint tape, thermal tape, or chemiluminescent lights is an important consideration. These can be used to mark the FLOT, casualties, cleared buildings and rooms, weapons positions, and individual Marines. Their use must be clearly addressed in the unit's SOP. When markers are used for extended periods, their meanings should change because the enemy may be able to capture or manufacture and use these marking devices to their advantage.

b. The use of tracer and incendiary ammunition may be restricted to prevent fires. The light of fires whites out some night vision devices and may interfere with or confuse thermal devices.

c. The control of power stations may be essential to operations during periods of limited visibility. Control of power stations enables friendly forces to control, to a degree, background illumination. Shutting off the power to the street lights is much easier than shooting out the lights. Commanders must balance the tradeoff between force protection and maintaining law and order after the battle is over. During cold weather, the control of power stations may be critical to the welfare of the civilian population.

d. Identifying friendly units, noncombatant civilians, and enemy troops becomes more difficult during limited-visibility operations.

e. Locating the source of sounds becomes more difficult because of the natural echoing in built-up areas and the tendency of sounds to carry farther at night.

f. Locating boobytraps and obstacles also becomes more difficult at night. Movement rates are normally slower than during periods of normal visibility.

6. Special Equipment. Fighting during periods of limited visibility requires some specialized equipment to maximize advantages.

a. As a rule, thermal imaging devices such as the AN/PAS-7 infrared viewer and the AN/TAS-5 Dragon infrared sight are better for limited-visibility operations than are light intensification devices such as the AN/PVS-7B. Light intensification devices are more easily washed out by background light, weapons flashes in enclosed areas, and fires. Thermal devices, while also affected by fires, are not as easily washed out.

b. The AN/PAQ-4A infrared aiming device is similar to its off-the-shelf laser aiming sight counterparts, except it is not visible to the naked eye. Infrared pen lights can also be attached to weapons and, when used in conjunction with night vision devices, can provide a quick sight picture, illuminate rooms and hallways, identify obstacles and boobytraps, and identify friendly forces.

c. Other night sights for weapons include the AN/TVS-5 crew-served weapon night vision sight, the AN/PVS-4 individual weapons night vision sight, and the AN/UAS-12 night vision sight, which mounts to the modular universal laser equipment (MULE) and the TOW missile system.

d. Trip flares, flares, illumination from mortars, and artillery and spotlights (visible light or infrared radiation) can be used to blind enemy night vision devices or to artificially illuminate the battlefield (Figure H-1). (See FMFM 6-8, *Supporting Arms Observer, Spotter, and Controller*, for more information on illumination from mortars and artillery.)

e. Spare batteries for the night vision devices should be carried to keep the devices operational. Soft, clean rags should be used to clean the lenses.

7. Combat Support. Coordinating employment of forces and fire support in limited-visibility conditions is a major concern to leaders during MOUT. The concentration of forces and fires at the point of decision is facilitated by the technological edge that night vision devices provide.

a. Any degradation of accuracy in artillery fire will likely be a result of the limitations of target acquisition assets. While FOs and FACs may have thermal sights and laser range finders, most Marines on the battlefield do not yet have devices that will enable them to acquire targets accurately. The following are some devices and techniques to improve target acquisition for indirect fires:

(1) If the target is within LOS, tanks and LAVs can rapidly identify the target and provide an accurate range.

(2) Preregistered targets are effective if the target reference point can be observed and the observer has clear communications with the firing unit.

(3) Fixed-wing and rotary-wing aircraft can be used to identify targets and adjust supporting arms.

Figure H-1. Use of Indirect-Fire Illumination During MOUT

8. Combat Service Support. Maneuver unit commanders and their Marines are not the only individuals who must adjust to combat under limited-visibility conditions in MOUT. Logisticians at every level must anticipate requirements for these conditions.

a. Units conducting resupply operations during periods of limited visibility should consider:

- Issuing drivers and vehicle commanders night vision devices so that the vehicles going to and from logistic release points do not require illumination (This also prevents the enemy from acquiring resupply locations by following vehicles with blackout lights on.)

- Maintaining strict noise and light discipline

- Following a clearly marked route to avoid obstacles and prevent the resupply vehicle(s) from becoming disoriented

- Providing radios to resupply vehicles

- Issuing each vehicle a map of the AO (preferably a city map with the street names).

b. Combat units operating for an extended time under conditions of limited visibility should have enough batteries to keep the night vision devices functioning at optimum power and sensitivity.

c. A large operational readiness float of night vision devices should be maintained.

d. Casualty collection during periods of limited visibility is more difficult. Clear methods for marking any casualties must be established before the operation begins.

e. CSS operations located in existing structures should not be visible from a distance. This includes limiting vehicle traffic to an absolute minimum, sealing doors and windows to prevent light leakage, and dispersing assets as much as possible.

9. Operational Considerations. Marine units conduct attacks during periods of limited visibility to gain or sustain the momentum of the attack. Before conducting a limited-visibility attack, the commander must balance the risks and ensure that every Marine understands the mission, intent, and control measures. Rehearsals and strict command and control reduce casualties and greatly enhance the chances for mission accomplishment.

a. To reduce confusion, Marines should clear buildings and rooms using the same techniques they use during the day.

b. Movement rates are slower in the dark. Each Marine must remain alert for mines, boobytraps, and enemy positions.

c. Rifle squads and fire teams should be equipped with a mixture of both thermal imaging and light intensification devices whenever possible. This enables the squads and fire teams to obtain a better picture of the night environment and enables Marines to balance the strengths and weaknesses of each type of night vision device for maximum results.

d. When moving through buildings, assault forces mark cleared rooms and buildings and communicate with the support forces. Marking cleared rooms is especially critical if more than one assault force or element is in the same building. Communications with supporting forces are imperative to avoid fratricide.

e. If flashlights or chemiluminescent lights are used, they should be held away from the head or chest area. This will make it harder for enemy soldiers firing at the light to kill the Marine holding the flashlight or chemiluminescent light.

f. Units must know where everyone is during offensive operations. This not only reduces the risk of fratricide, but also decreases the time spent identifying, locating, and treating casualties. Also, it greatly reduces the chance of Marines becoming disoriented and separated from the unit.

g. Assault units should be aware of adjacent fires that may diminish the effectiveness of night vision devices. Weapons flashes within small rooms cause Marines to lose their night vision and can wash out light intensification devices. Also, enemy soldiers may use flares inside and outside of buildings to deliberately render night vision devices ineffective.

h. Leaders must ensure that all Marines follow the ROE and the laws of land warfare. This is critical if the enemy is intermixed with the local civilian population. Also, leaders must follow all control measures.

i. Enemy forces can be expected to use periods of limited visibility to the same advantages as Marine forces do.

j. Enemy forces may have access to sophisticated night vision devices manufactured in Europe, the United States, Japan, Korea, and the former Soviet Union.

Appendix I

Urban Building Analysis

As in other types of operations, success in urban combat depends largely on the ability to analyze the military aspects of terrain. Marines must be able to recognize certain terrain features when evaluating urbanized terrain.

1. Types of Mass-Construction Buildings. Mass-construction buildings are those in which the outside walls support the weight of the building and its contents. Additional support, especially in wide buildings, comes from using load-bearing interior walls, strongpoints (called pilasters) on the exterior walls, cast-iron interior columns, and arches or braces over the windows and doors (Figure I-1). Modern types of mass-construction buildings are wall and slab structures, such as many modern apartments and hotels, and "tilt-up" structures commonly used for industry or storage. Mass-construction buildings are built in many ways:

- The walls can be built in place by using brick, block, or poured-in-place concrete.
- The walls can be prefabricated and then tilted up, or they can be reinforced-concrete panels.
- The walls can be prefabricated and assembled like boxes.

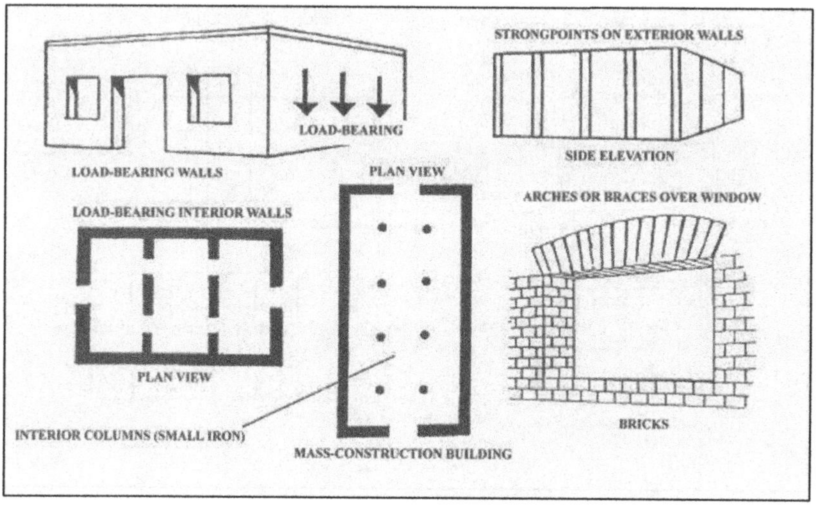

Figure I-1. Mass-Construction Buildings

.3

a. Brick buildings are, in some regions, the most common and most important of the mass-construction buildings. In Europe, brick buildings are commonly covered with a stucco veneer so that bricks do not show (Figure I-2). One of the most common uses of brick buildings is the small store. These buildings are found in all built-up areas but are most common in the core periphery (Figure I-3).

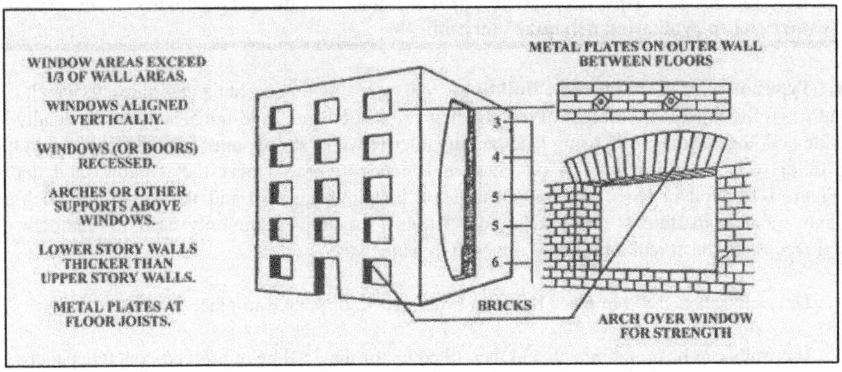

Figure I-2. Brick Buildings

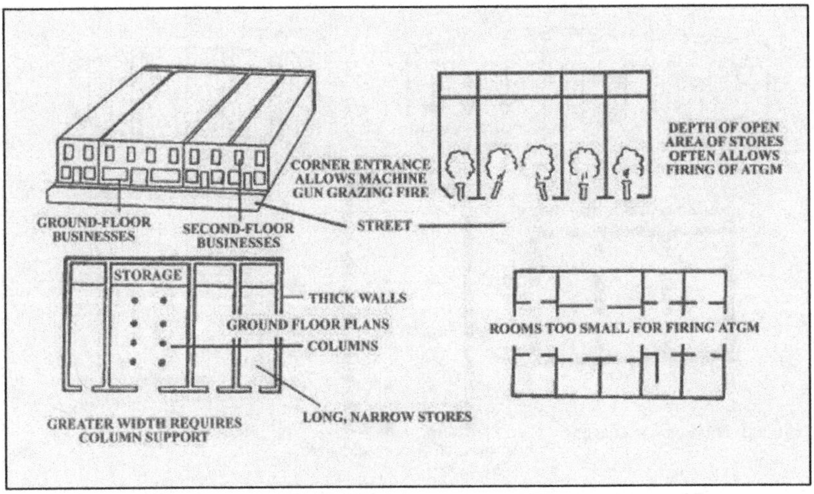

Figure I-3. Brick Store

b. Another common mass-construction building found mainly in industrial areas and along commercial ribbons is the warehouse. It is built of poured-in-place concrete reinforced with steel bars or of prefabricated tilt-up walls. The walls of warehouses provide good cover, although the roofs are generally light and vulnerable. The warehouses' large open bays permit firing of ATGMs or surface-to-air missiles and, because they are normally found in outlying areas, often afford adequate fields of fire for ATGMs. These buildings are built on slabs that can normally support the weight of vehicles and can provide excellent cover and concealment for tanks (Figure I-4).

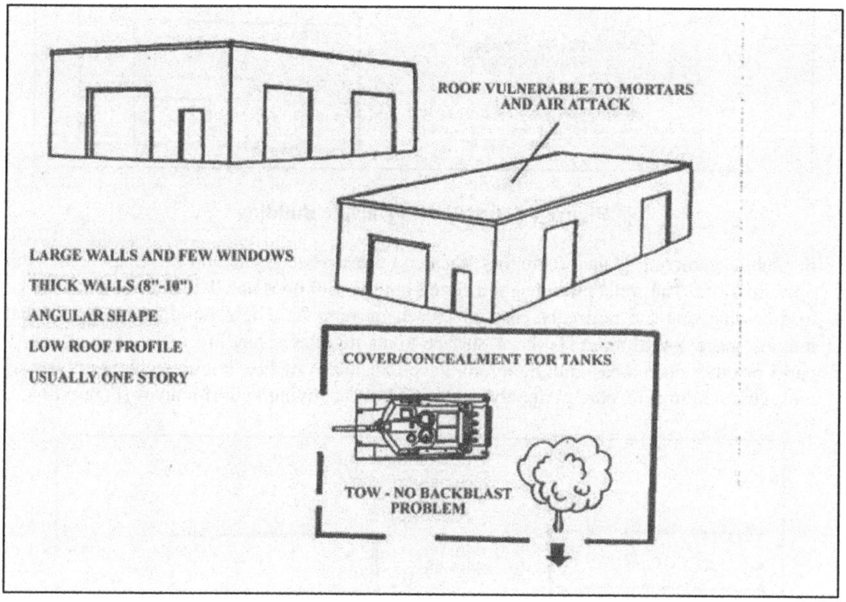

Figure I-4. Warehouse

c. Another type of mass-construction building is the box-wall-principle type. It is made from prefabricated panels of 6- to 8-inch-thick reinforced concrete. The outside wall is often glass. The box-wall-principle building provides good cover, except at the glass wall. The rooms are normally too small for ATGMs to be fired. A good circulation pattern exists from room to room and from floor to floor. These buildings are commonly used as hotels or apartments and are located mainly in residential and outlying areas (Figure I-5).

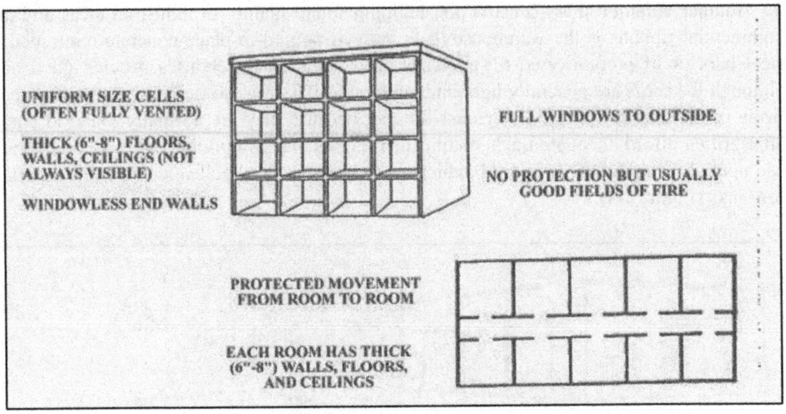

Figure I-5. Box-Wall-Principle Building

d. Public gathering places (churches, theaters) are mass-construction buildings with large, open interiors. The walls provide good cover, but the roof does not. The interior walls are not load-bearing and are normally easy to breach or remove. These buildings have adequate interior space for firing ATGMs or surface-to-air missiles. They are often located next to parks or other open areas and, therefore, have long fields of fire. Public gathering places are most common in core, core periphery, residential, and outlying high-rise areas (Figure I-6).

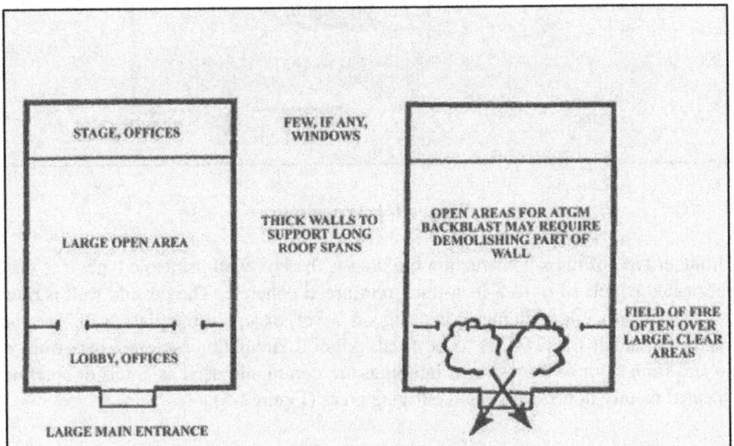

Figure I-6. Public Gathering Places

2. Types of Framed Buildings. Framed buildings are supported by a skeleton of columns and beams and are usually taller than frameless buildings (Figure I-7). The exterior walls are not load-bearing and are referred to as either heavy clad or light clad. Another type of framed building often found in cities is the garage, which generally has no cladding.

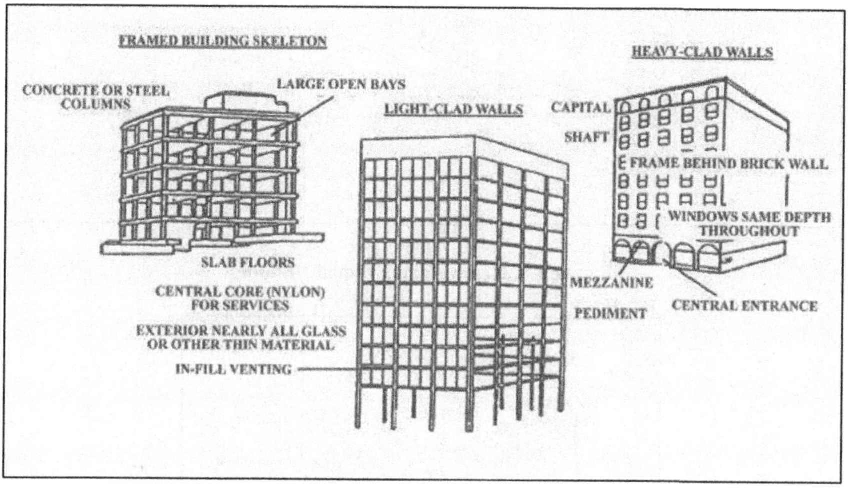

Figure I-7. Framed Buildings

a. Heavy-clad buildings were common when framed buildings were first introduced. Their walls are made of brick and block that are sometimes almost as thick as frameless brick walls, although not as protective. Heavy-clad framed buildings are found in core and core periphery areas. They can be recognized by a classic style or architecture in which each building is designed with three sections: the pediment, shaft, and capital. Unlike the brick building, the walls are the same thickness on all floors, and the windows are set at the same depth throughout. Often the frame members (the columns) can be seen, especially at the ground floor. The cladding, consisting of layers of terra cotta blocks, brick, and stone veneer, does not provide as good a cover as the walls of brick buildings. It protects against small-arms fire and light shrapnel but does not provide much cover against heavy weapons (Figure I-8).

(1) The floor plans of these buildings depend on their functions. Office buildings normally have small offices surrounding an interior hall. These offices have the same dimensions as the distance between columns (some large offices are as large as two times the distance between columns). These rooms are too small to permit firing of ATGMs but do provide some cover for snipers or machine gunners (Figures I-9 and I-10).

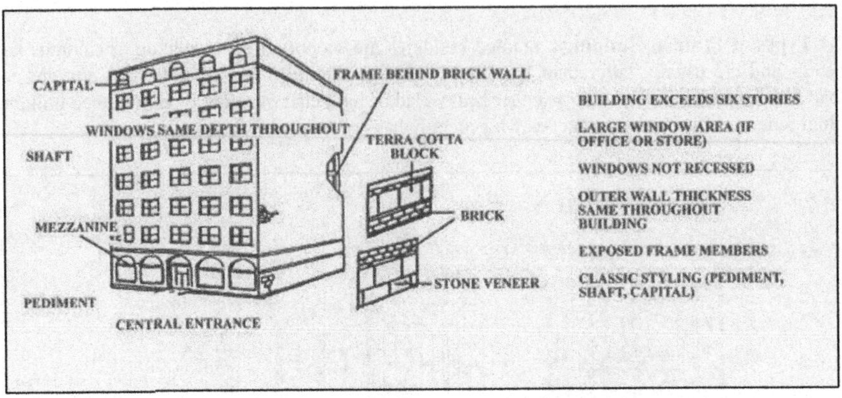

Figure I-8. Heavy-Clad Framed Building

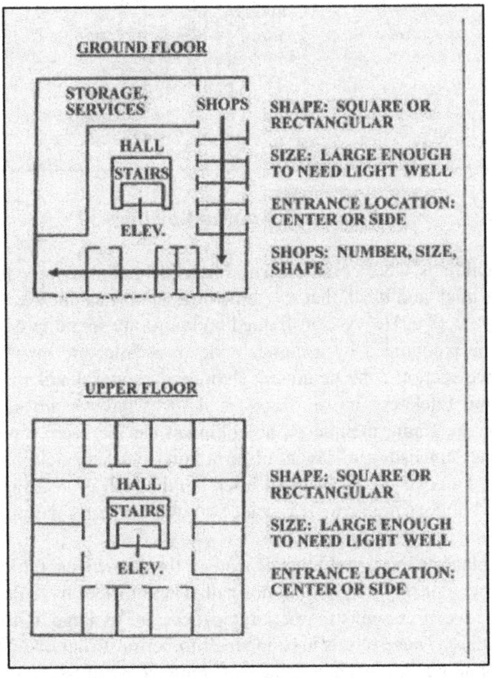

Figure I-9. Floor Plan of Heavy-Clad Framed Office Building

(2) Department stores normally have large, open interiors (Figure I-11). Such areas permit firing of ATGMs or man-portable surface-to-air missiles (if there are adequate fields of fire). Often a mezzanine level with a large backblast area permits firing down onto tanks. Steel fire doors often exist between sections of the store. The steel fire doors are activated by heat. Once closed, they are difficult to breach or force open, but they effectively divide the store into sections (Figure I-12).

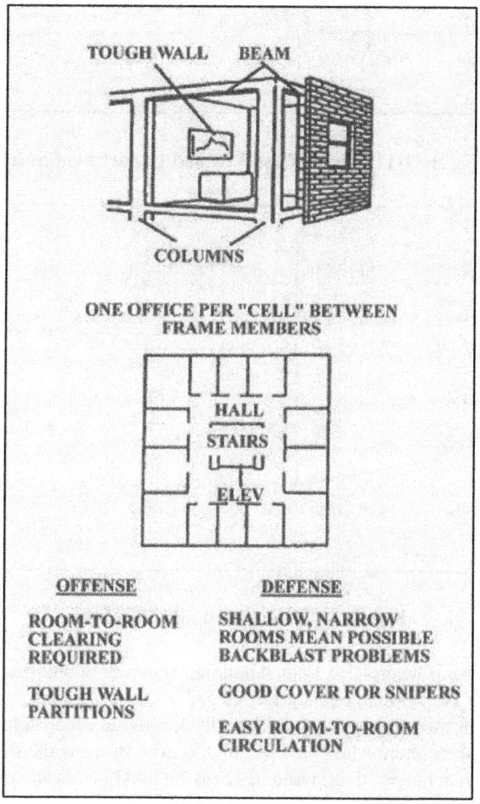

Figure I-10. Heavy-Clad Framed Office

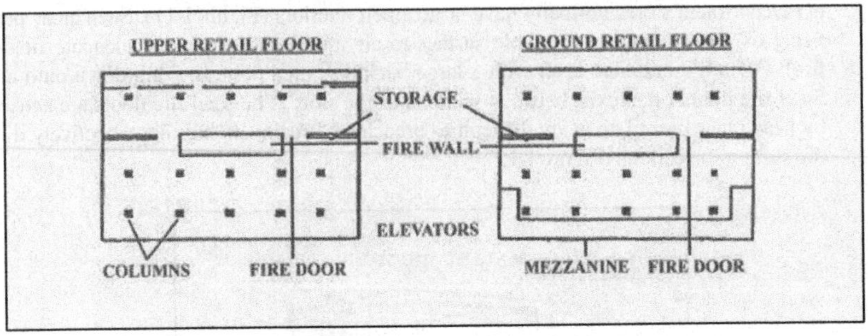

Figure I-11. Heavy-Clad Framed Department Store

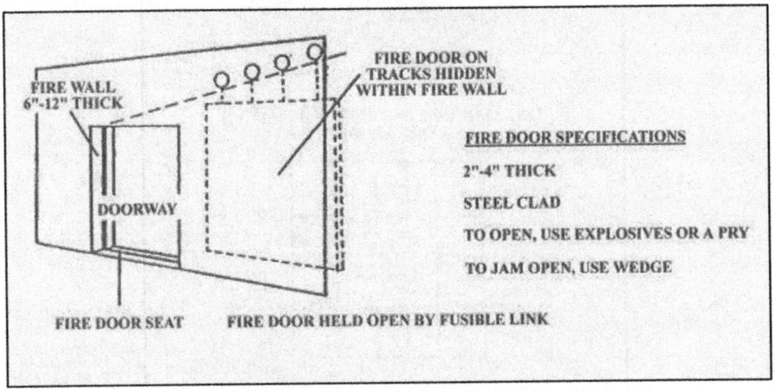

Figure I-12. Fire Wall and Fire Door

(3) Another type of heavy-clad framed building is used as a high-rise factory (Figure I-13). Such buildings are normally easily recognized because the concrete beams and columns are visible from the outside. They are usually located in older industrial areas. The large windows and open interior favor the use of ATGMs. Because the floors are often made to support heavy machinery, these buildings provide good overhead cover.

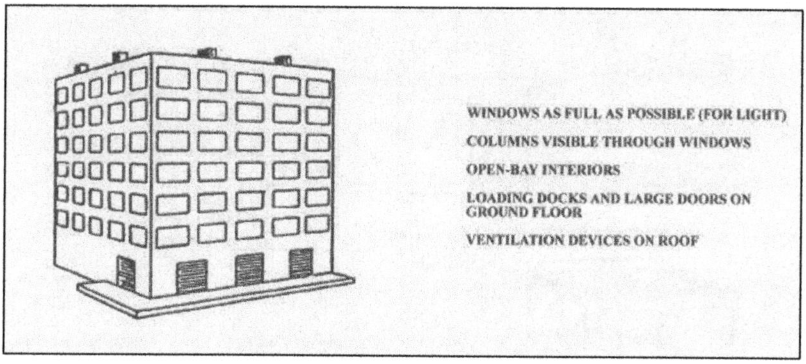

Figure I-13. High-Rise Factory

b. Light-clad buildings are more modern and may be constructed mostly of glass (Figure I-14). Most framed buildings built since World War II are light-clad buildings. They are found in both core and outlying high-rise regions. Their walls consist of a thin layer of brick, lightweight concrete, or glass. Such materials provide minimal protection against any weapon. However, the floors of the buildings are much heavier and provide moderate overhead cover. The rooms in light-clad framed buildings are much bigger than those in heavy-clad buildings. This feature, along with the fact that the buildings usually stand detached from other buildings, favors the employment of ATGMs. The interior partitions are thin, light, and easy to breach (Figure I-15).

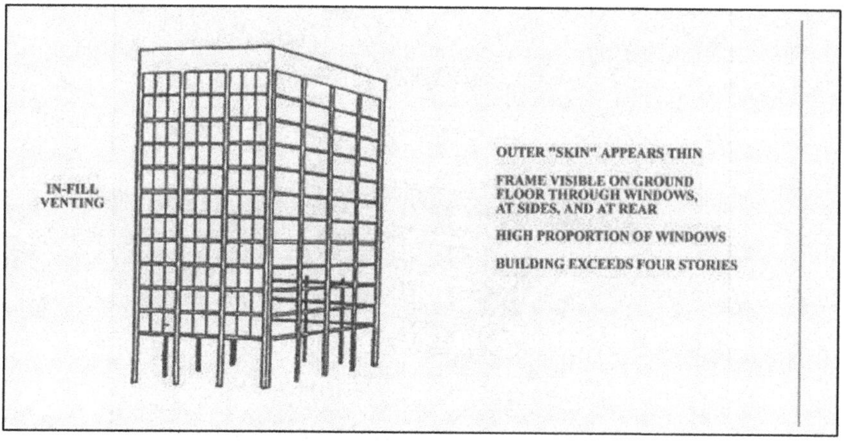

Figure I-14. Light-Clad Framed Building

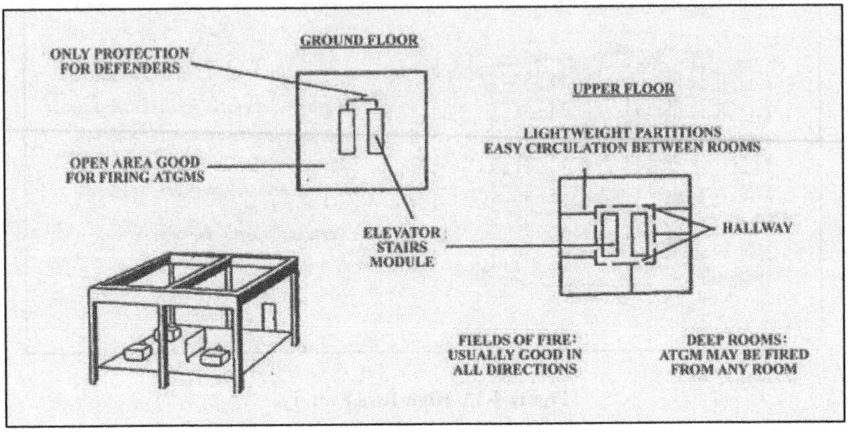

Figure I-15. Light-Clad Framed Room

c. The garage is one of the few buildings in an urban area in which all floors support vehicles. It provides a means to elevate vehicle-mounted TOWs, and the open interiors permit firing of ATGMs. Garages are normally high enough to provide a 360-degree field of fire for antiaircraft weapons. For example, a Stinger could hide under the top floor of the garage, come out to engage an aircraft, and then take cover again (Figure I-16).

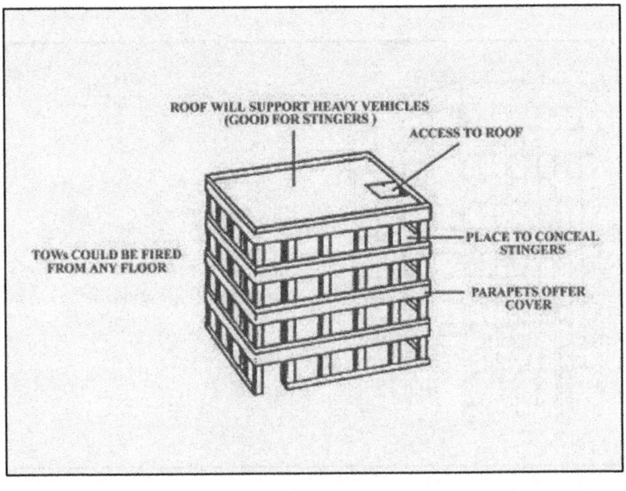

Figure I-16. Garage

3. Floor Plans. Floor plans in buildings follow predictable patterns. One of the factors that determines floor plans is building shape (Figure I-17). The basic principle governing building shape is that rooms normally have access to outside light. This principle helps Marines to analyze and determine the floor plans of large buildings.

4. Residential Areas. The two basic types of houses in the western world are located in and around cities and in rural areas. City-type houses are normally mass-construction brick buildings. Rural-type buildings in the continental United States, South America, and Southeast Asia are commonly made of wood, while in continental Europe, Southwest Asia, and sub-Saharan Africa, where wood is extremely scarce, rural buildings are normally constructed of concrete blocks. Rural-type buildings can also be found within cities (Figure I-18).

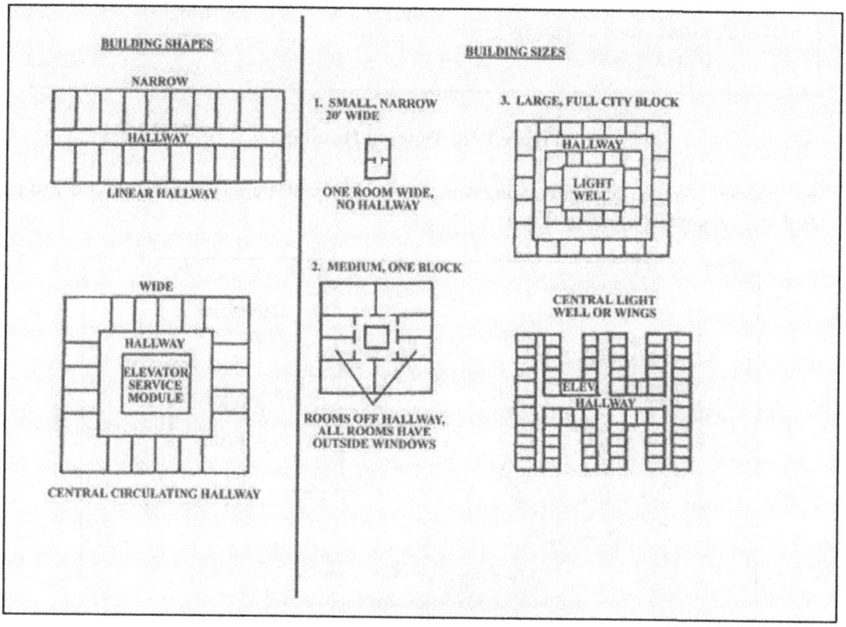

Figure I-17. Building Shapes and Sizes

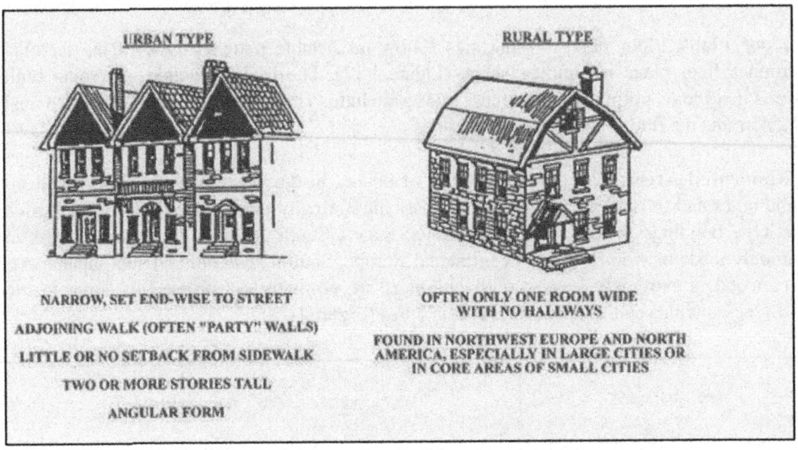

Figure I-18. Types of Housing

a. Another common type of building structure in cities with European influence is called the Hof-style apartment building (Figure I-19).

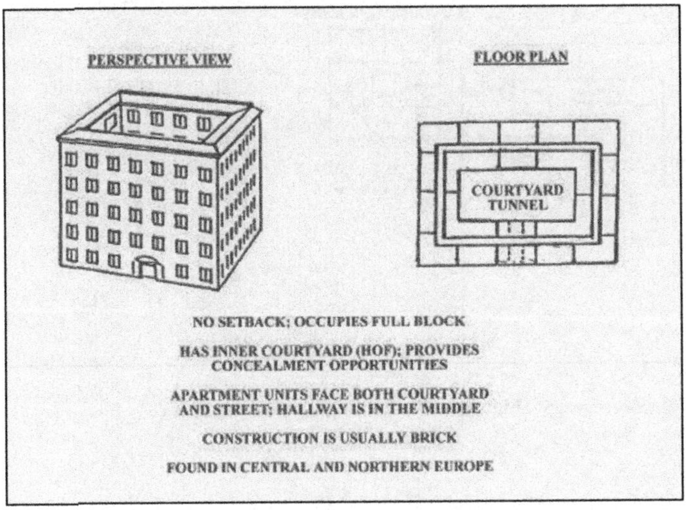

Figure I-19. Hof-Style Apartment Building

b. In the Middle East and tropical regions, the most common type of housing is the enclosed courtyard. Houses are added to one another with little regard to the street pattern. The result is a crooked, narrow maze that is harder to move through or fire in than dense European areas. (Figure I-20).

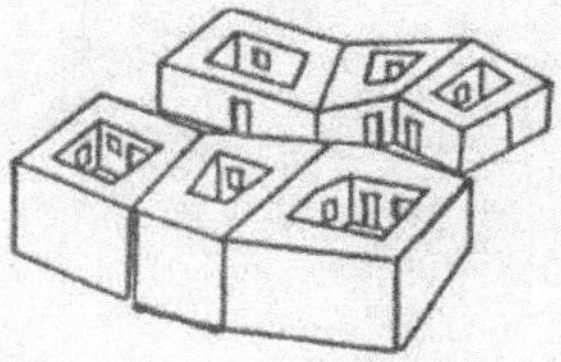

Figure I-20. Enclosed Courtyard

5. Characteristics of Buildings. Certain characteristics of both mass-construction and framed buildings can be helpful in analyzing a built-up area. Leaders can use Table I-1 to determine how to defend or attack a certain building given the unit's available weapons systems.

Type of Construction	Building Material	Height (Stories)	Average Wall Thickness (cm)
Mass	Stone	1 to 10	75
Mass	Brick	1 to 3	22
Mass	Brick	3 to 6	38
Mass	Concrete block	1 to 5	20
Mass	Concrete wall and slab	1 to 10	22 to 38
Mass	Concrete tilt-ups	1 to 3	18
Framed	Wood	1 to 5	3
Framed	Steel (heavy cladding)	3 to 50	30
Framed	Concrete/steel (light cladding)	3 to 100	2 to 8

Table I-1. Characteristics of Buildings

6. Distribution of Building Types. Certain types of buildings dominate certain parts of a city; this establishes patterns within a city. Analysis of the distribution and nature of these patterns has a direct bearing on military planning and weapon selection (Figure I-21).

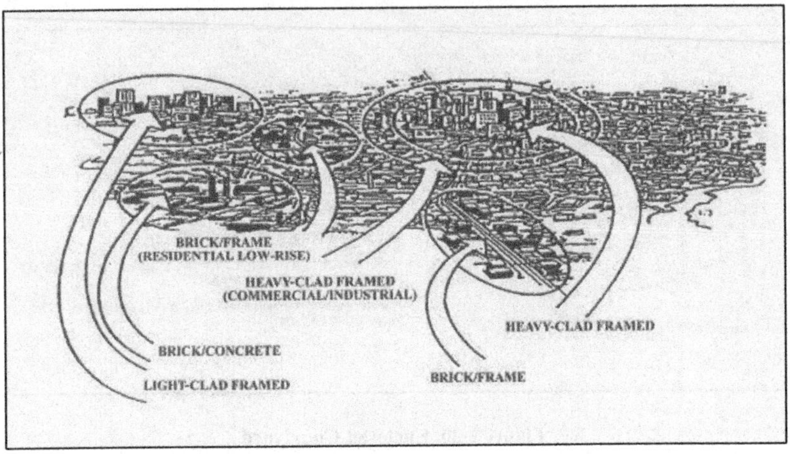

Figure I-21. Distribution of Building Types

a. Mass-construction buildings are the most common structures in built-up areas, forming about two-thirds of all building types. Brick structures account for nearly 60 percent of all buildings, especially in Europe.

b. Steel- and concrete-framed multistory buildings have an importance far beyond their one-third contribution to total ground floor area. They occupy core areas—a city's most valuable land—and as centers of economic and political power, they have a high potential military significance.

c. Open space accounts for about 15 percent of an average city's area. Many open spaces are grass-covered and are used for parks, athletic fields, and golf courses; some are broad, paved areas. The largest open spaces are associated with suburban housing developments where large tracts of land are recreation areas.

d. Streets serving areas consisting of mostly one type of building normally have a common pattern. In downtown areas, for example, high land values result in narrow streets. Street widths are grouped into three major classes: 7 to 15 meters, located in medieval sections of European cities; 15 to 25 meters, located in newer, planned sections of most cities; and 25 to 50 meters, located along broad boulevards or set far apart on large parcels of land. When a street is narrow, observing or firing into windows of a building across the street can be difficult because an observer is forced to look along the building rather than into windows.

When the street is wider, the observer has a better chance to look and fire into the window openings (Figure I-22).

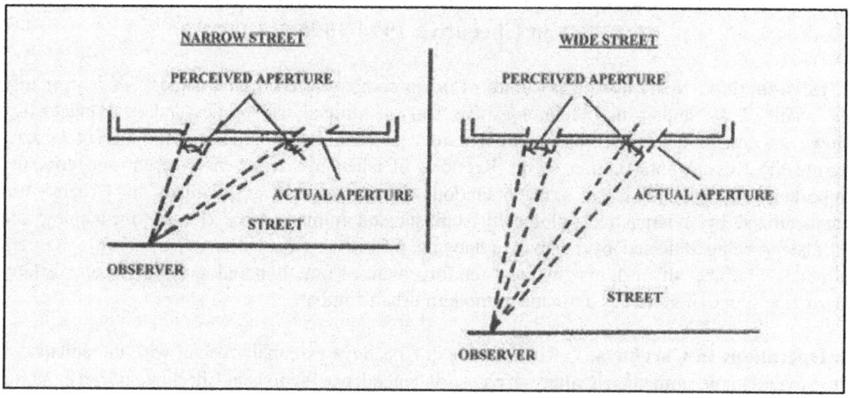

Figure I-22. LOS Distances and Angles of Obliquity

Appendix J

Lessons Learned from Russian Military Operations in Chechnya 1994-1996

1. Introduction. Many historical studies of urban combat have become dated. This is primarily the result of the impact that technology has had on military capabilities and evolving tactics. However, certain tenets of urban combat remain constant. The battles for the city of Grozny during the Russian intervention in the Republic of Chechnya represent a recent and critically important example of large scale operations in urban combat. Combat in Grozny was characterized by a large, technologically sophisticated military force (Russian) engaging and ultimately being defeated by a small, relatively primitive irregular force (Chechen). Grozny provides a number of fresh insights, and reinforcement of time honored tenets of urban warfare, across the scope of activities germane to modern urban combat.

2. Operations in Chechnya. Russia's war in Chechnya essentially began with the collapse of the Soviet Union, immediately after which pro-independence factions in Chechnya began agitating to free the Republic from Russian domination. This eventually led to civil war with pro-Russian factions within the Republic. Russia committed troops to Chechnya on 11 December 1994, for what many in the Russian government and military thought would be a quick campaign to restore Russian primacy. What followed was a bitter, costly, and protracted struggle that ended with the complete withdrawal of Russian military forces from Chechnya.

A primary focus of Russian operations in Chechnya was the capture of the capital city, Grozny. The initial attack on Grozny was conducted by a mechanized force, consisting of 6,000 troops mounted in tanks, BMP infantry fighting vehicles, and BTR armored personnel carriers. The Russians anticipated light opposition, but instead met a determined and heavy resistance from Chechens armed with a large number of antitank weapons. The attack was repulsed with significant Russian losses, both in personnel and armored vehicles (105 of 120 vehicles were lost). A second mechanized attack conducted on New Years Eve, 1994, was also repulsed with the loss of 140 of 200 tanks employed. The Russians captured Grozny some two months later, primarily through the use of excessive, overwhelming firepower, and at great cost to themselves and the local Chechen population.

The capture of Grozny marked the end of the first phase of Russian combat operations in Chechnya. Beginning in March 1995, the Russians became heavily engaged in anti-partisan operations as they tried to gain control of the country. Though the Russians controlled several major Chechen cities and a portion of the countryside by May 1995, they were never able to fully isolate the Republic. Chechen separatist fighters continued to receive a steady flow of arms and supplies from neighboring countries. Consequently, the Chechen fighters were able to maintain the initiative throughout this period, engaging Russian forces whenever and wherever they chose.

Even late in the campaign, Russian intelligence failed to detect Chechen fighters infiltrating Grozny, which enabled the Chechens to launch a major attack against the Russian-controlled city in early August 1996. The Chechens succeeded in capturing the city less than two weeks later.

.3

The recapture of Grozny was a significant loss for the Russians, precipitating a general cease-fire and the withdrawal of Russian forces from the Republic.

Besides the military aspects of combat in Chechnya, other issues such as ethnic, cultural, and religious divides impacted upon the nature of the conflict. The Russians failed to culturally orient their troops, which resulted in many serious cultural mistakes when dealing with the Chechen civilians. Once insulted or mistreated, the Chechen civilians became active participants in the struggle against Russia. The war eventually degenerated into one of abject ferocity and brutality on both sides. The Chechens are reported to have executed Russian prisoners, decapitated bodies, and boobytrapped the dead. Russian wounded and bodies were hung upside down outside of Chechen defensive positions in buildings, causing the Russians to have to shoot through their own people to engage the Chechens. The Russians, initially hesitant to cause civilian casualties, increasingly engaged Chechen units with a growing disregard for the safety of the local populace. As Russian discipline degraded, the incidences of Russian troops engaging in looting, arson, torture, even the summary execution of civilians, rose significantly. The result was the deaths of tens of thousands of civilians.

3. Lessons Learned. The intervention in Chechnya and the battles for Grozny generated many important lessons across the warfighting functions of combat in urbanized terrain. The following information is primarily from the Russian point of view, though some Chechen capabilities are included. The lessons learned are arrayed as warfighting functional considerations.

a. Command and Control.

(1) Russian military commanders lacked clear policy guidance. The Russian military's ability to attain and maintain steady, logical operational progress was hampered, if not negated, by a lack of clear governmental policy. The Russian government was split at the highest levels over the wisdom of committing forces to Chechnya, and the conduct of the intervention.

(2) Poor Russian command and control structure throughout all levels. There was no direct, unified chain of command for the operations in Chechnya. Operational directives often flowed from a number of governmental agencies directly to local unit commanders, often bypassing intermediate military commanders. This resulted in a lack of overall situational awareness, and exceptionally poor coordination among tactical forces.

(3) Russians were unable to contain/prohibit Chechen operations against Russia. When the Russian forces finally began to achieve progress in late May 1995, the Chechens initiated attacks against targets inside Russia. The result was a major propaganda victory for the Chechens.

(4) Excessive senior officer (flag level) interference contributed to operational confusion. There were eight major changes at the senior command level during the course of the intervention. Additionally, there were approximately one hundred general officers on the operational scene, about one general officer per 3,000 troops, which resulted in confusion as they got in each others way.

(5) Poor small unit leadership, particularly at the NCO level, was a primary cause of Russian tactical failures in Grozny. While the Russians recognize that urban combat is small unit intensive, they do not possess a professional NCO corps. Additionally, Russian forces lacked some 12,000 platoon leaders when the intervention began. This had a crippling effect on small unit operations.

(6) Russian inability to isolate Chechnya/Grozny allowed a constant flow of reinforcements and supplies into the country/city. Russian control of any given area was at best, porous. The Chechens continued to receive outside support throughout the campaign, as well as, a large supply of arms and equipment from poorly disciplined Russian troops inside Chechnya.

(7) Both sides used commercial off-the-shelf communication assets. The information war was as vitally important as any other aspect of the campaign. The Chechens had to make do with commercially available communication equipment. They fully exploited the use of cellular telephones, Motorola radios, improvised television stations, video cameras, and the internet. The Russians lacked sufficient military communication equipment, and had to supplement with commercial assets.

(8) Russian tactical communication was very difficult in Grozny. Urban structures in Grozny interfered with Russian military radios, severely hindering tactical communications. The Russians had to establish ground-based and aircraft-based relay stations to overcome the interference.

(9) Boundaries between units were tactical weak points. Units had to think of boundaries in both a horizontal and vertical context. There were many instances when the Chechens held the third floor and above, while the Russians held the first two floors and maybe the roof. This caused a lack of coordination, and mutual support. Additionally, due to poor command and control and coordination between units, boundaries became seams through which the Chechens could pass unopposed.

b. Intelligence.

(1) Lack of high quality intelligence made operations much more difficult and dangerous. Russian intelligence collection and dissemination capabilities were poor throughout the intervention. In March 1996, the Russian Minister of the Interior complained that poor reconnaissance and intelligence allowed Chechen fighters to move in and out of Grozny without opposition, which contributed to the heavy Russian losses.

(2) Directives to minimize civilian casualties ultimately put the Russians at a disadvantage. An advantage accrues to the side with less concern for the safety of the civilian population. Initially, the Russians attempted to limit civilian casualties. The Chechen fighters took advantage of this reticence, often using civilians, schools, and apartment buildings as a shield. While this may seem to be cold-blooded of the Chechen fighters, what must be understood is that the majority of the populace in Grozny were actually ethnic Russians, and that the ethnic Chechens had no remorse in using them in this manner.

(3) Russian concerns for the safety of the Chechen populace and property declined. As the Russian troops became increasingly frustrated in their attempts to separate Chechen fighters from the local populace, and their casualties began to mount, concerns for civilian safety waned, resulting in greater civilian casualties and further alienating the Chechen people.

(4) Rigorous communications security is essential, even against less sophisticated enemies.Much of the Russian tactical radio traffic was broadcast in the clear, which allowed the Chechens to intercept it. The Chechens were able to redirect Russian airstrikes onto Russian targets, and engage Russian artillery forward observers who broadcast their own position coordinates.

(5) The Russians made extensive use of RPVs for reconnaissance. The Chechen campaign saw the first Russian use of unmanned aerial vehicles for reconnaissance and intelligence gathering. There were primarily used by Russian airborne forces.

c. Maneuver.

(1) Inadequate training in the most basic maneuver and combat skills inhibited Russian operations. Overall, poor Russian combat performance could be traced to a lack of training in fundamental military skills, a situation that was then exacerbated by an additional lack of training in specific urban combat skills. Individual skills training was so inadequate that some units were entirely incapable of undertaking combat operations. The Russian's ability to conduct large scale operations had been seriously eroded by a complete cessation of exercises at the division level in the 2 years prior to the Chechen campaign. Even regimental and battalion level exercises had been curtailed by 75 percent or more. Additionally, The air forces were not any better trained or prepared for their operational role in Chechnya.

(2) Lack of cohesiveness among the Russian forces led to confusion and antagonism. The Russian ground forces consisted primarily of regular army units and Ministry of Interior Internal Troops, which were essentially a police force. With no real unified command and control, the two groups were usually conducting uncoordinated missions, which led to confusion and fratricide.

(3) The nature of cities channels combat operations along narrow lanes of activity. Combat conditions in Grozny were characterized by narrow fields of view, limited fields of fire, and constricted avenues of approach. For the Russians, this meant movement along approaches well known and heavily defended by the Chechens.

(4) Night fighting was the most difficult activity for Russian infantry. The Russian infantry, at all levels, was inadequately trained for night operations, and lacked night vision equipment.

(5) Armored vehicles cannot operate in cities without extensive dismounted infantry support. The Chechens formed antitank hunter-killer teams equipped with RPG-7 and RPG-18 antitank missiles. They engaged Russian tanks with volley fire from above, behind, and the sides. Armor columns not supported by dismounted infantry suffered great losses.

(6) Armored engineer vehicles are indispensable for removing obstacles. The Russians found that armored engineer vehicles were critical to restoring or improving mobility in Grozny. Eventually, the Russians formed special assault groups that included armored engineer vehicles in each group.

(7) The Chechens preferred small unit, hit-and-run ambush tactics. The Chechens relied on small units (15 to 20 troops) to conduct quick, sharp ambushes, from which they would disengage before the Russians could bring significant firepower to bear. Ambushes were often conducted in three tiers, with Chechens in the underground, on the ground floor, and on the roof, with each element having a different task in the ambush. They also used ambush tactics against Russian helicopters.

(8) Urban combat is extremely manpower intensive, and results in significant attrition of personnel and materiel for the attacker. The Russians discovered that a 5 to 1 manpower ratio was often not enough, due to the high casualty rates consistent with urban combat and the requirement to guard virtually every building taken.

(10) Tracked vehicles are preferable to wheeled vehicles in the urban environment. Tracked vehicles provided better mobility than wheeled vehicles due to their significantly greater ability to negotiate obstacles and rubble.

(11) Armored vehicles required additional protection (i.e., reactive armor, sandbags, wire mesh, etc.) to negate attacks from the sides, rear, and top. Russian tank design emphasized greater armor protection on the frontal arc, and less on other areas of the vehicle. This proved inadequate for urban combat, where attacks can come from any direction. The majority of lethal hits came from above, easily penetrating turrets and engine decks, and from the rear.

(12) The spatial qualities and perspectives of urban combat are more "vertical" than non-urban combat. The Chechens regularly fought from basements, rooftops, and upper-floor rooms, making them difficult to locate and engage. Russians soldiers trained in open areas, had to reorient to a confined, highly vertical environment.

(13) Standard Russian military unit configurations were inappropriate for urban combat. Typical Russian army unit organizations, oriented to open spaces and mechanized operations, proved to be inadequate for urban combat. They eventually had to adopt specially configured assault teams that included, among other assets, significantly greater firepower and engineer assets. Essentially, they moved towards a better combined arms team.

(14) The Russians found their helicopters to be far too vulnerable to rooftop snipers and ambushes in the urban setting. While Russian doctrine had long specified the capture of a building from the top-down, it became apparent in Grozny that this tactic was too expensive in terms of helicopter losses. The Chechens made effective use of snipers and RPG-equipped ambush teams to damage or destroy helicopters supporting Russian ground forces.

d. Fires.

(1) Artillery was invaluable in the direct fire mode. The Russians found artillery, to include rocket launchers and mortars, to be very effective for direct fire against point targets , usually at ranges of 150 to 200 meters.

(2) Obscurants (smoke, white phosphorous, tear gas) are especially useful in urban combat. The Russians used significant amounts of smoke and white phosphorus to mask the movement of forces. Every fourth or fifth artillery or mortar round was either smoke or WP. WP also had a toxic effect, easily penetrating Chechen protective masks.

(3) The Chechens effectively used RPGs against helicopters. The Chechens made effect use of ambush teams equipped with RPGs to engage Russian helicopters, either by damaging or destroying the helicopter or simply prohibiting helicopterborne operations.

(4) Air defense guns (i.e., ZSU-23-4) were valuable in suppressing ground targets. The Russians found air defense guns effective against multi-story buildings because they had sufficient elevation to hit targets in the upper stories. Air defense guns became a primary asset in the Russian special assault groups.

(5) The Russians determined that overwhelming firepower can make-up for organizational and tactical deficiencies in the short run if collateral damage is disregarded. Whenever Russian military units proved incapable of effective urban operations, the Russians simply fell back on massive, overwhelming firepower to reduce a strong point. It was through the use of excessive, unrestrained firepower, and a complete lack of regard for collateral damage, that finally enabled the Russians to gain control of Grozny.

(6) The Chechens negated Russian supporting fires by "hugging" Russian units. Whenever possible, Chechen fighters adopted the tactic of staying as physically close to Russian units as possible, so that the Russians could not employ supporting arms without the risk of injuring their own troops.

(7) The Russians found "bunker busting" weapons to be invaluable. The Russians came to rely heavily on "bunker busting" weapons for engaging Chechens, primarily snipers, entrenched in buildings.

(8) Russian helicopters needed stand-off weapons, that would allow them to engage targets while outside the range of Chechen weapons. The Chechens employed 23mm cannons and heavy machine guns against Russian helicopters, causing the Russian's helicopter gunships to fire from ranges of 3,000 meters or more. This required the use of wire-guided munitions and large caliber (240mm) rockets.

(9) The Russians made extensive use of precision-guided weapons. When the weather allowed, the Russians were able to use precision guided weapons, such as laser-guided bombs and missiles. They had great effect against priority targets such as bridges, major road intersections, and buildings.

e. Logistics.

(1) Deficiencies in the Russian logistics system degraded combat effectiveness. Significant inadequacies in the Russian logistics apparatus effected virtually all combat forces and operations. Some soldiers actually entered combat in Grozny without weapons or ammunition. The Russian transportation system failed. Supply officers were unprepared for the high volume demands for ammunition, hand grenades, smoke grenades, demolition charges, and one-shot antitank weapons.

(2) Russian logistics personnel were ill-prepared in basic soldier skills. Russian infantry often had to be used to conduct logistics operations into Grozny, because the logistics personnel lack the basic skills to defend themselves against the Chechens.

(3) The Russian ground forces required special equipment not normally found in unit T/Es. The T/E for the typical Russian infantry unit did not include such items as ropes, grappling hooks, and ladders, all of which proved essential for urban combat.

(4) Armored vehicle recovery is extremely difficult and dangerous. The Russians found the recovery of vehicles to be extremely difficult due to the terrain (i.e., narrow streets, rubble and debris), and Chechen opposition.

(5) Some Russian equipment was modified in the field to counter Chechen tactics and equipment. The Russians had to adopt add-on armor for their vehicles. They fabricated metal screens and wire mesh guards, which helped to defeat RPG and Molotov cocktail attacks.

(6) Medical support was of great importance. Russian casualty evacuation was difficult due to the lack of vehicular mobility, and the vulnerability of helicopters to Chechen fire. Many casualties were psychological, due in great part to the intense, violent nature of close quarter combat. The Russians were ill-prepared for the level of mental health problems they encountered. Additionally, logistics units were often unable to provide fresh drinking water, which caused Russian troops to consume contaminated water that resulted in other health problems.

f. Force Protection.

(1) Operations took much longer than expected, and the cost in terms of casualties was much greater than anticipated. The Russians committed unprepared, poorly trained and poorly equipped forces to Chechnya, based on an unrealistic operational time frame. Ultimately, this cost the Russians significant casualties and the war.

(2) Fratricide was a serious and continuing problem throughout the campaign. The situation facing Russian combat forces was ripe for fratricide. Poorly trained units, operating in a confused and uncertain urban environment, often unable to tell friend from foe, and lacking quality leadership and inter-unit coordination, were often as dangerous to themselves as they were to the Chechens .

(3) Chechen use of adhoc air defense systems eliminated Russian air support, particularly helicopters, in Grozny. The Chechens were able to improvise effective anti-air defenses using a mixture of anti-air and ground (i.e., heavy machine guns, RPGs) weapons, and ambush tactics specifically adapted to the urban environment.

(4) The Chechens made extensive and effective use of snipers. Snipers fired from well inside rooms versus near window openings, as well as, from rooftops and basements. The Russians lacked an effective sniper and counter-sniper capability of their own.

(5) The Chechens improvised crude chemical weapons. The Chechens, lacking access to military chemical weapons, Used off-the-shelf materials and equipment to make chemical mines (i.e., chlorine gas), detonated remotely by radio.

(6) The Chechens made extensive and effective use of boobytraps. It seemed that the Chechens mined and boobytrapped 7-8everything. They had a good understanding of the average Russian soldier's actions and reactions, and boobytrapped accordingly. Additionally, the Russians found it difficult to maintain boobytrap awareness among their troops.

Appendix K

Glossary

I. Acronyms

AAA .... antiaircraft artillery
AAAV .... advanced assault amphibious vehicle
AAV .... assault amphibious vehicle
ACE .... aviation combat element
AO .... area of operations
APC .... armored personnel carrier
APDS-T .... armor-piercing, discarding sabot with tracer
APFSDS .... armor-piercing, fin-stabilized, discarding sabot
API .... armor-piercing incendiary
ARVN .... Army of the Republic of Vietnam
ATF .... amphibious task force
ATGM .... antitank guided missile
BDA .... battle damage assessment
BMNT .... .beginning of morning nautical twilight
CA .... civil affairs
CAG .... civil affairs group
CE .... command element
CI .... counterintelligence
COC .... combat operations center
CONAD .... consolidated administration center
CP .... command post
CSS .... combat service support
CSSE .... combat service support element
DOD .... Department of Defense
DS .... direct support
EOD .... explosive ordnance disposal
EPW .... enemy prisoner of war
EW .... electronic warfare
FAC(A) .... forward air controller (airborne)
FASCAM .... family of scatterable mines
FDC .... fire direction center
FEBA .... forward edge of the battle area
FLOT .... forward line of own troops
FM .... U.S. Army field manual
FM .... Frequency modulated
FMFM .... Fleet Marine Force manual
FMFRP .... Fleet Marine Force reference publication
FO .... forward observer
FPL .... final protective line
FRAGO .... fragmentary order
FSC .... fire support coordinator
GCE .... ground combat element
GPS .... global positioning system
GS .... general support
HE .... high explosive
HEAT .... high-explosive antitank
HEAT-MP .... HEAT-multipurpose
HEDP .... high-explosive dual-purpose
HEI-T .... high-explosive incendiary with tracer
HLZ .... helicopter landing zone
HMMWV .... high-mobility, multipurpose wheeled vehicle
HSS .... health services support
HUMINT .... human intelligence
IDF .... Israeli Defense Forces
IFV .... infantry fighting vehicle
IPB . . intelligence preparation of the battlefield/battlespace
ITOW .... improved TOW
JTF .... joint task force
KOCOA .... key terrain, observation and fields of fire, cover and concealment, obstacles, and avenues of approach
LAV .... light armored vehicle
LCAC .... landing craft air cushion
LD .... line of departure
LOS .... line of sight
LZ .... landing zone
MAGTF .... Marine air-ground task force
MCDP .... Marine Corps doctrinal publication
MCRP .... Marine Corps reference publication
MCWP .... Marine Corps warfighting publication
MEDEVAC .... medical evacuation
MEF .... Marine expeditionary force
MET .... meteorological
METT-T ... mission, enemy, terrain and weather, troops and support available-time available
MEU .... Marine expeditionary unit
MOOTW .... military operations other than war
MOPP .... mission-oriented protective posture
MOUT .... military operations on urbanized terrain
MP .... military police
MPF .... maritime prepositioning force
MULE .... modular universal laser equipment
NBC .... nuclear, biological, and chemical
NCA .... National Command Authorities
NCF .... naval construction force
NEF .... naval expeditionary force
NEO .... noncombatant evacuation operation
NGF .... naval gunfire
NSFS .... naval surface fire support
NVA .... North Vietnamese Army
NWP .... naval warfare publication

Appendix J

Glossary

I. Acronyms

OP .... observation post
OPLAN .... operation plan

OPORD
operation order
OPSEC
operations security
PDDA
power-driven decontamination apparatus
PGM
precision-guided munition
PLO
Palestine Liberation Organization
PLRS
position location reporting system
POL
petroleum, oils, and lubricants
PSYOP
psychological operations
ROE
rules of engagement
ROK
Republic of Korea
RP
red phosphorus
RSOP . . . . reconnaissance, selection, and occupation of position
SASR
special application sniper rifle
SMAW
shoulder-launched multipurpose assault weapon
SOP
standing operating procedure
STANAG
NATO standardization agreement
STB
supertropical bleach
TOW
tube-launched, optically tracked, wire-guided missile
UAV
unmanned aerial vehicle
UCMJ
Uniform Code of Military Justice
U.N.
United Nations
VC
Viet Cong
VHF
very high frequency
WP . . . . white phosphorus

II. Definitions

A

assault—1. The climax of an attack, closing with the enemy in hand-to-hand fighting. 2. To make a short, violent, but well-ordered attack against a local objective, such as a gun emplacement, a fort, or a machine gun nest. (Joint Pub 1-02)

assault—The culmination of an attack which closes with the enemy. (MCRP 5-2A)

attack—An offensive action characterized by movement supported by fire with the objective of defeating or destroying the enemy. (FMFRP 0-14)

B

battlespace—All aspects of air, surface, subsurface, land, space, and electromagnetic spectrum which encompass the area of influence and area of interest. (FMFRP 0-14)

block—1. A tactical task assigned to a unit that requires it to deny the enemy access to a given area or to prevent enemy advance in a given direction or an avenue of approach. It may be for a specified time. Units assigned this mission may have to retain terrain and accept decisive engagement. 2. An obstacle effect that integrates fire planning and obstacle effort to stop an attacker on a specific avenue of approach or to prevent an enemy from exiting an engagement area. (MCRP 5-2A)

breach—A tactical task where any means available are employed to break through or secure a passage through an enemy defense, obstacle, minefield, or fortification. (MCRP 5-2A)

built-up area—A concentration of structures, facilities, and populations, such as villages, cities, and towns, that forms the economic and cultural focus for the surrounding area. (MCRP 5-2A, modified)

bypass—A tactical task that involves maneuvering around an obstacle, position, or enemy force to maintain the momentum of advance. Bypassed obstacles and enemy forces are reported to higher headquarters. (MCRP 5-2A)

C

clear—A tactical task to remove all enemy forces and eliminate organized resistance in an assigned zone, area, or location by destroying, capturing, or forcing the withdrawal of enemy forces such that they cannot interfere with the friendly unit's ability to accomplish its mission. (MCRP 5-2A)

combat arms—Units who close with and destroy enemy forces or provide firepower and destructive capabilities on the battlefield. (MCRP 5-2A, modified)

combat operations center (COC)—The primary operational agency required to control the tactical operations of a command that employs ground and aviation combat, combat support, and combat service support elements or portions thereof. The combat operations center continually monitors, records, and supervises operations in the name of the commander and includes the necessary personnel and communications to do the same. (FMFRP 0-14)

combat service support (CSS)—The essential capabilities, functions, activities, and tasks necessary to sustain all elements of operating forces in theater at all levels of war. Within the national and theater logistic systems, it includes but is not limited to that support rendered by service forces in ensuring the aspects of supply, maintenance, transportation, health services, and other services required by aviation and ground combat troops to permit those units to accomplish their missions in combat. Combat service support encompasses those activities at all levels of war that produce sustainment to all operating forces on the battlefield. (Joint Pub 1-02)

combat support (CS)—Fire support and operational assistance provided to combat elements. (Joint Pub 1-02)

combat support (CS)—Units that provide critical combat functions in conjunction with combat arms units to secure victory (MCRP 5-2A, modified).

command post (CP)—A unit's or subunit's headquarters where the commander and the staff perform their activities. In combat, a unit's or subunit's headquarters is often divided into echelons; the echelon in which the unit or subunit commander is located or from which such commander operates is called a command post. (Joint Pub 1-02)

contain—A tactical task to restrict enemy movement. (MCRP 5-2A)

cover—1. Shelter or protection from enemy observation that reduces the effects of enemy direct and indirect fire. 2. A type of security operation that protects the force from surprise, develops the situation, and gives commanders time and space in which to respond to the enemy's actions. (MCRP 5-2A)

D

defend—A combat operation designed to defeat an attacker and prevent him from achieving his objectives. It employs all means and methods available to prevent, resist, or destroy an enemy attack. Forms of defensive operations are area and mobile. Choices of defensive operations are in-depth and forward. The defensive techniques are defend in sector, defend a battle position, and defend a strong point. (MCRP 5-2A)

destroy—1. A tactical task to physically render an enemy force combat-ineffective unless it is reconstituted. 2. To render a target so damaged that it cannot function as intended nor be restored to a usable condition without being entirely rebuilt. (MCRP 5-2A)

disrupt—A tactical task or obstacle effect that integrates fire planning and obstacle effort to break apart an enemy's formation and tempo, interrupt the enemy's timetable, or cause premature commitment of enemy forces, or the piecemealing of his attack. (MCRP 5-2A)

II. Definitions

F

fire support—Assistance to elements of the Marine air-ground task force engaged with the enemy rendered by other firing units, including (but not limited to) artillery, mortars, naval surface fire support, and offensive air support. (FMFRP 0-14)

fix—1. A tactical task in which actions are taken to prevent the enemy from moving any part of his forces either from a specific location or for a specific period of time by holding or surrounding them to prevent their withdrawal for use elsewhere. 2. A tactical obstacle effect that integrates fire planning and obstacle effort to slow an attacker within a specified area—normally an engagement area. (MCRP 5-2A)

H

helicopterborne operation—A military action in which combat forces and their equipment maneuver about the battlefield by helicopters or vertical-landed aircraft. Aviation activities are under the control of the aviation combat element commander who is assigned in direct or general support of one or more combat element(s). (FMFRP 0-14)

hold—1. To maintain or retain possession of by force, as a position or an area. 2. In an attack, to exert sufficient pressure to prevent movement or redisposition of enemy forces. (Joint Pub 1-02)

I

infiltration—The movement through or into an area or territory occupied by either friendly or enemy troops or organizations. The movement is made, either by small groups or by individuals, at extended or irregular intervals. When used in connection with the enemy, it infers that contact is avoided. (Joint Pub 1-02)

isolate—A tactical task given to a unit to seal off (both physically and psychologically) an enemy from his sources of support, to deny an enemy freedom of movement, and prevent an enemy unit from having contact with other enemy forces. An enemy must not be allowed sanctuary within his present position. (MCRP 5-2A)

L

linkup—An operation wherein two friendly ground forces join together in a hostile area. (FMFRP 0-14)

M

Marine air-ground task force (MAGTF)—A task organization of Marine forces (division, aircraft wing, and service support groups) under a single command and structured to accomplish a specific mission. (Joint Pub 1-02)

military operations on urbanized terrain (MOUT)—All military actions that are planned and conducted on a topographical complex and its adjacent natural terrain where manmade construction is the dominant feature. It includes combat in cities, which is that portion of MOUT involving house-to-house and street-by-street fighting in towns and cities (MCRP 5-2A, modified)

military operations other than war (MOOTW)—Operations that encompass the use of military capabilities across the range of military operations short of war. These military actions can be applied to complement any combination of the other instruments of national power and occur before, during, and after war. (Joint Pub 1-02)

N

neutralize—1. To render enemy personnel or material incapable of interfering with a particular operation. 2. To render safe mines, bombs, missiles, and boobytraps. (MCRP 5-2A)

O

occupy—A tactical task in which a force moves onto an objective, key terrain, or other man-made or natural terrain area without opposition, and controls that entire area. (MCRP 5-2A)

P

peace enforcement—Application of military force, or the threat of its use, normally pursuant to international authorization, to compel compliance with resolutions or sanctions designed to maintain or restore peace and order. (Joint Pub 1-02)

peacekeeping—Military operations undertaken with the consent of all major parties to a dispute, designed to monitor and facilitate implementation of an agreement (ceasefire, truce, or other such agreement) and support diplomatic efforts to reach a long-term political settlement. (Joint Pub 1-02)

protect—A tactical task to prevent observation of or engagement or interference with, a force or location. (MCRP 5-2A)

pursuit—An offensive operation designed to catch or cut off a hostile force attempting to escape, with the aim of destroying it. (Joint Pub 1-02)

R

reduce—1. A tactical task to gain control over an enemy position or objective. 2. A task to create lanes through or over an obstacle sufficient to allow the attacking force to accomplish its mission. (MCRP 5-2A)

retain—A tactical task to occupy and hold a terrain feature to ensure it is free of enemy occupation or use. (MCRP 5-2A)

rules of engagement (ROE)—Directives issued by competent military authority which delineate the circumstances and limitations under which United States forces will initiate and/or continue combat engagement with other forces encountered. (Joint Pub 1-02)

S

screen—A task to maintain surveillance, provide early warning to the main body, or

II. Definitions

impede, destroy, and harass enemy reconnaissance within its capability without becoming decisively engaged. (MCRP 5-2A)

search—A systematic reconnaissance of a defined area, so that all parts of the area have passed within visibility. (Joint Pub 1-02)

secure—A tactical task to gain possession of a position or terrain feature, with or without force, and to deploy in a manner which prevents its destruction or loss to enemy action. The attacking force may or may not have to physically occupy the area. (MCRP 5-2A)

seize—To clear a designated area and obtain control of it. (FMFRP 0-14)

suppression—A tactical task to employ direct or indirect fires, electronic attack, or smoke on enemy personnel, weapons, or equipment to prevent or degrade enemy fires and observation of the friendly forces. (FMFRP 0-14)

T

terrorism—The calculated use of violence or threat of violence to inculcate fear; intended to coerce or to intimidate governments or societies in the pursuit of goals that are generally political, religious, or ideological. (Joint Pub 1-02)

Appendix L

References and Related Publications

Field Manuals (FMs)

3-11	Flame, Riot Control Agent, and Herbicide Operations
5-34	Engineer Field Data
5-102	Countermobility
5-250	Explosives and Demolitions
7-98	Operations in a Low-Intensity Conflict
9-15	Explosive Ordnance Disposal Service and Unit Operations
20-32	Mine/Countermine Operations
21-75	Combat Skills of the Soldier
3-30	Grenades and Pyrotechnic Signals
23-34	TOW Weapon System
27-10	Law of Land Warfare
33-1	Psychological Operations
34-130	Intelligence Preparation of the Battlefield
41-5	Joint Manual for Civil Affairs
41-10	Civil Affairs Operations
90-8	Counterguerrilla Operations
100-20	Military Operations In Low Intensity Conflict

Technical Manual (TM)

9-2350-264-10-2	Operator's Manual for Operation Under Usual and Unusual Conditions for Tank, Combat, Full-Tracked: 120mm Gun, M1A1 General Abrams, with Change 4

Marine Corps Concept Paper

A Concept for Future Military Operations on Urbanized Terrain

Marine Corps Doctrinal Publications (MCDPs)

1	Warfighting
1-3	Tactics

Marine Corps Reference Publications (MCRPs)

5-2A	Operational Terms and Graphics

Marine Corps Warfighting Publications (MCWPs)

3-15.3	Scout Sniping (under development)
3-37.1	Chemical Operations
3-37.4	NBC Reconnaissance

Fleet Marine Force Manuals (FMFMs)

0-9	Field Firing for the M16A2
0-25	The Law of Land Warfare
1-3	Tactics
2-6	MAGTF Rear Area Security
2-11	Antiarmor Operations
3-5	Employment of Military Police in Combat
3-30	Communications
6	Ground Combat Operations
6-7	Scouting and Patrolling for Infantry Units
6-8	Supporting Arms Observer, Spotter, and Controller
6-9	Marine Artillery Support
6-21	Tactical Fundamentals of Helicopterborne Operations
7-14	Combating Terrorism
7-34	MAGTF Civil Affairs
7-34-1	Civil Affairs Guidance for the Small-Unit Leader
7-40	Helicopter Insertion/Extraction
8-2	Counterinsurgency Operations
11-9	NBC Protection
11-10	NBC Decontamination
11-11	Treatment of Chemical Agent Casualties and Conventional Military Chemical Injuries
11-12	Chemical Hazard Assessment Guide
11-13	NBC Handbook
11-17	Chemical and Biological Contamination Avoidance
11-18	Nuclear Contamination Avoidance
11-20	NBC Reconnaissance
13-7	MAGTF Breaching Operations

Fleet Marine Force Reference Publications (FMFRPs)

0-6	Marine Troop Leader's Guide
0-14	Marine Corps Supplement to DOD Dictionary of Military and Associated Terms
4-26	Enemy Prisoners of War and Civilian Internees
7-14A	The Individual's Guide for Understanding and Surviving Terrorism
7-37	Vehicle Bomb Search
12-15	Small Wars Manual

12-18	Mao Tse-Tung on Guerrilla Warfare
12-25	The Guerrilla and How to Fight Him
13-5	Engineer Field Data

Joint Publications

1-02	DOD Dictionary of Military and Associated Terms
3-07	Joint Doctrine for Military Operations Other Than War
3-11	Joint Doctrine for Nuclear, Biological, and Chemical (NBC) Defense
3-15	Joint Doctrine for Barriers, Obstacles, and Mine Warfare
3-53	Doctrine for Joint Psychological Operations
4-06	JTTP for Mortuary Affairs in Joint Operations

Standard NATO Agreements (STANAGs)

2003	Patrol Reports by Army Forces
2014	Operations Orders
2029	Method of Describing Ground Locations, Areas, and Boundaries
2070	Emergency War Burial Procedures
2088	Battlefield Illumination
2101	Establishing Liaison
2868	Land Force Tactical Doctrine (ATP-35(B))

Miscellaneous

Fratricide: Reducing Self-Inflicted Losses, Newsletter No. 92-4, Center for Army Lessons Learned (CALL), U.S. Army Combined Arms Command, Fort Leavenworth, Kansas, April 1992.

HR, 25, Annex to Hague Convention No. IV, *Embodying Regulations Respecting the Laws and Customs of War on Land*, 18 October 1907.

Modern Experience in City Combat, U.S. Army Human Engineering Laboratory, Aberdeen Proving Ground, Maryland, 1987.

Table of Contents

Part I. Scouting

Chapter 7. Scouting Fire Team

PART 2. INFANTRY PATROLLING

Chapter 8. Fundamentals of Infantry Patrolling

Chapter 9. Patrol Organization

Chapter 10. Patrol Preparation

Chapter 11. Movement to and Return from the Objective Area

Chapter 12. Reconnaissance Patrols

PART I. SCOUTING

CHAPTER 1. FUNDAMENTALS OF SCOUTING

Scouting involves observing terrain and/or the enemy, and accurately reporting those observations. Scouting requires proficiency in the use of weapons, cover and concealment, route selection, and skill in unobserved day or night movement.

1001. PURPOSE

When an infantry unit is not actively fighting the enemy, it should be actively searching for the enemy, which is the primary purpose of scouting. The unit attempts to keep the enemy off balance while making preparations for further attacks. Physically locating and keeping the enemy off balance are normally accomplished by small units ranging from a two-person scouting party to a squad-size patrol.

Infantrymen are sent out as scouts or as members of a patrol because the commander needs information about the enemy, terrain, and the location of friendly troops. The lives of the entire unit may depend upon the success or failure of a scout or patrol and the accuracy and timeliness of the report. The success of the scout or patrol will depend upon their training, preparation by the commander, and understanding their mission and the commander's requirements.

To wage combat successfully, a commander must have accurate, detailed, and timely information about the enemy, the terrain, and adjacent friendly units. Well-trained scouts and capably led patrols are among the most effective means the commander has for acquiring the information necessary to plan tactical actions and make decisions in execution.

1002. REQUIRED SCOUTING SKILLS

To be effective, a scout must be able to—

- Recognize terrain features.
- Read a map and determine direction.
- Practice and implement the principles of cover and concealment.
- Fully utilize movement and route selection.
- Know the enemy (estimate enemy unit composition and strength).
- Observe and report information accurately.
- Select routes and move through numerous types of terrain.

CHAPTER 2. TERRAIN, MAPS, AND DIRECTION

A scout must understand map symbols, identify elevations from contour interval lines, scale distance on a map, relate natural and man-made features shown on the map to the actual features on the ground, plot a course from one point to another, and locate his current position. To relate a map to the actual terrain and its features, a scout must be able to orient it to the ground using a compass, two points, a watch, and the sun or the stars.

2001. TERRAIN FEATURES

Since the infantry works and fights on the ground, terrain ashore information that scouts gather and report on is of great importance. Hills, valleys, woods, and streams are the forms and growths commonly referred to as natural land features. Artificial or man-made features include houses, bridges, and railroads. Figure 2-1 shows some important terrain features.

2002. THE LENSATIC COMPASS

The best method of finding direction, during both day and night, is with a compass. The lensatic and M2 are the two types of compasses issued to infantry units. Both work on the same principle.

The standard compass for general use in the Marine Corps is the pivot-mounted lensatic compass, so called because azimuths are read through a magnifying lens in the eyepiece. Figure 2-2, on page 2-2, shows the lensatic compass and its nomenclature. The meter graphic scale on the side of the compass is 1:50,000, which is the most used scale in military mapping. The graphic scale is useful in the field as a straightedge, an aid in orienting the map, and a means of reading map grid coordinates. The plastic dial is graduated in both degrees and mils. Numbers on the dial are printed in black. There are luminous markings on the bezel, floating dial, and on both ends of the sighting wire, plus a 3-degree bezel serration and clicking device that permit reading azimuths at night. The compass is carried in a nylon case that may be attached to the cartridge belt.

Magnetic compasses are affected by the presence of iron or magnetic fields. Consequently, the scout should not be within the influence of local magnetic

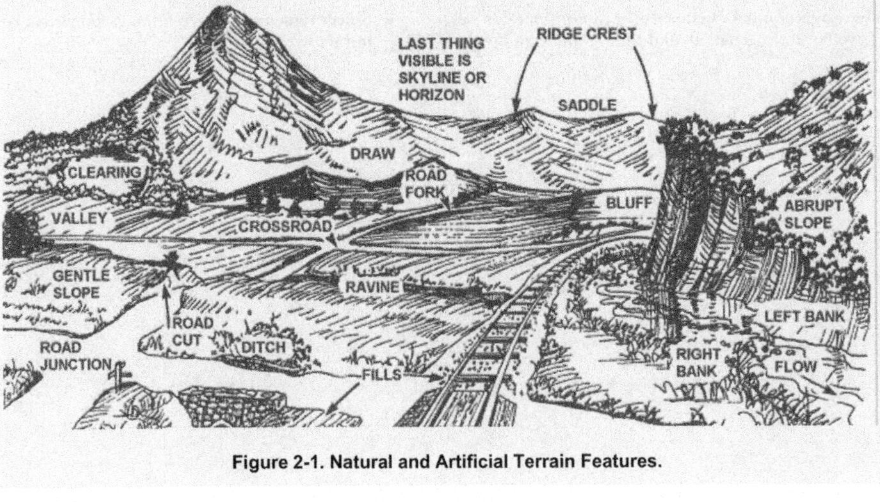

Figure 2-1. Natural and Artificial Terrain Features.

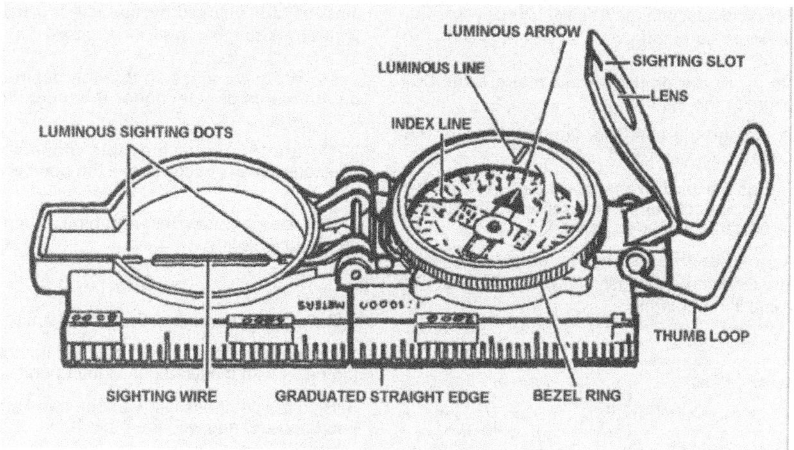

Figure 2-2. Lensatic Compass.

attraction while using a compass to determine direction. The rifle, pistol, and other metal objects must be laid aside when reading the compass. The minimum distances the scout should be from visible masses of iron and electrical fields of magnetism for the compass to provide accurate readings are provided below.

Magnetic Fields	Minimum Distances (meters)
High tension powerlines	60
Field guns	20
Vehicles (wheeled or tracked)	20
Telephone and telegraph wires	10
Barbed wire	10
Machine gun	3
Rifle, pistol	1

Reading an Azimuth

To read an azimuth to any point, the cover of the compass is raised to an angle of 90 degrees in relation to the index face, and the eyepiece is lifted to a 45-degree angle in relation to the bezel, or so the numbers on the dial can be seen. The thumb of either hand is placed in the thumb loop, the index finger extended along the side of the compass case, and the remainder of the hand closed. The closed hand and wrist are grasped with the other hand. The elbows are drawn in close to the body, forming a firm foundation for the compass. The eye is placed to the lens on the eyepiece. The compass is pointed at the object or point to which the azimuth is to be read. A sighting is taken through the sighting slot in the eyepiece, and the point is lined up with the sighting wire in the cover. The compass is held until the dial steadies; then the reading is taken through the lens of the eyepiece. This reading is the magnetic azimuth of the line from the observer to the point.

Reading a Back Azimuth

A back azimuth is the direction opposite the line of sight. If the azimuth is less than 180 degrees, the back azimuth is obtained by adding 180 degrees. If the azimuth is greater than 180 degrees, the back azimuth is obtained by subtracting 180 degrees. Back azimuths are used to determine a return route or to resection to determine a current position.

Circumventing Obstacles

When a scout is traveling on an azimuth and comes upon an obstacle—such as a contaminated area, minefield or swamp—the following steps (sometimes referred to as the *90-degree offset* method) is employed to go around or circumvent the obstacle and resume movement along the original azimuth (see fig. 2-3). The steps are as follows:

1. Move up to the obstacle and make a full 90-degree turn to the right (or left).

2. Walk beyond the obstacle, keeping track of the distance in paces or meters.

3. Stand at the end of the obstacle, face in the original direction of march, and follow that azimuth until the obstacle has been passed.

4. Make a 90-degree turn to the left (or right) and move the distance previously measured to return to the original line of march.

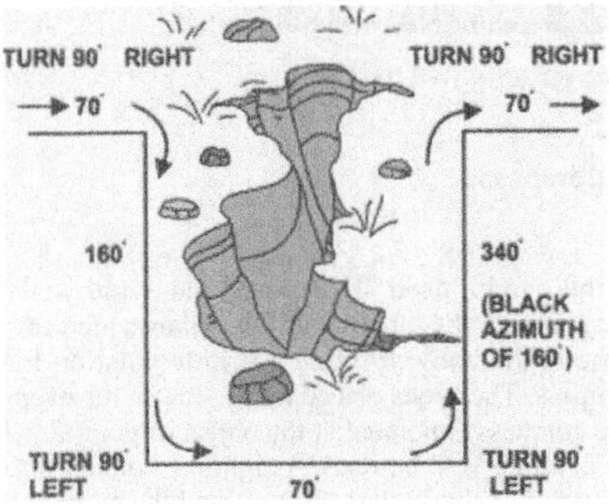

Figure 2-3. Circumventing an Obstacle.

Following an Azimuth During the Day

The eye is placed to the lens in the eyepiece and the compass moved until the desired azimuth reading is visible beneath the fixed index. Without moving the compass, the vision is shifted from the lens through the sighting slot in the eyepiece, and a sighting is taken out beyond the sighting wire in the cover. A prominent terrain feature on this line of sight is selected, the compass closed, and the landmark approached. When the scout reaches the landmark, the procedure is repeated.

Following an Azimuth at Night

It is necessary to prepare and set the compass before departing on a night movement because at night only the luminous parts of the compass can be seen. To prepare the compass for night use, the luminous parts must be fully charged by sunlight or artificial light, such as a flashlight. To set a compass—

1. Move the compass so that the desired azimuth on the dial is directly under the index line on the lower glass.

2. Rotate the upper movable glass so that the luminous line is directly above the north arrow of the dial.

3. Set the compass for marching at night on the specified azimuth.

—OR—

1. Face the general direction of movement.

2. Line up the north arrow and the luminous line on the bezel with the luminous sighting dots.

3. Hold the compass still with one hand and grip the knurled bezel ring with the other hand.

4. Turn the bezel ring the prescribed number of clicks in the proper direction, remembering that each click equals 3 degrees. For example, to set an azimuth of 21 degrees, the bezel ring would be turned seven clicks to the left.

5. Turn the whole compass until the north needle lines up with the luminous line. The compass is then set on the desired azimuth. The azimuth is the line formed by the two luminous sighting dots on the inside of the cover.

To march on a preset azimuth during night movement, open the compass and move it so the north arrow is directly below the luminous line (see fig. 2-4). Move

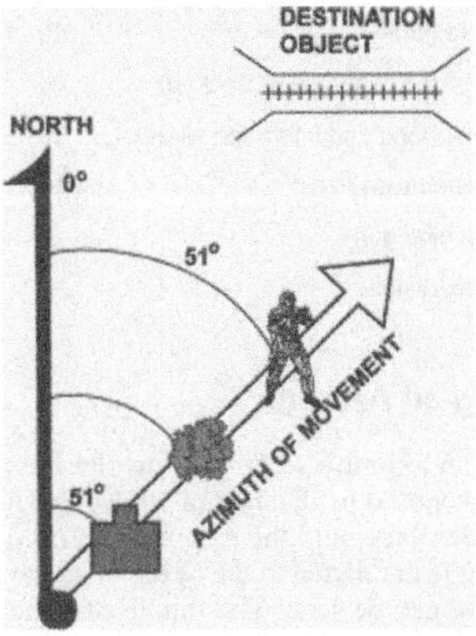

Figure 2-4. Following a Night Azimuth.

in the direction of the line formed by the two luminous sighting dots. It is necessary to refer to the compass more frequently at night than during the day. If stars are visible, find a prominent star along the azimuth of movement to use as a reference point. When the view of the sky is restricted by overcast conditions or vegetation, send a scout forward along the azimuth of movement to the limit of visibility. This scout is guided along the azimuth of movement by a stationary navigator. When the scout reaches the limit of visibility, the navigator moves to the scout's location. This process is repeated until the destination is reached.

A more rapid method for reaching the scout's destination is to equip the navigator with a compass. The navigator can set the compass as explained earlier and the scout proceeds providing security 180 degrees to the front on the specified azimuth, receiving right and left corrections from the navigator while both are on the move. The point scout must stay within visual range of the navigator. If available, a strip of white or luminous tape on the back of the point scout's helmet will assist.

Intersection

Intersection is the location of an unknown point by successively occupying at least two, preferably three known positions and sightings on the unknown point. It is used to locate features not depicted on the map or not readily identifiable. To determine an intersection, perform the following steps (see fig. 2-5):

1. Orient the map using the compass.
2. Locate and mark your position on the map.
3. Measure the magnetic azimuth to the unknown position; then convert to grid azimuth.
4. Draw a line on the map from your position on this grid azimuth.
5. Move to a second known position from which unknown point is visible. Locate this position on the map and again orient the map using the compass. The second unknown position should be a minimum of 30 degrees offset from the first position.
6. Repeat steps 4 and 5.

Figure 2-5. Intersection.

To check accuracy, move to a third position and repeat steps 1 thorough 4. Where the lines cross is the location of the unknown position. Using three lines, a triangle is sometimes formed—called the *triangle of error*—instead of an intersection. If the triangle is large, recheck your work to find the error. Do not assume that the position is at the center of the triangle.

Resection

Resection is the location of the user's unknown position by sighting on two or three known features that are identifiable on the map. To determine a resection, perform the following steps (see fig. 2-6):

1. Orient the map using the compass.
2. Locate two or three known positions on the ground and mark them on the map.
3. Measure the magnetic azimuth to a known position then convert to grid azimuth.

4. Change the grid azimuth to a back azimuth and draw a line on the map from the known position back toward the unknown position.

5. Repeat step 3 and step 4 to determine a second known position.

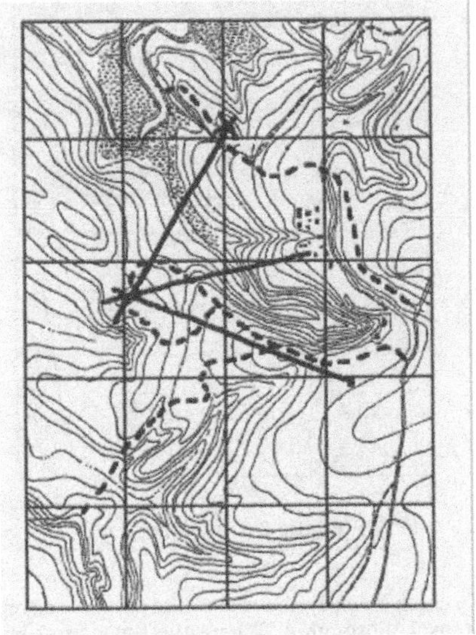

Figure 2-6. Resection.

To check accuracy, repeat the steps above for a third known position. The intersection of the lines is your location. Using three lines, a triangle of error may be formed. If the triangle is large, recheck.

2003. STARS, SUN, AND OTHER FEATURES

In rare cases when a scout is without a compass, the following examples are alternate means to determine direction. When using constellations to determine direction, identify your location's Temperate Zone. A Temperate Zone is the area between the tropics and the polar circles.

At night, the stars provide an excellent means of maintaining a line of march. In the North Temperate Zone (north of the equator), the Big Dipper constellation is one key to determining direction of true north. It is made up of seven fairly bright stars in the shape of a dipper with a long curved handle (see fig. 2-7). The two stars that form the side of the cup farthest from the handle, used as pointers, are situated in the direction of a bright star that is about five times the distance between the two stars of the dipper cup. This bright star is the North Star and is directly over the North Pole. The pointers always designate the North Star, which is the direction of true north.

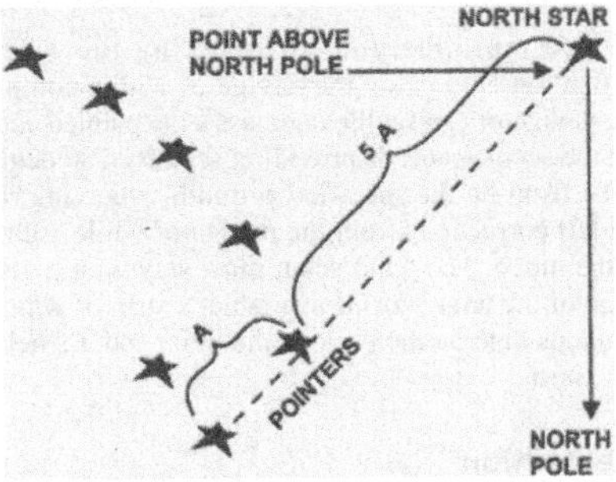

Figure 2-7. Locating the North Pole.

In the Southern Hemisphere, true south is determined in relation to the Southern Cross, a constellation composed of five stars. Two bright pointer stars in the vicinity of the Southern Cross serve as locators to help locate true south (see fig. 2-8). The outer four stars are

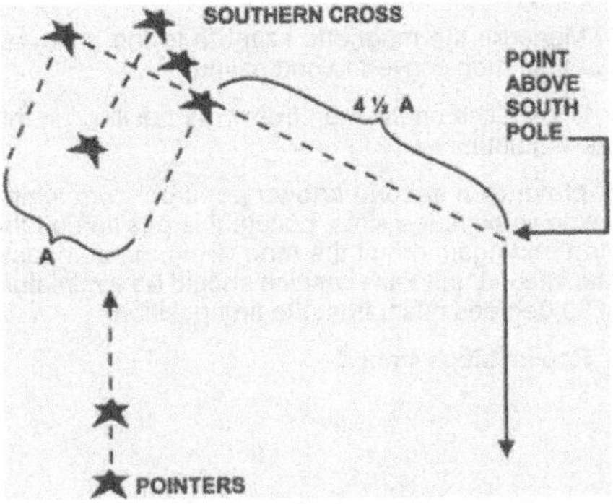

Figure 2-8. Locating the South Pole.

fairly bright and form a cross. This cross is imagined as the frame of a kite. A straight tail, four and one half times as long as the length of the kite itself, is put on the kite using finger widths for a measuring stick. The end of this tail will be close to a position directly over the South Pole. Usually, it will not be possible to see a star in the immediate vicinity, because there is no bright star visible directly above the South Pole.

During daylight hours, a watch and the sun can be used to determine direction within 8 degrees. In the North Temperate Zone, the watch is held horizontally, face up, and the hour hand pointed at the sun (see fig. 2-9). The north-south line and the direction of south can be found midway between the hour hand and the number 12, if the watch is set on standard time. If in daylight savings time, the direction of south is found midway between the hour hand and the number 1.

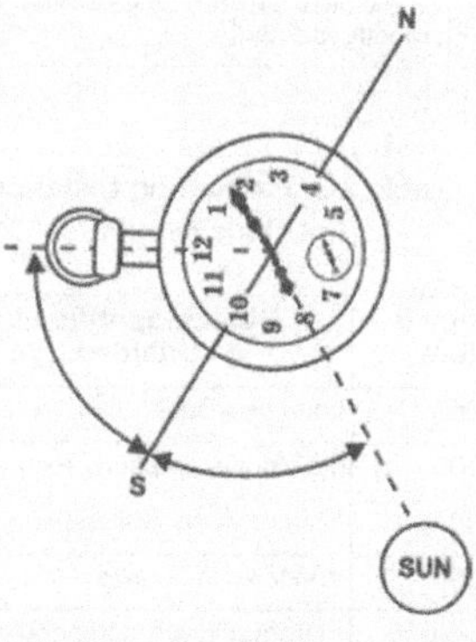

Figure 2-9. Determining Direction by Watch and Sun (North Temperate Zone).

In the South Temperate Zone, if the watch is set on standard time, the number 12 on the watch is pointed at the sun; if the watch is set on daylight savings time, the number 1 is pointed at the sun. North is midway between 12 (or 1) and the hour hand (see fig. 2-10).

When laying in a north-south line, if any doubt exists as to which end of the line is north, remember that the sun is in the east before noon and in the west in the afternoon.

In addition to the sun and stars, other methods a scout without a compass can use to determine direction include determining prevailing wind direction and

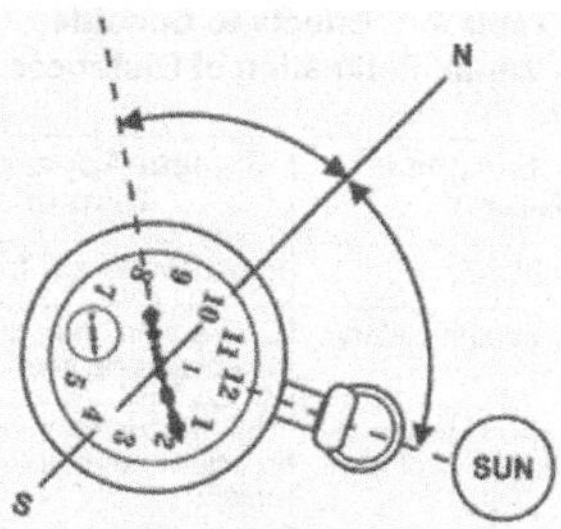

Figure 2-10. Determining Direction by Watch and Sun (South Temperate Zone).

using a mountain for orientation. By previous study of maps and photographs, a scout can keep informed of location and direction by using a distinctive edge of woods, a deep ravine or the direction of a stream's flow. A scout should constantly evaluate and memorize both the immediate terrain and general area for prominent features and landmarks.

2004. RANGE DETERMINATION

Range determination is the method of finding the distance between an observer and an enemy target or an object. By accurate range determination, the members of a given unit can set their sights correctly and place effective fire on enemy targets. The degree of accuracy is dependent on several factors, such as terrain relief, time available, and experience of the observer.

Mental Estimation

A mental distance estimate is made using a known unit of measure. Distance is estimated to the nearest 100 meters by determining the number of known units of measure between the observer's position and a target. For example, a football field, which is 100 yards, can be used as a known unit of measure for determining the distance between an observer's position and a target. For longer distances, progressive estimation may be necessary. To do this, the observer determines the number of units of measure to an intermediate point and doubles the value. The observer should consider the effects in table 2-1 in estimating distances.

Table 2-1. Effects to Consider in Mental Estimation of Distances.

Objects Appear Nearer	Objects Appear More Distant
In bright light.	In poor light or in fog.
In clear air at high altitude.	Only a small part of the object can be seen.
The background is in contrast with the color of the object.	The background is similar in color to that of the object.
The observer is looking down from a height.	The observer is looking over a depression, most of which is visible.
The observer is looking over a depression, most of which is hidden.	The observer is kneeling or sitting, especially on a hot day, when the ground is moist.
The observer is looking down a straight feature such as a road.	
The observer is looking over water, snow, or a uniform surface such as a cultivated field or desert.	

Table 2-2. Estimating Distance in Wooded Terrain.

Distance in Meters	Tree Description
1,000	Trunk and main branches are visible. Foliage appears in cluster-like shape. Daylight may be seen through the foliage.
2,000	Trunk visible, main branches distinguishable, foliage appears as smooth surface. Outline of foliage of separate trees distinguishable.
3,000	Lower half of trunk visible. Branches blend with foliage. Foliage blends with adjoining trees.
4,000	Trunk and branches blend with foliage and appears as a continuous cluster, smooth in appearance. Movement of foliage due to wind cannot be deleted.
5,000 and beyond	Whole area covered by trees and appears smooth and dark.

Table 2-3. Estimating Distance in Urban Terrain.

Distance in Meters	Object Identified by the Unaided Eye
1,000	Lone tree trunk
1,500	Individuals and horsemen
3,000	Chimneys on rooftops
4,000	Windows in houses
4,000–5,000	Individual houses in populated area
8,000–9,000	Villages and individual houses
15,000–18,000	Large houses, towers, and steeples

Estimating in Good Visibility

When visibility is good, distances can be estimated by using the appearance of tree trunks, branches, and foliage (as seen by the naked eye) in comparison with map data. Table 2-2 is a guide for wooded terrain. Table 2-3 is a guide for urban environments.

Estimating From a Terrain Study

The Marine should always use terrain/map analysis to assist in estimating distances. When the Marine is looking in a specific direction, the estimation of distance can be enhanced by studying the terrain and comparing it with the map. Particular emphasis should be given to color contrasts of terrain features seen along the observer-target line (OTL). For example, the distance across successive ridge lines or depressions in the distance may be identifiable by only slight changes of color to the eye. Different colors of grass might reveal a hidden terrain feature such as a stream.

CHAPTER 3. ENEMY ACTIVITY

A commander often acts on information furnished by scouts. Therefore, scouts must aim at absolute accuracy in reporting enemy activity. This chapter discusses estimating enemy strengths, interpreting signs and tracks, and knowing the enemy.

3001. ESTIMATING ENEMY STRENGTHS

If troops cannot be counted, their strength may be estimated by: noting the length of time it takes various types of moving columns to pass given point, the area required of a unit in camp or bivouac, or the front on which they are deployed. When the ground is dry, infantry on the march raise a low, thick cloud of dust, and motor vehicles or mechanized units raise a thick, rapidly moving cloud. Additionally, through practice, a scout may gain information as to the strength and composition of enemy forces by listening to noises and observing lights, fires, and smoke. A scout gains valuable experience in estimating enemy strengths by observing friendly forces in camp, on the march, and deployed. The knowledge scouts gain during field exercises of the appearance and tactical dispositions of squads, platoons, companies, and larger units will be of great assistance in estimating the strength and composition of enemy units observed under various conditions.

3002. INTERPRETING SIGNS AND TRACKS

In addition to estimates made through direct observation, a scout may often be able to estimate size, composition, direction, rate of movement, condition, discipline, state of training, and morale of enemy forces through signs and tracks left behind.

Signs

The examination of vacated enemy positions provides valuable information. The size of a bivouac or defense area ordinarily indicates the number of enemy occupants. Clothing, ration containers, dumps, etc., further indicate the quantity of the departed enemy force. The condition of the bivouac area and amount of material abandoned give an indication of the enemy morale, training, and discipline. A well-policed area indicates good discipline. Rubbish, ration and smoking residue, and nonessential personal items of equipment adrift indicate a lower state of morale, training, and discipline. Stores and material left behind in good condition may indicate a hasty movement or withdrawal. Burned or destroyed materials indicate a deliberate, orderly withdrawal or movement. Letters, insignia, and other articles may reveal the identity of the enemy unit.

In the case of a moving enemy, the distance between periodic halts indicates the rate of march if enemy habits relative to marches and halts are known. Condition of the halt areas indicates the state of morale, training, and discipline.

The physical condition of enemy dead and wounded and their personal equipment and weapons are reported. The general condition and state of maintenance of destroyed or abandoned vehicles should also be reported.

Tracks

A track is a mark left on the ground by the passage of a person or object. Examination of tracks reveals information about the enemy.

Troops

A few tracks overlapping each other on both sides of a road or trail may indicate a patrol in staggered formation. A large number of tracks indicates troops in column formation. A large column will wear a dry road smooth and flat. In damp terrain, a freshly made track will have sharp edges; ordinarily, signs of moisture will disappear in about 15 minutes. A runner's toes are dug into the ground; a walker's footprint is fairly even.

Vehicles

The type of track indicates whether the vehicle is wheeled or tracked. A scout acquires the necessary experience to make the proper determination by observing vehicle tracks during training.

The direction of travel can be determined by the way tracks pass across ruts, by impressions on the edges of holes in the ground, how water is splashed from puddles, or by the way grass, twigs, and branches are bent; for example—

- A vehicle (wheeled or tracked) entering a rut pushes dirt into the rut and leaves an indentation on the exit side of the rut.
- A wheel going over holes in the ground leaves a deeper impression on the edge toward the direction of travel.
- The side of a puddle with the greater splash indicates the direction of travel of the vehicle.
- When traveling cross-country, the direction in which grass is bent and/or twigs, branches, and bushes are broken indicates direction of travel.

A general rate of speed can be estimated by the amount of water or mud splattered. A fast-moving vehicle will throw larger amounts of water or mud a greater distance to the front and sides than a slow-moving vehicle, and it leaves a deeper impression on the exit edges of holes. The faster the travel, the deeper the impression.

3003. KNOWING THE ENEMY

A scout should learn as much as possible about enemy psychology, habits, organization, and tactics. The more knowledge gained about the enemy, particularly the enemy's normal security measures, the better the scout's chances are to observe and obtain accurate information with minimum risk to the mission's success. Scouts gain much of this knowledge through experience, but they also gain a great deal of their preliminary information, particularly that pertaining to enemy organization and tactics, during training and may be updated by unit commanders and intelligence officers.

CHAPTER 4. DAYLIGHT SCOUTING

A scout must be able to operate in all types of terrain and under all conditions of visibility. He must be thoroughly familiar with the principles for using cover and concealment, camouflage, individual movement, and route selection, both to and from the objective.

4001. COVER AND CONCEALMENT

Cover is protection from the fire of hostile weapons. Concealment is protection from observation or surveillance from hostile air and ground observation, but not from hostile fire. Both cover and concealment are divided into two main categories: natural and artificial. Natural cover includes small hills, ditches, rocks or vegetation. Fighting holes, bunkers, and brick walls are examples of artificial cover. Some features, such as buildings, provide both cover and concealment. In deciding whether to seek cover or concealment, a scout must make the best choice to complete the mission (see fig. 4-1).

Concealment Principles

Concealment principles are as follows:

- Remain motionless while observing. Anything in motion attracts the eye.
- Use all available concealment.
- Observe from the prone position (it offers a low silhouette and makes detection by the enemy difficult.
- Expose nothing that reflects light.
- Blend with the background because contrasting colors are noticeable.
- Remain in the shade because moving shadows attract attention.
- Distort or change the regular outline of objects. Most military objects have distinctive shapes that make obvious shadows and silhouettes.
- Avoid the skyline. Figures on the skyline can be seen from great distances and are easily identified by their outlines.

Concealment Techniques

Concealment techniques are as follows:

- When observing, the scout looks around an object's side (unless it is transparent) and prepares to fire, if necessary, around the side of or, if possible, through an object.
- Looking or firing over an object can make the scout an easily visible target for the enemy. If the scout must fire over the top of concealment or cover, the outline of the head or helmet should be broken or distorted.
- Upon the approach of an airplane, the scout takes a prone position, turns face-down, and remains motionless. If surprised by an airplane, the scout remains in place and does not look up.

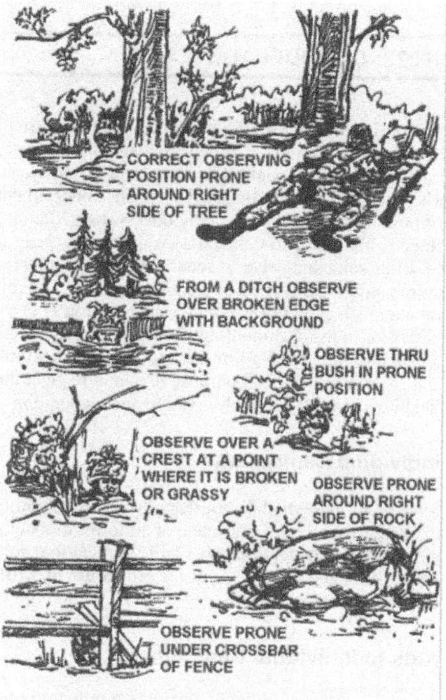

Figure 4-1. Correct Use of Cover.

- The scout covers exposed body parts such as the face, back of the neck, and hands with grease paint, mud or other materials to reduce sun reflection.
- Camouflage for equipment can be improvised from garnishing or sandbags to prevent sun reflection.
- In snowy terrain, white overgarments are worn.
- The helmet cover outline should be distorted.

4002. CAMOUFLAGE

Camouflage is the use of concealment and disguise to minimize the possibility of detection and/or identification of troops, material, equipment, and installations. The purpose of camouflage is to provide concealment of military objects from enemy observation. Camouflage is also used to conceal an object by making it look like something else. A scout's mission usually requires individual and equipment camouflage. If natural camouflage is not adequate, the position is camouflaged. In using camouflage, remember that objects are identified by their form (outline), shadow, texture, and color. The principal purpose of camouflage in the field is to prevent direct observation and recognition.

Individual Camouflage

Successful individual camouflage involves the ability to recognize and take advantage of all forms of natural and artificial concealment available (vegetation, soil, debris, etc.) and knowledge of the proper use of artificial camouflage materials.

Aids to Individual Camouflage

A scout must recognize the terrain's dominant color and pattern and must change the appearance of clothing and equipment accordingly in order to blend and not contrast with the terrain (see fig. 4-2).

The helmet is camouflaged by breaking up its shape, smooth surface, and shadow. Use of a helmet cover works best. In the absence of a helmet cover, mud can be irregularly blotched on the helmet to disguise its form and dull the surface. A helmet cover may be improvised from irregularly colored cloth or burlap to blend with the background. Foliage can be draped to prevent the visor of the helmet from casting a dark shadow across the face. Foliage should not stick up like plumes because any head movement will give away the position.

A small, thin bush in the shadow of a large bush makes a good observation point. Lone trees, rocks, fence corners, and outstanding landmarks are easily picked up by the enemy as obvious observation posts.

If camouflage clothing is not available, other available clothing can be attached in irregular splotches of appropriate colors.

Exposed skin reflects light and attracts the enemy's attention. Even very dark skin will reflect light because of its natural oil. The buddy system is recom-mended when applying camouflage. Standard Marine Corps issue camouflage face paint sticks are two toned:

- Loam and light green for light-skinned troops, in all but snow regions.
- Sand and light green for dark-skinned troops.
- Loam and white for troops in snow-covered terrain.

Shiny areas (forehead, cheekbones, nose, and chin) are painted with a dark color. Shadow areas (around the eyes, under the nose, and under the chin) are painted

Figure 4-2. Avoid Contrasting Backgrounds.

with a light color. Skin that is exposed on the back of the neck and hands is painted with a two-color combination in an irregular pattern (see fig. 4-3).

When standard issue face paint sticks are not available, burnt cork, charcoal or lamp black can be used to tone down exposed areas of skin.

Mud is used only in an emergency because it changes color as it dries and may peel off, leaving the skin exposed. Since mud may contain harmful bacteria, mud should be washed off as soon as possible.

Any equipment that reflects light should be covered with a nonreflective material that aids in the concealment of the weapon (for example, black electrical tape or mud). The straight line of the rifle or other infantry weapons may be very conspicuous to an enemy observer. The barrel and hand guard should be wrapped with strips of contrasting colored cloth or tape to break the regular outline. Mud or dirt dulls the reflecting surface of the stock, barrel, and bayonet where coloring has been worn. Lamp black may also be used on metal parts. The function of the weapon must not impaired.

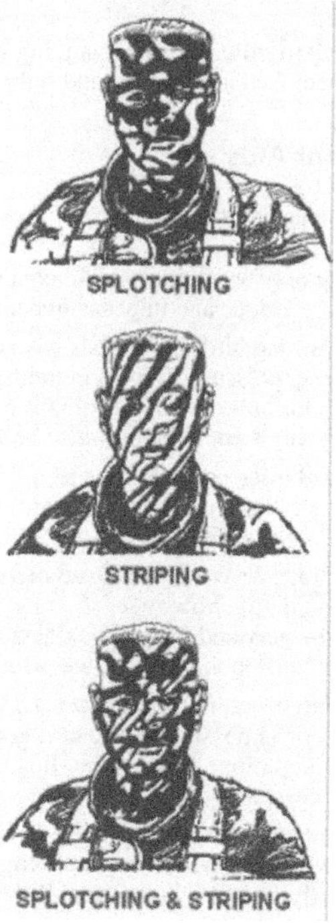

Figure 4-3. Face Camouflage.

If time, material, and surroundings permit, a ghillie suit should be constructed. (Refer to MCWP 3-15.3, *Scout Sniping*.)

Aids to Camouflage a Position

To successfully camouflage a position, the scout must remember to—

- Camouflage the position as soon as it is occupied.
- Avoid using too much material for camouflage. Even though natural materials are used, too much may make the object and its shadow stand out from its surroundings, thus attracting the attention of a hostile observer.
- Inspect completed camouflage work from the enemy's point of view to check effectiveness.

Continuous Camouflage

Camouflage around and on the scout's position must be maintained in a fresh condition as wilted and dead foliage can give the position away. If the mission dictates that the position should be occupied for longer periods, wilted foliage should be replaced during periods of reduced visibility.

4003. INDIVIDUAL MOVEMENT

Principles

The principles of individual movement are as follows:

- Scouts move from one concealed position to another. When not changing positions, they remain motionless.
- The scout's head is lifted slowly but steadily, without abrupt movements, to search for a new position.
- Scouts select the next stopping place before moving and ensures it is not contained by the enemy.
- Scouts change position on the run: springs up, runs with the body bent low, zigzags, quickly drops to the ground slightly to the right or left of the objective, then rolls or crawls to the desired position.

(Remember the phrase, "I'm up—he sees me—I'm down.")

Rushing

When starting from the prone position—

- Raise the head slowly and steadily and select a new position.
- Lower the head slowly, draw arms inward, cock right leg forward, and prepare to rush.
- Use one movement to raise the body by straightening both arms.
- Spring to your feet, stepping off with the left foot.
- Bend forward as low as possible when running. Never advance directly to the next position; always zigzag.

When hitting the deck—

- Stop.
- Plant both feet in place.
- Drop quickly to the knees and slide the hand to the heel of the rifle.
- Fall forward, breaking your fall with the butt of the rifle. (To confuse the enemy, roll over after hitting the deck and roll into firing position with feet, knees, and stomach flat on the ground.)
- Keep head down if you do not intend to fire.

When rolling over—

- Hit the deck and assume the prone position.
- Bring the rifle in close to the body, placing the rifle butt in the crotch.
- Roll over swiftly to confuse any enemy observers as to final intended location. Never reappear at the same place you went down.

Low Crawl

The low crawl is used when—

- Cover and concealment are scarce.
- The enemy has good observation over the area in which the scout is moving.
- Speed is not essential.

To perform the low crawl, keep the body as flat as possible against the ground. Grasp the rifle sling at the upper sling swivel. Let the balance of the rifle rest on the forearm and let the butt of the rifle drag on the ground. Keep the muzzle off the ground.

To start forward, push arms forward and pull right leg forward. To move forward, pull with arms and push with right leg. Change the pushing leg frequently to avoid fatigue.

High Crawl

The high crawl is used when—

- Cover and/or concealment are available.
- Poor visibility reduces enemy observation
- Greater speed of movement is required.

To perform the high crawl, keep body off the ground. Rest weight on forearms and lower legs. Cradle rifle in arms, keeping the muzzle off the ground. Keep knees well behind the buttocks to stay low.

Move forward, alternately advancing right forearm and left knee; then left forearm and right knee.

Movement Aids

Aids to movement include—

- Carrying only necessities. Additional weight causes premature fatigue and impedes free movement.
- Not disturbing birds or animals whose flight would betray your presence. If birds or animals are alerted, remain motionless under cover for a few minutes, as the enemy's attention may also be attracted.
- Moving during an incident that diverts attention, such as an airplane flight, a distant disturbance or sudden bursts of fire.
- Fog, smoke, or even light haze offer concealment for movement; however, the enemy may have thermoimagery and night vision devices. Therefore, darkness and smoke cannot be used as easily.
- Following a stream or road by staying as far away from them as possible while still keeping them in sight. Keep close to the dune line when moving along a beach.
- When moving through tall grass or similar growth, move when the wind blows, changing direction frequently. A straight route will be more readily noticed.
- Whenever possible, avoid areas of soft ground so as not to leave tracks.

- When crossing a road or water obstacle, choose crossing sites where the enemy's observation is restricted (an area in shadows or near a bend) and cross rapidly.

4004. ROUTE SELECTION

Prior to Movement

A scout and the immediate commander conduct a map reconnaissance before starting on a mission. This assists them in selecting the route according to available cover and concealment and any indicated enemy activity.

Prior to and during the course of the mission, move to an observation point to visually reconnoiter the terrain for movement and select the tentative route. It may be necessary to make wide detours around open spaces or those containing enemy patrols or other enemy activity.

Carefully study the country to be traversed and pay close attention to the general features, streams, ridges to be crossed, and their relation to the general direction to be taken (see fig. 4-4).

Make notes of terrain features and landmarks along the proposed route and rely on notes for guidance (see fig. 4-5 on page 4-6). Additionally, determine the compass direction and readings for each change of direction at the start. Finally, learn the location of unit boundaries and observation/listening posts as well as general location of other friendly or scouting parties. Be sure to avoid man-made and natural obstacles as they will slow progress and overall success of the mission. If possible, use the local populous as a source of intelligence. When returning to friendly lines, avoid using the same route.

En Route

En route, the actual advance will be a series of movements from one observation point to the next. The distance and route will depend on cover and terrain. Assess the cover, terrain, and any enemy or civilian activity to determine whether or not to modify the approach or return routes. Unless the mission requires it, avoid danger areas (for example, houses, villages, potential assembly or bivouac areas, roads, and streams) that may give away your position by being

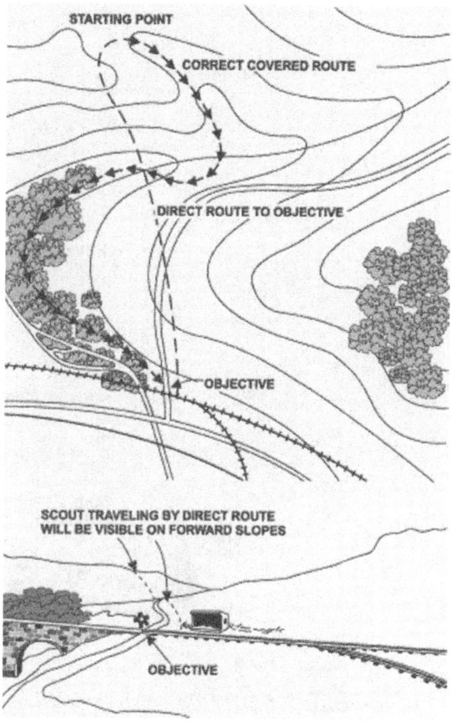

Figure 4-4. Choosing a Concealed Route of Advance from a Map.

observed by the enemy. When required to reconnoiter danger areas, choose a covered approach and return, and make entry or passage as quietly and quickly as possible. If part of a larger effort, the approach and return should be covered by observation and fires of the other members of the scouting party or patrol.

Stream Crossings

When the crossing does not appear to be held by the enemy, advance upon it rapidly. If there are two or more scouts, one crosses while the other(s) provide protection. Note the length, width, depth, and approaches to a crossing. Observe the condition of the road or trail that crosses the stream, and report on the suitability of the crossing for use by tracked and wheeled vehicles. If the crossing is under observation by enemy, seek another crossing site or dash across to avoid detection.

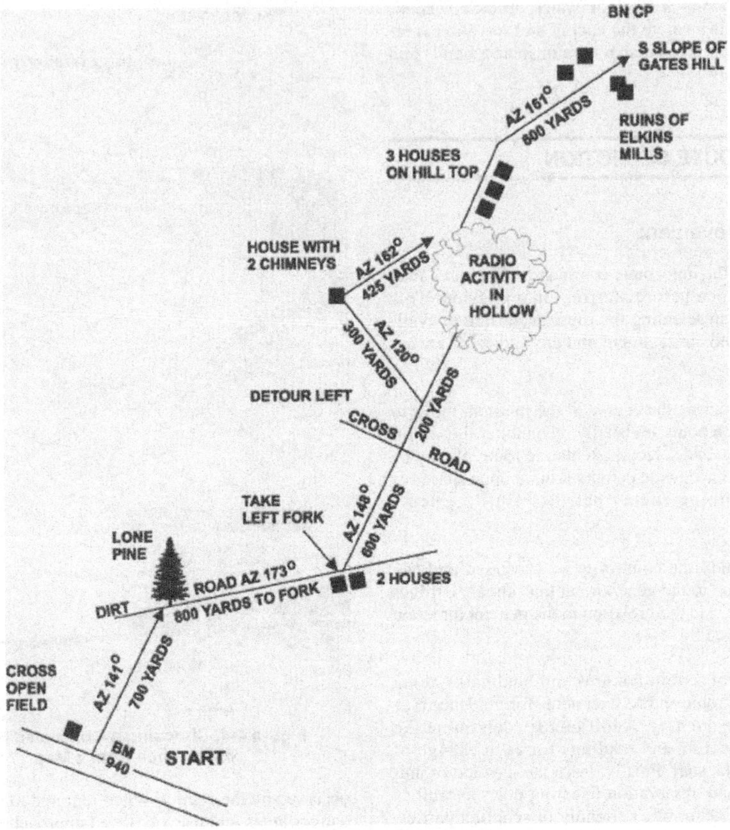

Figure 4-5. Proposed Route Sketch.

CHAPTER 5. NIGHT SCOUTING

Night scouting presents many of the same problems encountered in day operations-such as cover, concealment, movement, and camouflage-as well as additional considerations. Knowledge of human eye construction and operation will enable maximum advantage under night conditions or poor visibility.

5001. NIGHT VISION

Certain parts of the eye correspond to parts of a simple camera (see fig. 5-1). The lens focuses light entering the eye similar to a camera lens. The iris (colored part of eye) corresponds to the diaphragm of a camera, opening and closing to regulate the amount of light entering the eye through the pupil. The retina corresponds to camera film. Light rays strike the retina, form an image, and cause an impression to be transmitted to the brain through the optic nerve. In a camera, the image is formed and fixed on film.

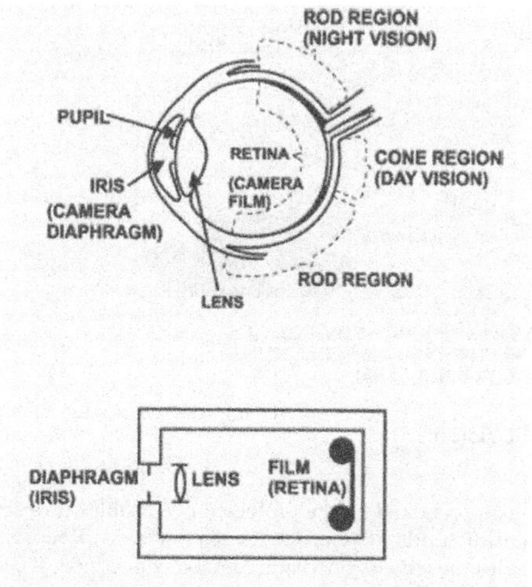

Figure 5-1. The Eye is Like a Camera.

The retina is composed of cone cells and rod cells, so-called because of their shapes. Cone cells distinguish color, shape, and sharp contrast. Because they are activated by light conditions, they are blind during periods of low illumination. Rod cells produce a chemical substance called visual purple that makes them active in darkness, low illumination or night conditions. Rod vision distinguishes black, white, shades of gray, and general outlines.

Principles

To effectively "see" at night, the principles of night vision dark adaptation, off-center vision, and scanning are applied.

Dark Adaptation

Allowing the eyes to become accustomed to low levels of illumination is called dark adaptation. It takes the rod cells about 30 minutes to produce enough visual purple to activate them and enable the eye to distinguish objects in dim light. This may also be accomplished by staying in a red-lighted area, or by wearing red goggles for 20 minutes, followed by 10 minutes in darkness (which allows the pupils to open wide). This method saves valuable time by allowing Marines to be in a lighted area to receive orders, check equipment, or perform some other function before moving into darkness.

Off-Center Vision

The technique of focusing on an object without looking directly at it is called off-center vision. When looking directly at an object, the image is formed on the cone region, which is not sensitive at night (see fig. 5-2 on page 5-2). When looking slightly to the left, right, above or below an object, the image is formed on the area of the retina containing rod cells, which are sensitive in darkness. The most sensitive area varies in individuals, but is usually found by looking 6 to 10 degrees away from an object; in effect, out of the corner of the eye (see fig. 5-3 on page 5-2).

Scanning

Off-center vision used to observe an area or an object is called scanning. When using rod vision, the visual purple in the rod cells bleaches or blacks out in 4 to 10 seconds and the object observed disappears. As the visual purple in the rod cells in one area bleaches out, the eyes must slightly shift to use fresh rod cells. Eyes

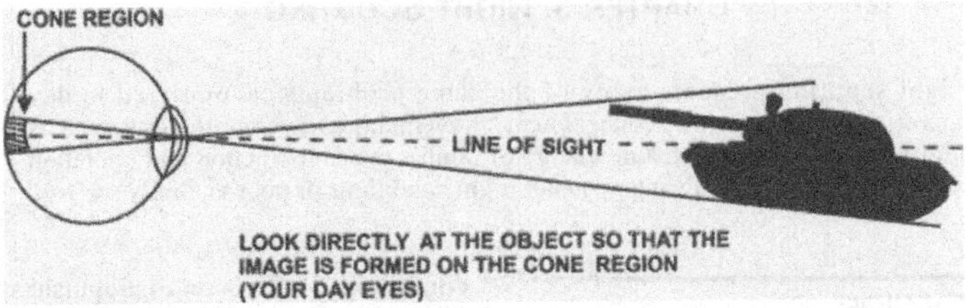

Figure 5-2. Day Vision.

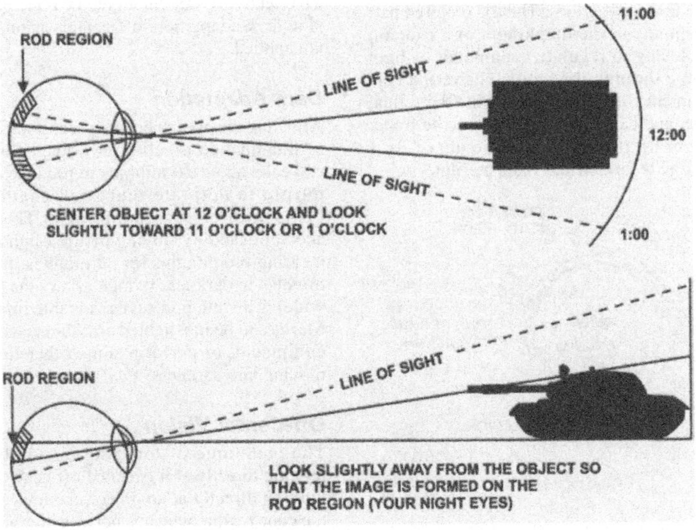

Figure 5-3. Night Vision.

should be moved in short, abrupt, irregular movements over and around the target (see fig. 5-4).

Preserving Night Vision

Night vision is quickly destroyed if bright light is allowed to enter the eye. When entering a lighted area or when observing in a temporarily lighted area (illumination, flares), one eye should be closed and covered to preserve its night vision. When the light goes off, fades or the lighted area is exited, the night vision retained by the protected eye enables it to see until the other eye adapts to the darkness. Red light helps preserve night vision, but like white light, it can be observed at long distances.

Factors that decrease night visual acuity include fatigue, lack of oxygen, long exposure to sunlight, alcohol, nicotine (within the past 48 hours), and age. When night vision has been attained, straining will not improve effectiveness; however, practice identifying objects at night will improve perception.

5002. APPEARANCE OF OBJECTS

Darkness not only makes objects difficult to see but also changes their appearance, distorts size, and blots out details. A tree visible against the night sky appears smaller than in the daytime because the twigs at the end of branches cannot be seen. A scout must train to identify objects by block outlines at night and cannot rely on details visible in daylight. Binoculars enlarge objects or parts of objects otherwise too small to be seen and help identify objects already spotted. Night observation devices increase night visibility and should be used whenever possible.

5003. SOUNDS

At night, sounds become very important. By listening, a scout gains information about the enemy and by exercising care, keeps information from the enemy. A scout stops frequently to listen. Scouts must listen for long periods in perfect silence. Hearing is amplified with the mouth open; removing the helmet will reduce sound distortion. Sounds are transmitted a greater distance in wet weather and at night than in dry weather and in the daytime. By holding the ear close to the ground sounds of people walking and vehicles moving can be heard. Sound travels approximately 370 meters a second. When a flash from a fired weapon is observed, the range to the weapons can be easily estimated by counting the time interval between the flash and hearing the report. For example, counting to three (one thousand one, one thousand two, one thousand three), indicates the distance is 1,110 meters. The cadence is determined by actual practice at known ranges.

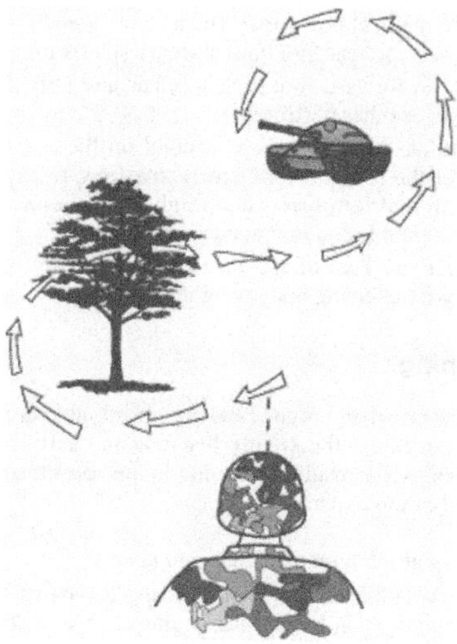

Figure 5-4. Scanning

5004. SMELLS AND TOUCH

A scout's sense of smell can warn of enemy fires, cooking, motor parks, gasoline and diesel engines, and bodies of water. A scout must feel and recognize objects in the dark, and adjust and operate equipment quietly by sense of touch.

5005. CLOTHING AND WEAPONS

All loose clothing must be secured (string or tape can be used) to prevent snagging on barbed wire, brambles, and brush. Helmet covers are worn to muffle sounds made by low branches.

The belt buckle should be turned around to the side in order to move in a prone position without scraping the buckle against stones or hard surfaces. Identification tags can be taped together to prevent rattling. Hands, face, and neck can be blackened so skin does not reflect light or appear as white spots in the darkness. (Refer to para. 4002.)

Scouts are normally armed with rifles. Rifle slings should be taped to prevent rattling. All weapons parts should be checked for glare elimination measures.

5006. CONCEALMENT

Although total darkness provides concealment, scouts must observe the same principles of concealment during moonlight conditions as in the daytime. Scouts should assume enemy employment of night observation devices and observe the principles of night movement such that presence will not be disclosed by noise when close to the enemy.

5007. AIDS TO NIGHT SCOUTING

Aids to night scouting include the following:

- Carry out scouting missions close to or within hostile positions on dark or rainy nights.
- Stifle a threatening sneeze by pressing fingers upward against the nostrils.
- Stifle a threatening cough by applying slight pressure with the finger on the Adam's apple.
- Stop a ringing sound that interferes with hearing by yawning.
- Speak softly rather than whispering if voice communication is necessary.
- Move boldly and rapidly when taking advantage of any sound-such as shelling, rustling wind or distant firing-to push forward if firing is taking place.
- Avoid shell craters and depressions in damp weather conditions if the enemy has employed chemical munitions.
- Move the eyes constantly; concentrating on one object too long will strain them.
- Take notice of the enemy's use of flares. When the enemy employs flares, few enemy patrols are apt to be out; when flares are not employed, the enemy's patrols are likely to be numerous.
- Drop to a prone position upon hearing a flare being fired and before it illuminates. Remain motionless while it is burning. If open or moving when a flare bursts in the air, freeze or drop quickly in the split second after the flare illuminates while the enemy is blinded. You are an easy target for the enemy if the flare bursts in the air or on the ground behind you. Never look at a flare. If you activate a trip flare, drop to the ground and crawl away from the illuminated area.
- Consider all patrols or individuals encountered as hostile until proven friendly. If encountering someone, crouch low, silhouetting the approaching person against the sky. At the same time, make yourself an indistinct target in case the person encountered is an enemy.
- Return fire only to avoid capture if fired on when close to enemy positions.

5008. AIDS TO NIGHT MOVEMENT

- Aids to night movement include the following:
- Move silently.
- Advance in stealthy legs. Each leg should follow some terrain feature that serves as a guide. When there are no terrain features to serve as guides, move in a straight or nearly straight line from one defined point to another, or maintain direction by using a compass.
- Avoid running, except in an emergency.
- Take advantage of sounds that may distract the enemy.
- Fall silently without making an outcry.

Walking

When walking at night—

- Place the heel down first. Balance the weight of the body on the rear foot until a secure spot is found.
- Lift the forward foot high to clear any stiff grass, brush, or other obstruction.
- Continue to balance body weight on the rear foot, lower the forward foot gently, toe first, to explore the ground for objects that might make noise. Step over fallen logs and branches, not on them.
- Lower the heel of the forward foot slowly; gradually transferring body weight to that foot.

Creeping

The low crawl and high crawl are not suitable at night when very near the enemy because an easily heard shuffling noise results. Creeping is the recommended method of movement:

- Creep at night on the hands and knees.
- Use your hands to feel for twigs, leaves or other substances that might make a noise. Clear a spot to place your knee. Keeping your hand at that spot,

bring your knee forward until it meets your hand. Then place your knee on the ground and repeat the action with the other hand and knee.

- Lay the rifle on the ground at your side and clear an area for it. Lift the rifle up and move it forward. Movement is slow and tedious, since it must be done silently.

Hitting the Deck at Night (Right-Handed Shooter)

To safely hit the deck at night from the standing position—

- Advance your left leg, place the butt of the rifle in your right armpit with the hand remaining on the pistol grip, and grasp it with the right hand at the balance.
- Quietly drop down on the right knee and left hand.

- Move the left leg carefully to the rear, and then move the right leg to the rear.

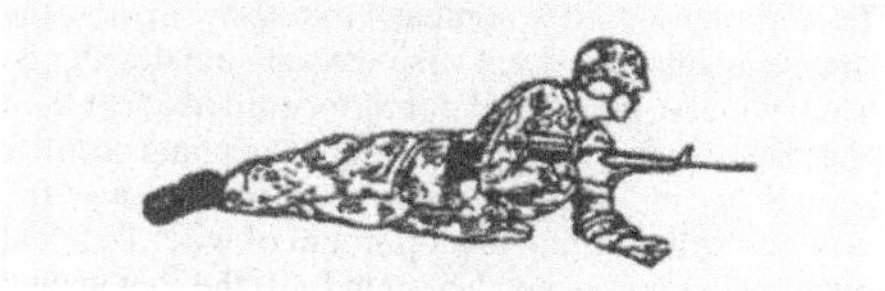

- Lie flat on the ground, or take up a firing position if necessary.

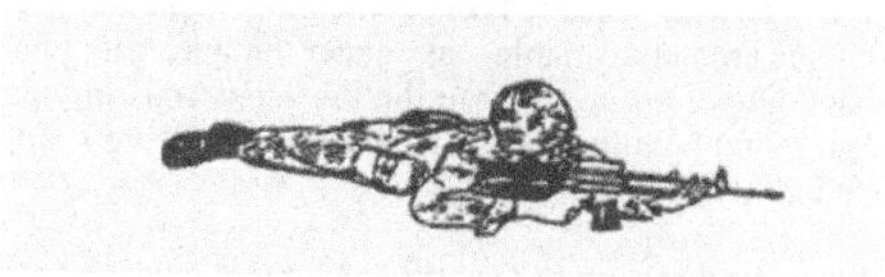

Wire Obstacles

A mission often requires a scout to pass through and work behind enemy positions. To accomplish this, the scout must be able to quietly pass through enemy wire obstacles and cross trenches. Cutting a gap in wire is time-consuming. If possible, walk over the low bands of enemy wire and crawl under the high bands (see fig. 5-5). Avoid movement along wire barriers, as enemy

Figure 5-5. Crossing Wire Silently at Night.

covering fires are generally planned parallel to them to take advantage of canalization and enfilade fire.

To step over low wire at night, crouch low to view the strands against the sky. Grasp the top strand with one hand; with the other hand, reach forward and feel for a clear spot for foot placement without stepping on other strands or any object apt to make a noise. Raise the body up, still grasping the top strand of wire. To avoid catching the foot in another strand, lift the foot up and over, passing it close to the hand grasping the wire.

If a high wire obstacle is encountered at night and wire cutters are not available, pass under the wire with your back on the ground. Grasp the lowest strands in your hands and hold them clear of the body while you slide under them.

When cutting wire and working solo, cut a wire near a post (see fig. 5-6), then dispose of all but one loose end. Grasp the wire close to a post and cut between your hand and the post, muffling the sound and keeping the loose wire in your grasp. When cutting wire in tandem, one firmly holds the wire with the hands positioned close to the cutters, in order to muffle the sound and prevent the loose ends from flying back, while the other one cuts. In both instances, the loose ends of the wire are bent back to form a passage.

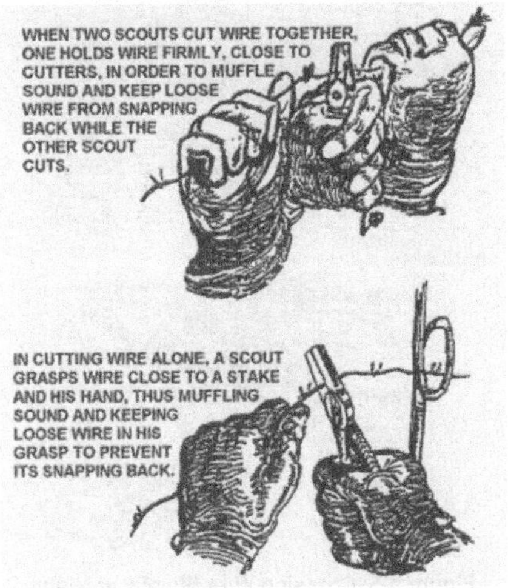

Figure 5-6. Cutting Wire Silently at Night.

Wrap a sandbag around the wire cutters and wire to deaden the sound.

Do not cut a complete gap in the wire; cut only the bottom wire(s). Leave the top wire(s) intact to lessen the chance of discovery by the enemy.

Crossing Trenches

Before approaching a trench, wait outside the trench for awhile and listen. Do not enter or cross a trench near its junction with a communication trench. Crawl silently up to the edge of the trench and look into it. Remove all loose dirt and rocks from the edge. If it is a narrow trench, spring up and jump across, sinking quietly to the ground on the other side and remaining there a moment to listen before proceeding. If the trench is wide, climb silently and slowly down into it and out the other side, using the revetment for support (see fig. 5-7). Do not enter enemy trenches unless it is absolutely necessary in order to accomplish the mission. Ordinarily, work is better accomplished from outside the trench. Sentries usually pay more attention to sounds in front of them; therefore, if it is necessary to enter a trench, cross it first at the place where enemy observation is restricted, then approach from the rear.

5009. LOCATING AND PLOTTING THE ENEMY AT NIGHT

For night work, a scout must understand the use of a lensatic compass. (Refer to para. 2002.) Using the lensatic compass, the scout can guide the platoon into position, locate adjoining elements of the command, keep direction when on patrol, determine the location of gaps in the enemy wire and the position of enemy out guards.

Locating Gaps in Enemy Wire

When searching for gaps in enemy wire, at least two lensatic compasses are needed: one to register the gap in the wire and the other for navigation. When a gap in enemy wire is located, lie outside the gap, keeping a distance of 10 meters from the barbed wire. Sight with the lensatic compass on a prominent point on the skyline in line with the gap. In selecting the prominent point in the skyline, pick one that appears on the map (i.e., hill mass, house, road junction). If the only prominent point available is one not identifiable on a map (i.e., a tree, destroyed vehicle, enemy position),

Figure 5-7. Crossing Trenches Silently at Night.

use it. The exact location of this point can be fixed the next day by visual reconnaissance of the area from an observation point. When the needle rests—

- Clamp it in place by lowering the eyepiece to the closed position.
- Rotate the movable bezel ring until the luminous line is directly over the north end of the needle. The azimuth of the gap from the prominent point is now registered.

The compass is carried back without further adjustment. The azimuth setting can be recorded later on a map (see fig. 5-8).

Locating Enemy Out Guards

At night, locate enemy guards by their sounds and failure to maintain light discipline. As sounds of the enemy are heard, and/or observations of the enemy made, shoot an azimuth with one compass. When the needle comes to rest, clamp the compass by lowering the eyepiece to the closed position.

Figure 5-8. Locating a Gap in Enemy Wire at Night.

Note the time and nature of each sound, the estimated distance, and which compass was used to fix the location. Plot this data on a map then wait until the debrief to turn in the notes and compasses. See figure 5-9.

Figure 5-9. Locating Enemy Outguards at Night.

50010. ROUTES OF MOVEMENT

Prior to beginning a night missions, a scout studies the ground in detail from an observation point, air photos, and a map during daylight. The route of advance should be below the skyline. Avoid becoming a silhouette (see fig. 5-10).

Unless the moon is bright, avoid, if possible, passing through woods, ditches, ravines, and brush, because noises of movement may lead to discovery. If the enemy is known to have night observation device capability, avoiding these kinds of terrain may not be possible. To avoid enemy ambushes, return by a different route from the advance and change routes on successive nights.

Figure 5-10. Difference Between Correct Routes by Day or Night.

CHAPTER 6. OBSERVING AND REPORTING

When sent out on a mission, a scout's duties are to observe and report. the first five chapters covered scout protection measures and movement in enemy territory. This chapter discusses performing the assigned mission and reporting the mission after its completion.

6001. OBSERVATION POSTS

Positions

The following guidelines apply to observation posts. See figure 6-1.

- When selecting observation posts, scouts should chose the least prominent position.
- Scouts may occupy one or more observation posts.
- An observation post should not be manned for more than 24 hours.
- A selected observation post should be observed for 10 to 15 minutes to ensure it is not occupied.
- Scouts move to the chosen observation post by a concealed route.
- If the post is located on a hill, crawl to a position where the skyline is broken.
- If a tree is used, the position should have a background so as not to be silhouetted against the sky while climbing or observing.
- When leaving the observation post, a different route from that of the approach should be used.
- If a radio is used, its antenna should be located to provide clear communication to the controlling commander but masked from enemy observation and direction-finding equipment. Upon departure, scouts should remove the antennae from the observation post so as not to give away the position.

Observing

Using all senses available, be particularly alert for movement, objects, sounds, and smells inappropriate to the surroundings. While observing, avoid all unnecessary movement. If observing from a building, keep back from doors and windows.

In daylight, look first at the ground nearest you. Begin observing close to your post and search a narrow strip 50 meters or less deep, going from right to left parallel to your front; then search from left to right a second and similar strip farther away but overlapping the first. Continue to observe until the entire field of view has been searched (see fig. 6-2 on page 6-2).

At night, use a night observation device. If one is not available, search the horizon with short, jerky movements, and short pauses. Look a little to one side of an object and then to the other. Lower the head close to the ground to view the object more clearly. Use low-powered field glasses to increase sight range.

Figure 6-1. Method of Approaching an Observation Post.

Figure 6-2. Method of Searching Ground.

6002. REPORTING

It is imperative that the scout accurately and completely report who, where, when, and what was observed to the proper authority upon mission completion. In most cases, a scout will not be equipped with a radio.

Verbal Reports

Verbal reports should be made when writing is impractical, when the information is not complicated or when the enemy is likely to intercept a messenger carrying a written message.

Written Messages

Written messages, preferred to verbal reports, are recorded in message book blanks issued for that purpose (see fig. 6-3) and delivered to a higher authority as soon as possible. The NATO spot report (SPOTREP) should be reviewed as the written message is a NATO format.

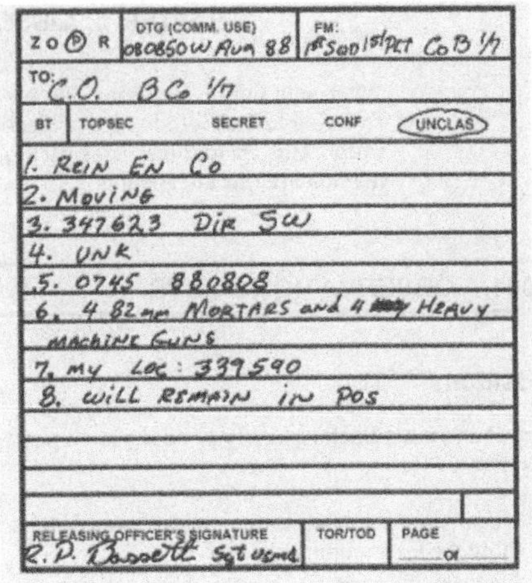

Z O P R	DTG (COMM. USE) 080850W AUG 88	FM: 1st SQD 1st PLT Co B 1/7
TO: C.O. B Co 1/7		
BT	TOPSEC SECRET CONF UNCLAS	

1. REIN EN CO
2. MOVING
3. 347623 DIR SW
4. UNK
5. 0745 880808
6. 4 82mm MORTARS and 4 HEAVY MACHINE GUNS
7. MY LOC: 339590
8. WILL REMAIN IN POS

RELEASING OFFICER'S SIGNATURE	TOR/TOD	PAGE
R.P. Bassett Sgt USMC		of

Figure 6-3. Sample Field Message.

The message body is brief, accurate, and clear; facts and opinions are distinguished. If secondhand information is reported, its source is included. Reports include all information of value, first about the enemy, and then about the originator. Information about the enemy should cover—

- Size and/or strength.
- Actions or activity.
- Location and direction of movement.
- Unit identification. (The designation of the enemy unit may be derived from unit markings, uniforms worn or through prisoner interrogation.)
- Time of observation.
- Equipment and weapons.

Messages are printed in block letters. Individual items of information are numbered and separated into paragraphs. If doubt exists as to message receipt by the commander, a summary of its contents is included in the next message. Information about the originator or writer should cover—

- Location at the time of enemy observation (reference to an important terrain feature, by map coordinates, by the back azimuth from each of two definitely located points, or the back azimuth and distance from one known point).
- Intentions. (Remain in position? Continue on the mission? Take other action?)

The message is carefully reread and if possible read by another person to ensure understandability. If a messenger is used, the messenger must read and understand the message in order to answer any questions the commander might have.

Sketch

Information difficult to describe may be given accurately on a simple sketch. The sketch may give all the necessary information or it may be used to supplement a written message. A military sketch is generally one of two types: simple or panoramic. The simple sketch is easily made and read.

Figure 6-4 shows a simple sketch that has been included in the message itself. The sketch may be on a separate sheet of paper, but all of the necessary information must be contained in either the sketch, the message or both.

The panoramic sketch is a picture of the terrain's elevation in perspective, as seen from one point of observation. Although a panoramic sketch is not difficult to create, skill and training are necessary to enhance usefulness. Figure 6-5 on page 6-4 illustrates how to make a panoramic sketch.

Overlay

The same information sent back to higher headquarters on the sketch may be sent on an overlay, if the sender and the person to whom the message is to be sent have copies of the same map. Figure 6-6 on page 6-6 illustrates a simple overlay. The overlay is drawn on transparent paper as follows:

- Orient the map and place it on a hard, flat surface.
- Place the transparent paper over the part of the map of the object or information to be transmitted and hold the paper in this position.
- Orient the overlay to the map by tracing in the intersecting grid lines at two opposite corners of the overlay. Write the correct number designation on the overlay. The cross made by the intersection is called a *tick mark* and enables the receiver to locate the exact area on the map covered by the overlay.
- Sketch the object seen or the information to be transmitted on the tracing paper (the sheet on top of the map) in the exact location it would appear on the map (the sheet underneath the tracing paper).

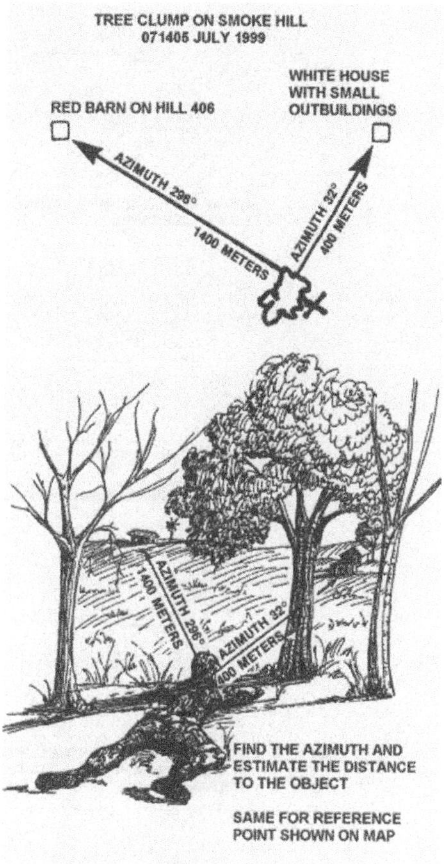

Figure 6-4. Making a Simple Sketch.

Explanatory notes are annotated in the overlay's margin, arrows point to the objectives.

- Indicate with an "X" and an appropriate explanation the position from which the observer saw the object or obtained information.
- Include title and scale of the map from which the overlay was made, date and hour the information was obtained, and signature of the observer on the overlay in the lower right-hand corner.

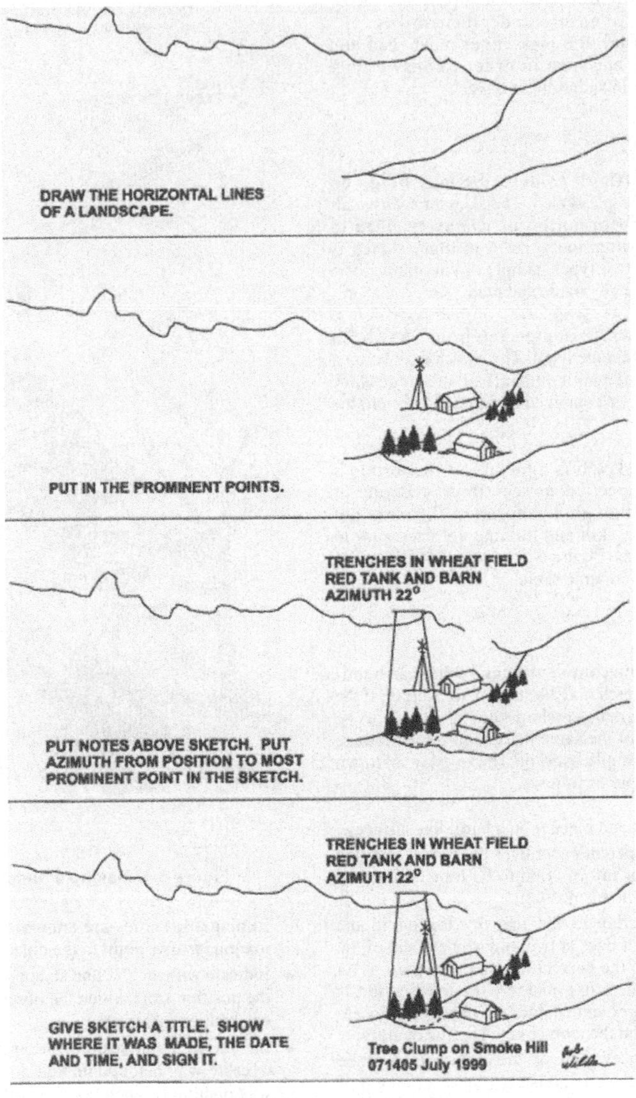

Figure 6-5. Panoramic Sketch.

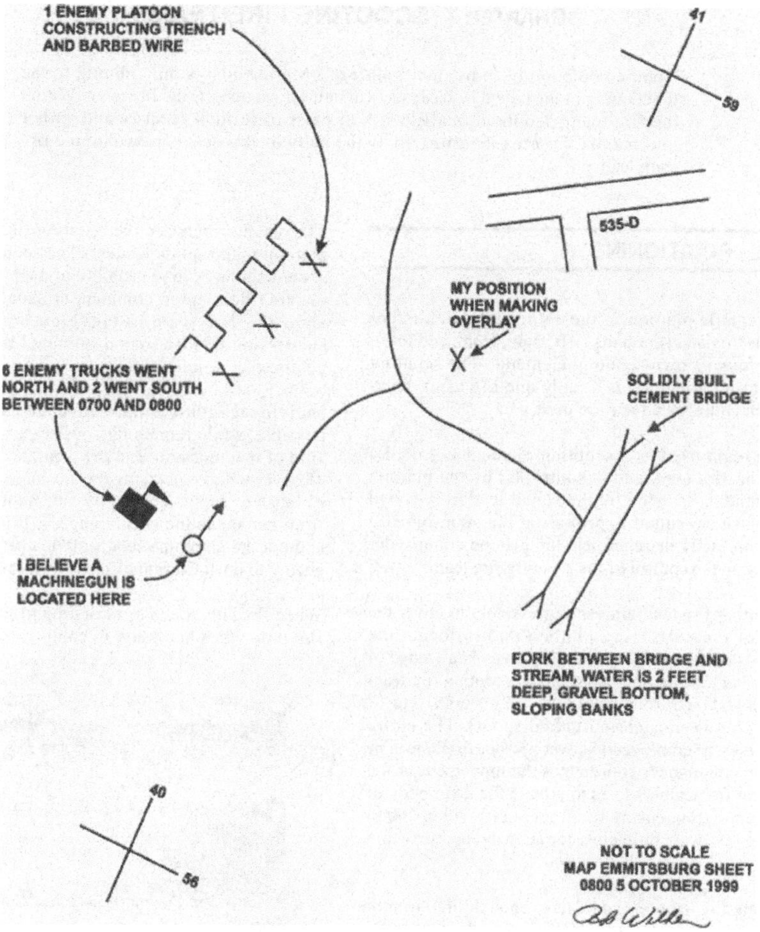

Figure 6-6. Simple Overlay.

CHAPTER 7. SCOUTING FIRE TEAMS

Upon completion of individual training, a Marine adapts this training to the functioning of the unit. The basic tactical unit for scouting is the fire team. Within the fire team, scouts normally work in pairs to furnish security and gather information. Information gathered by the individual scouts is passed to the fire team leader.

7001. POSITIONING

When a rifle platoon in the approach march is not preceded by adjacent units (left, right, front, and rear), it employs its own scouting elements. The scouting element for a platoon is usually one fire team; however, an entire squad may be used.

A fire team used as a scouting element is called a scouting fire team and is controlled by the platoon commander, assisted by the squad leader. A squad leader whose squad is providing the scouting fire teams normally marches near the platoon commander to assist in the control of the scouting fire teams.

A scouting fire team moves aggressively to cover the front of the advancing platoon and to locate the enemy's position(s). It generally moves in a wedge or skirmisher's formation. Normally, a scouting fire team is deployed on a frontage of 50 to 75 meters (10 to 17 meters between each individual scout). The entire squad may be employed to cover a wider frontage. The platoon commander coordinates the movement of the scouting fire team(s) so as to protect the main body of the platoon from enemy fire from points within 400 to 600 meters away, or in close terrain from points within the limits of enemy observation.

Scouting fire team(s) should have enough firepower to overcome resistance from small enemy advanced posts and patrols; the intent is to make enemy riflemen and machine gunners open fire and disclose their positions. Without scouts in advance, the platoon may move into areas where enemy fire may prevent further advance or maneuver and inflict heavy casualties (see fig. 7-1).

Scouting fire teams are covered by the platoon or, when the platoon is masked, the fire team leader ensures individual scouts are maneuvered and coordinated so that the fire team covers its own advance. The fire team leader constantly watches for signals from the platoon commander and remains in visual contact at all times.

The distance between the scouting fire team and the platoon is terrain-dependent. The scouting fire team should not be beyond visibility of the platoon. In open terrain, the platoon commander usually directs the scouting fire team to move by bounds along a succession of locations designated by the platoon commander as intermediate objectives.

Individual scouts should advance as stealthily as possible, while remaining consistent with their mission of reconnaissance to the front, taking advantage of cover without delaying the advance. An occasional glimpse of scouts constantly advancing over a wide front can make the enemy uneasy. It is this activity, and not the target the scouts offer, that may cause the enemy to open fire and disclose its location.

When fired upon, scouts must drop to cover and return fire only when necessary to complete the mission. If

Figure 7-1. Position of Scouts Preceding an Attacking Platoon.

necessary, one scout reports back to the fire team leader who informs the platoon commander.

7002. LOCATING ENEMY POSITIONS

The key terrain of defensive positions are those points that afford extended observation over the ground where the attack must advance. The enemy will place machine guns and infantry to defend critical points.

Members of a scouting fire team preceding an attacking platoon identify the probable enemy infantry and machine gun positions (see fig. 7-2). They use concealment and cover to conduct their advance in order to discover the exact location of enemy positions.

7003. ACTION WITH AN ATTACKING PLATOON

The scouting fire team reconnoiters to the front of the advancing platoon. As soon as the scouting fire team leader indicates the area is secure, the platoon advances and the scouting fire team moves forward. Squads within the platoon advance by bounds; at least one squad is positioned to support the other(s) by fire. Successive positions along the line of advance are selected and designated by the platoon commander as intermediate objectives, and reconnoitered by the scouting fire team before occupation. By conducting proper reconnaissance, surprise by the enemy or movement in the wrong direction may be prevented.

Movement

The distance between the scouting fire team and the front of the main body of the platoon is dependent on the mission, enemy, terrain and weather, troops and support available, and time available (METT-T). In close terrain, such as dense woods, the scouting fire team's movements closely resemble those used for night operations. In approaching houses, woods, and villages, one scout of each pair covers the other while the latter reconnoiters (see fig. 7-3).

A scouting fire team moves forward aggressively to cover the front of the advancing platoon, usually adopting either skirmisher's or wedge formation in order to be prepared to go into action immediately and to cover a wide frontage of 50 to 75 meters.

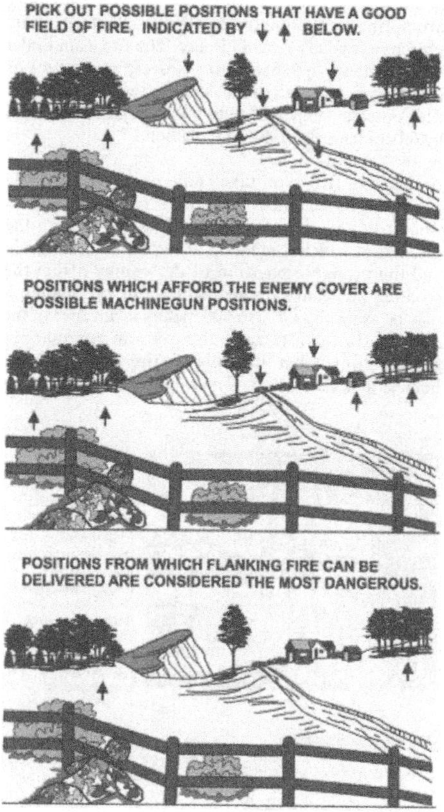

Figure 7-2. Assessing Probable Enemy Machine Gun Positions.

As a scouting fire team advances in open terrain, it is supported when possible by elements of the platoon; in close terrain, by mutual support within each fire team. Mutual support within the fire team is accomplished by the fire team leader and the automatic rifleman forming a team that supports by fire the advance of the rifleman and assistant automatic rifleman until they reach a location designated by the fire team leader. The rifleman and assistant automatic rifleman then support by fire the movement of the team leader and automatic rifleman. These successive points to which the sub elements of the fire team move are normally designated as fire team intermediate objectives by the fire team leader. The fire team leader sets as many fire team intermediate objectives as necessary to maintain mutual support within the team. This process is repeated until the team can be covered by other elements of the rifle platoon.

A scouting fire team takes advantage of available cover and concealment without delaying its advance. The orders of the platoon commander govern the distance at which it precedes the platoon. The terrain and the probable position of the enemy affect the scouting fire team's distance in front of the platoon. It may be as much as 400 to 600 meters in advance of the platoon. In open terrain, the platoon commander usually directs that the scouting fire team move by bounds to a succession of intermediate objectives. In close terrain or conditions of limited visibility, the scouting fire team is normally ordered to precede the platoon at the limit of visibility, maintaining visual contact with the platoon commander.

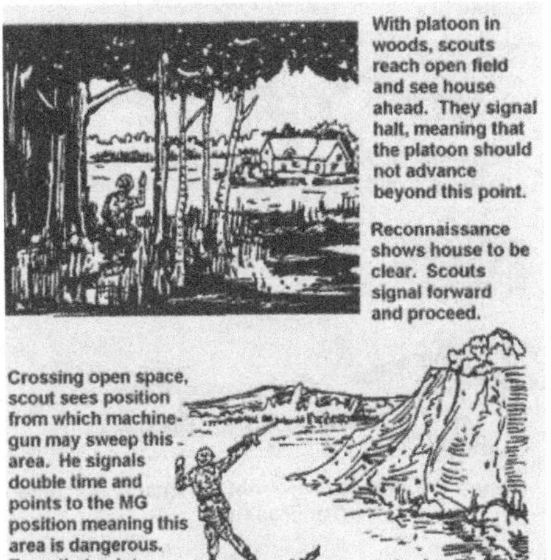

Figure 7-3. Conduct of Scouts Preceding a Platoon.

Action in Woods

When a scouting fire team is directed to advance over open ground to the edge of a woodline, two members of the team, preferably the rifleman and assistant automatic rifleman, reconnoiter inside the woodline while the remainder of the fire team covers them. It is not recommended they separate until finished with their reconnaissance of the far side of the danger area. Both members staying together can cover the same area using a zigzag reconnaissance and they are better equipped to overcome any opposition.

In heavy underbrush and/or poor visibility, the rifleman and assistant automatic rifleman proceed into the woodline together for 50 to 60 meters. The two then separate, searching out either flank to the first high ground or limits of observation, probably 50 to 100 meters (see fig. 7-4 on page 7-4). After the initial search and out posting the limit of advance, the remaining scout signals the fire team forward.

In light underbrush and/or good visibility, the assistant automatic rifleman remains at the edge of the woodline while the rifleman searches the woodline. (see fig. 7-5 on page 7-5). The rifleman searches the woodline in a zigzag pattern, reports back to the assistant automatic rifleman, then moves to an outpost position at the limit of advance. In turn, the fire team leader signals the platoon commander that it is safe for the platoon to move forward.

The fire team leader then moves the remainder of the fire team into the woods, joining up with the forward scout manning the outpost. The scouting fire team occupies and holds a line 50 to 75 meters within the woods and observes toward the direction of movement until the platoon closes up. The scouting fire team leader awaits further word from the platoon commander before moving the team further into the woods.

When directed, the scouting fire team leader moves the team forward until they reach the far edge of the woods. The team is held at the edge of the woods and the fire team leader notifies the platoon commander of the situation. The platoon commander moves the platoon to a position where it can cover the scouting

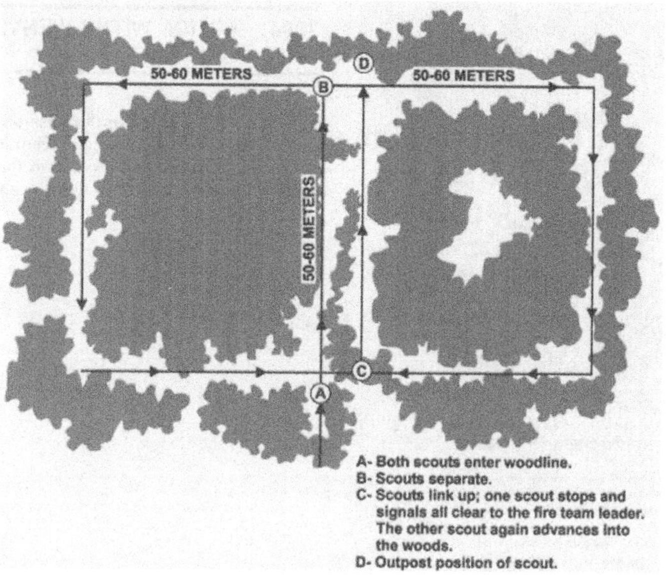

Figure 7-4. Searching Edge of a Woodline (Dense Underbrush/Poor Visibility).

fire team as it exits the woods and directs the team leader to move out and continue the scouting mission.

A scouting fire team passing through woods ahead of its unit maintains a distance allowing visual and oral communications. If an obstacle is encountered, reconnaissance to its front and flanks must be carried out. When advancing along a road or path, scouts precede the platoon to provide necessary protection and to prevent surprise fire on the platoon. When crossing a road or path, they reconnoiter well to the flanks before signaling "all clear" to the platoon.

The scouting fire team will not exit the woods until the arrival of the platoon commander, who will then be given an opportunity to alter the disposition or direction of march. The point where the platoon exits the woods is considered a danger area where the platoon is vulnerable to enemy fire. The scouting fire team is sent ahead to reconnoiter the danger area, as well as the next area to be occupied by the platoon. They signal back whether conditions require a halt, an advance or a quick rush across the open area. The scouting fire team leader must be continually on the lookout for signals from the rear.

Action Under Fire

When a scouting fire team is fired upon, they immediately take cover, locate targets, and return fire. The scouting fire team leader then determines—

- Location of enemy (range and reference points).
- Extent of position (location of flanks).
- Types of positions (obstacles, bunkders, fighting holes, etc.).
- Number of enemy.
- Enemy weapons (machine guns, mortars, tanks, etc.).

The platoon commander assesses the situation as quickly as possible based on the limited information obtained. Usually the platoon commander brings up the remaining squads, sets up a base of fire, and assaults the enemy position. Should the enemy position prove too strong for the platoon, the platoon remains engaged with the enemy as a base of fire until the remainder of the company is committed to clear the enemy resistance.

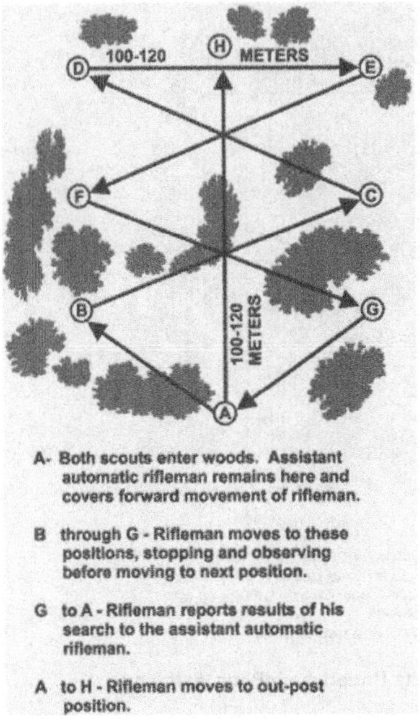

Figure 7-5. Searching Edge of a Woodline (Light Underbrush/Good Visibility).

7004. ACTION WITH AN ENVELOPING UNIT

When a platoon is given the mission to envelop an enemy position, a scouting fire team is employed for protection and reconnaissance in the same way as when the platoon is advancing in the approach march.

PART 2. INFANTRY PATROLLING

CHAPTER 8. FUNDAMENTALS OF INFANTRY PATROLLING

This chapter begins Part 2 and provides basic information about infantry patrols; specifically, their purpose, types, and missions. Infantry patrol training is also addressed and keys to successful patrolling are presented. Subsequent chapters of this part cover patrol organization, preparation, movement, and reconnaissance actions.

8001. DEFINITIONS

A patrol is a detachment of ground, sea or air forces sent out for the purpose of gathering information or carrying out a destructive, harassing, mopping-up or security mission (Joint Publication [JP] 1-02). The mission to conduct a patrol may be given to a fire team, squad, platoon or company.

8002. RELATION OF PATROLLING TO SCOUTING

Each patrol member must be knowledgeable in the principles of scouting and maintain membership of a larger team. To develop the teamwork skills required among the members of a patrol, additional training beyond the basic principles is necessary to become a well-trained scout. A patrol member must respond quickly to the decisions and orders of the patrol leader. There must be complete confidence among all members of the patrol and the confidence that they, as a team, will be successful in their mission.

8003. PURPOSE

A commander must have current information about the enemy and the terrain in order to employ the unit effectively. Patrols are an important means of gaining this information and are used to destroy enemy installations, capture enemy personnel, perform security missions or prevent the enemy from gaining information. Modern warfare places a high premium on effective patrolling because units have larger areas of operations and can be threatened from all directions. As distances between units increase, more patrolling becomes necessary to prevent infiltration by guerrillas or small enemy units, as well as to maintain contact with friendly adjacent units. Active patrolling by numerous small groups is needed to locate the enemy and gather information on the enemy's disposition, strength, morale, and weapons, as well as gather and confirm information about the terrain.

8004. TYPES OF PATROLS

Classification as to Mission

Reconnaissance Patrol

Reconnaissance patrols gather information about the enemy, terrain or resources. Relying on stealth rather than combat strength, they gather this information and fight only when necessary to complete the mission or to defend themselves. The distance covered by reconnaissance patrols varies based on the terrain and mission. The squad is ideally suited for reconnaissance patrol missions because of its relative small size and its experience of working together.

Combat Patrol

A combat patrol is a fighting patrol assigned missions that require engagement with the enemy in combat. Larger and more heavily armed than reconnaissance patrols, combat patrols have a mission to capture enemy documents, provide security, and capture or destroy enemy equipment and installations. Such action is ordinarily followed by a return to friendly positions. Regardless of the mission, the patrol reports any information concerning the enemy and terrain acquired during the accomplishment of the assigned mission. There are four types of combat patrols: raid, contact, ambush, and security (normally conducted by a Marine rifle platoon). A rifle platoon reinforced with crew-served weapons is normally considered the minimum size for contact, economy of force or ambush patrols. In some situations, such as the capture of a small enemy outpost, a rifle platoon could conduct a raid. However, a raid is a complex mission and, due to the organization of a raid force (command, reconnaissance, assault, support, security, and reserve elements), a rifle company is normally the smallest force assigned to a raid.

Classification as to Means of Movement

Foot Patrols

Movement by foot is the most common means; however, there are inherent disadvantages. Foot patrols travel slowly and carry limited quantity and types of equipment and supplies. Range and area coverage is relatively restricted. Foot patrols also have apparent advantages in that they have fewer restrictions as to terrain that can be covered; are more difficult for the enemy to detect; provide thorough coverage within limits of range; and are generally not inhibited by weather.

Motorized Patrols

Where terrain and road networks permit, a motorized patrol overcomes the inherent disadvantages of the foot patrol. Mechanized forces require patrolling units that can keep pace with them. However, motorized patrols are restricted to certain types of terrain, and tend to bypass areas that may be advantageous to and occupied by enemy infantry.

Waterborne Patrols

Waterborne patrols move over seas, lakes, rivers and streams, canals, and other inland waterways. The water is either used as a medium of entry to an objective area or is the actual patrol route. Waterborne patrols are limited by the location of water routes in the terrain and tend to bypass areas that may be advantageous to and occupied by the enemy.

Helicopterborne Patrols

Where terrain is extremely difficult or the enemy situation precludes the use of vehicle or motorized patrols, helicopterborne patrols are a method or means to conduct a patrol.

8005. TRAINING

Training is essential to successful patrolling. Premature and unordered actions by members of the patrol destroy coordination and control. Leaders are trained to issue their orders calmly to inspire confidence and discipline, and to avoid misunderstanding. Patrol members must work together and fight as a team. Training should develop the following skills:

- Expertise in handling individual and special weapons, and familiarity with enemy weapons that may be captured.
- Recognize camouflaged personnel, equipment, and defensive positions; ability to pick up fleeting targets fire the rifle from any firing position.
- Understand fire discipline and, after weapon firing, immediately change position (see fig. 8-1).
- Quick and accurate observation skills, and the ability to recall and transmit clearly and briefly, both orally and in writing.

Figure 8-1. Changing Position After Firing.

- Recognize and quickly respond to improvised signals, visually or by sound.
- Ability to swim with weapon and equipment.
- Use issued or improvised camouflage suits and garnish helmet in order to blend with the surroundings. Smudge face, hands, and any bright surfaces of weapons and equipment with some substance, such as mud or charcoal, to prevent the reflection of light.
- Silence self, equipment, and weapon.
- Use antimalarial and water purification tablets.
- Acclimation to temperature extremes.
- Develop a sense of direction and learn how to follow a course by compass, stars, sun, flow of streams, prominent terrain features, and by observing other natural phenomenon. Learn to determine the distance traveled from a known point and to keep a record of azimuths and the distance traveled on each azimuth (dead reckoning).
- Call for and adjust indirect fire assets.
- Familiarization with all communications assets and the use of field expedient antennae.

8006. KEYS TO SUCCESSFUL PATROLLING

Regardless of the category or means of conducting a patrol, the keys to successful patrolling are—

- Detailed planning. Every portion of the patrol must be planned, all possible contingencies considered.
- Productive, realistic rehearsals. Each phase of the patrol is rehearsed, beginning with actions in the objective area. Similar terrain and environmental conditions are used when conducting rehearsals.
- Thorough reconnaissance. Ideally, the patrol leader will physically conduct a reconnaissance of the route and objective. Photographs and/or maps will be used to supplement the reconnaissance.
- Positive control. The patrol leader must maintain positive control, this includes supervision during patrol preparations.
- All-around security. Security must be maintained at all times, particularly near the end of the patrol where there is a natural tendency to relax.

Chapter 9. Patrol Organization

Organizing a patrol is a two-step process: the general organization of the entire patrol and the task organization of various patrol elements. Normally, the nature of patrolling does not permit long preparation periods and rehearsals to specifically build a unit for each mission. Accordingly, the patrol leader must combine unit integrity considerations with proven concepts of patrol organization.

9001. GENERAL ORGANIZATION

The patrol leader establishes a patrol headquarters and elements to accomplish the mission.

The headquarters is composed of the patrol leader and the personnel who provide support for the entire patrol, such as a forward observer, corpsman, and radio operator.

The major subdivisions of reconnaissance and combat patrols are elements. The existing infantry structure (squads and fire teams) is reinforced as required. For example, a reinforced platoon tasked to conduct a combat patrol that will raid an enemy outpost could be organized as follows:

- Platoon headquarters (command element):
  - Patrol leader (platoon leader).
  - Assistant patrol leader (platoon sergeant).
  - Navigator.
  - Radio operator (company tactical net).
  - Radio operator (patrol tactical net).
  - Corpsman.
- The first squad (security element) provides security en route to the objective area (the point, flank security, and rear security) and at the objective area (the flanks and the objective rally point).
- The second squad (support element) supports by fire for the attack, covering fire for the withdrawal, and supporting fires to cover the crossing of danger areas.
- The third squad (assault element) provides the assault force to attack and seize the objective; searchers to clear the objective; pacers, compass man, navigator, and the assistant patrol leader en route and back from the objective area.

Any attachments a patrol may have (i.e., demolition team, scout snipers, and machine gun squad) will be added to the element that supports its function. For example, the demolition personnel should go with the unit conducting the attack, and scout snipers and machine gunners should stay with the support squad.

9002. TASK ORGANIZATION

The preceding paragraphs described the elements necessary for a patrol to accomplish its mission. These elements reflect the internal functions or tasks required for the patrol to succeed. Depending on the METT-T, there are various methods of grouping these elements together. Task-organization is the further subdivision of patrol elements into teams that are required to perform essential tasks. In creating teams, unit integrity of infantry units should be maintained.

The patrol is organized so each individual, team, and element is assigned a specific task, but capable and prepared to perform other tasks. This may not be possible for certain specialist tasks requiring a trained technician.

Chapter 10. Patrol Preparation

For a patrol to succeed, all members must be well trained, briefed, and rehearsed. The patrol leader must have a complete understanding of the mission and a thorough understanding of the enemy and friendly situations. The patrol leader should make a complete reconnaissance of the terrain to be covered (either visual or map), and must issue an order to the patrol, supervise preparations, and conduct rehearsals.

10001. MISSION

The mission assigned to a patrol must be clear and oriented toward one objective with a specific task and purpose. More than one primary objective or indefinite missions invites confusion, casualties, and failure.

10002. FACTORS INFLUENCING PATROL SIZE

The circumstances under which patrolling by infantry units is conducted make it necessary for combat patrols to be able to fight offensively, security patrols to defend themselves, and reconnaissance patrols to move quickly and only fight if necessary to break contact or defend themselves. The size of a patrol depends on METT-T.

Generally, a patrol should contain the least number of members needed to accomplish the mission. Combat missions ordinarily require larger patrols than reconnaissance missions.

10003. COMMANDER'S DUTIES

Determining Patrol Requirements

The need for conducting patrols derives from the commander's stated mission (issued by higher level commander) and other specified and implied tasks (secondary or supporting efforts necessary to accomplish the stated mission). This requires analyzing the unit's mission and determining the necessary reconnaissance and/or combat tasks that must be performed. By considering the mission, along with time available, the commander develops the overall concept of operations to include the patrol plan and the specific tasks for each patrol.

Assigning Units

When assigning patrol missions, maintaining the integrity of the existing unit while considering the skills and experience of the unit and its leader are critical factors to the infantry commander. To provide operational depth and equitable apportionment of hazardous assignments, the commander ensures that each of the subordinate leaders and units develops the skills and experience necessary to conduct successful patrols. Assignment of patrol units must consider the commander's concept of operations as a whole and the plans for subsequent employment of assigned forces after completion of the assigned patrolling mission.

Providing Adequate Time

The commander must allow the patrol sufficient preparation time by completing the mission analysis, estimate, and preparation of orders. The commander should use warning orders to alert subordinates to possible requirements and afford them the opportunity for concurrent planning.

Providing the Patrol Leader Information

The following information should be provided:

- A simple, straightforward explanation of the mission, particularly for night patrols.
- General routes (defined by checkpoints) or exact routes (defined by avenues of approach or other terrain features) to follow.
- Enemy composition, disposition, and strength.
- Location and activities of friendly troops.
- Outposts or other security elements through which the patrol is to pass.
- Terrain conditions.
- Missions and routes of other patrols.

- Time patrol is to depart and return.
- Method of reporting information while on patrol (radio, messenger), place where messengers are to be sent, and place where the patrol leader is to report upon completion of the patrol.
- The challenge and password to be used for exiting and reentering friendly lines.
- Special instructions such as locations to be avoided and essential elements of information higher headquarters is seeking.
- Report barrier/obstacle plan of friendly units if known; support available from friendly unit (e.g., medical evacuation [MEDEVAC], react force).
- Fire support available.

Providing Special Skills and Equipment

There will be situations when the unit assigned to conduct a patrol does not have the necessary technical skills or equipment organic to it to successfully accomplish the patrol mission. In these cases, additions (attachments) are made to the unit, such as—

- Machine gun and/or shoulder-launched, multipurpose assault weapon (SMAW) team and/or squad.
- Forward observers (mortar/artillery).
- Radio operator(s).
- Combat engineers.
- Tracked vehicle crewman to assess the ability to traverse the terrain.
- Corpsman.
- Other personnel (snipers, translators) as required.

Providing Miscellaneous Support

The commander must ensure that the patrol leader is provided with the food, water, ammunition, radios and batteries, maps, special clothing, and any other items required by the unit (includin attachments) needed for the mission. Post-patrol support such as debriefings by intelligence personnel must also be planned.

Reviewing the Patrol Leader's Plan and Preparations

Once the patrol leader receives the mission, conducts visual and/or map reconnaissance, and develops the plan, the commander confirms the patrol leader's understanding of the mission and plan for accomplishing it. This discussion between the patrol leader and commander ensures that the patrol leader understands the commander's intent and is properly prepared to provide the patrol leader an opportunity to ask for clarification or additional support, if required.

Debriefing the Patrol

Upon return of the patrol, the commander receives the patrol report at a debriefing attended by the patrol leader and all patrol members. The debriefing should be conducted as soon as possible following the patrol's return, while information is still fresh in the minds of the patrol members. A patrol report, based on the information collected during the debrief, is generated and forwarded to the next higher com-mander.

10004. PATROL LEADER DUTIES

The patrol leader organizes and prepares the patrol by using the six troop-leading steps to make the best use of resources available. These steps are to—

1. Begin planning.
2. Arrange for reconnaissance and coordination.
3. Make reconnaissance.
4. Complete the plan.
5. Issue the order.
6. Supervise.

Begin Planning

The patrol leader begins by evaluating all factors affecting the mission. He looks for possible courses of action that lead to a decision, and then transforms this decision into an order. The first step includes making an initial assessment and decision on using available time, issuing a warning order and initial preparatory tasks, and initiating his estimate.

The initial planning effort assesses the time, assistance, and information available, and plans the proper use of each. Time allowances include reconnaissance; completion of the estimate and order; troop preparation; and such briefings, rehearsals, and inspections as required before beginning the patrol.

The patrol leader reviews the mission and the attachments and/or support available and decides what preparatory efforts must begin immediately. Proper use of subordinates to manage these initial tasks during this period reduces preparation time and frees the patrol leader for proper planning and reconnaissance. A warning order is issued using a modified five-paragraph order format (situation, mission, execution, administration and logistics, and command and signal [SMEAC]) as a checklist. (Refer to app. A for specific information that a patrol warning order should contain within each checklist item.

Once the initial preparations are set in motion, the patrol leader begins estimating by analyzing the mission; considering the friendly, enemy, and operating environments; considering each course of action available against what the enemy might do; comparing the courses of action in terms of mission accomplishment, capabilities, and probable casualties; and choosing one that becomes the basis for concept and order. While the patrol leader should use and organize notes, estimates must be done quickly and accurately, particularly for immediate situations. To organize thinking, the patrol leader uses METT-T, which consists of:

- Mission—the mission assigned to the patrol and how it relates to the mission of the commander who is sending the patrol.
- Enemy—what is known or suspected of enemy presence and capabilities, habits and characteristics, and fighting techniques.
- Terrain and weather—including ground, vegetation, drainage, weather, and visibility.
- Troops and support available—friendly situation and support available.
- Time available—the constraints and impact of time on preparation and mission accomplishment.

The estimate begins with mission analysis, which is the most important part of the entire planning process. Here, the patrol leader considers the specified tasks of the mission assigned and identifies other significant actions (specified and implied tasks) that must be undertaken to accomplish the stated mission. These tasks are arranged in sequence of accomplishment. The tasks and sequence create the framework for developing courses of action for the patrol concept of operations.

Arrange for Reconnaissance and Coordination

The patrol leader arranges a personal reconnaissance to observe as far forward as possible and also coordinates with the appropriate commanders for the patrol's "passage of lines" (see paragraph 11001 for complete definition) and supporting fires. The patrol leader also coordinates with other patrol leaders who may be operating in the same or adjacent areas and requests that the commander assigning him the patrol mission coordinate the patrol action with adjacent commanders, local security, and night defensive fires, as appropriate. The patrol leader may delegate any or all of these arrangements to the assistant patrol leader if the patrol leader requires the time for planning.

Make Reconnaissance and Complete the Estimate

The patrol leader uses personal reconnaissance to answer questions that arise from the map reconnaissance and METT-T evaluation. Specific points include passage points, lanes through obstacles, locations of friendly listening posts and observation posts, possible approach and return routes, enemy positions (if any), and intermediate observation points on the way to the objective.

In selecting approach and return routes, the patrol leader chooses routes that best use concealment and avoid opposition and obstacles. To lessen the chances of ambush by the enemy, the return trip is planned along a different route. In addition to personal reconnaissance and review of the map and aerial photographs, the advice of other patrol leaders who already are familiar with the terrain and the objective area should be considered.

After compiling information about the situation and possible time constraint, the patrol leader completes an estimate. The first step is developing courses of action, each of which will provide for movement to the objective area, mission accomplishment, and the return, based on the tasks and their sequencing identified in mission analysis. While the eventual concept of operations is presented in order of occurrence, the patrol leader must develop the courses of action by either backward or forward planning. In situations where the objective is well defined and there is sufficient information to plan the action for mission accomplishment (reconnaissance or combat), the patrol leader begins the scheme for accomplishing the mission at the objective and then, planning backwards, considers the options for getting there and back.

The following sections on movement to and return from the objective area, reconnaissance missions, and combat missions discuss methods and options available to the patrol leader in developing the courses of action. The principal variables between courses of action will be who, where, and how in the following:

- Patrol task organization.
- Routes to the objective area.
- Observation point(s) (reconnaissance patrol), ambush site, form of maneuver, type of ambush (combat patrol), fire support plan.
- Return routes.

The patrol leader then mentally considers the progress of each course of action (a map or simple sketch is a useful aid) against expected and unexpected enemy action. By comparing the options against each other and prospective enemy opposition, the patrol leader chooses the course of action that has the best chance of success. Includes in this mental preview process is the time to determine the patrol plan for unexpected contingencies (enemy attack/counterattack, casualty handling). These contingency actions, together with the selected course of action, become the patrol concept of operations.

Once the patrol leader determines the scheme of maneuver, the fire support required to accomplish the mission is addressed including the fire power organic to the patrol and what additional indirect fire support will have to be provided by other units. When planning for indirect fire support, the patrol leader considers the following questions:

- Will artillery, mortar, or close air support be required at the objective area (combat patrols)?
- What artillery and mortar targets exist along the routes to and from the objective area that can be employed by the patrol if it encounters the enemy during movement (reconnaissance patrols and combat patrols)?
- What additional fire support will be required to cover the patrol's movement from the objective area back to the friendly area once the enemy is aware of the patrol's actions at the objective area (combat patrols)?

The effect that casualties have upon the patrol depends upon many factors. Generally, more casualties can be expected in a combat patrol than in a reconnaissance patrol. A patrol may continue on to the objective carrying its casualties, send them back with a detail of Marines, abort the mission and return the entire patrol with the casualties, or call their parent unit for assistance.

Some factors that determine what action the patrol leader takes are: patrol's mission; unit's standing operating procedure for handling wounded; number of casualties and nature of their injuries; availability of aid, helicopters or other means of casualty evacuation. Helicopter evacuation should only be used for the most serious casualties. For infantry units conducting patrols in proximity to the enemy, helicopter evacuation of casualties may compromise the patrol's mission and force the patrol to return to friendly positions before the mission is completed.

The patrol leader determines the requirement for nuclear, biological, and chemical (NBC) defense equipment. Gas masks should always be carried due to the availability of riot control agents (RCAs) to the enemy. If chemical or biological agents have been employed in the area that the patrol must pass through, protective garments will have to be worn by patrol members for part of or the entire patrol. Wearing extra clothing and carrying extra equipment affects the speed of the patrol's movement. A contingency plan for post patrol decontamination must be developed.

Complete the Plan

At this point, the patrol leader has completed the basic thinking necessary for accomplishing the assigned mission. The patrol leader prepares the patrol order to spell out the details, assign tasks to subordinates, and explain the entire endeavor for ease of understanding by the other members of the patrol.

Prepare the Order

The patrol leader's order contains more detailed information than discussed in the warning order. Orders follow the prescribed five-paragraph order format but contain greater detail. The patrol order is a modified 5-paragraph order; the major modifications are to paragraphs 3a and 3c. The format for the order is contained in appendix B.

Issue the Order

The patrol leader asks for a status report on the initial preparatory tasks assigned to subordinate leaders and specialists when the warning order was issued. When the patrol leader has completed planning and initial preparations have progressed to the point where the patrol order may be issued, the members of the patrol are assembled. Roll call is taken to ensure all patrol members are present, then the prepared order is issued.

This will be the only opportunity for the patrol leader to issue detailed instructions. The mission, in particular, must be unmistakably clear so that once the patrol is committed, all subordinate leaders can act with unity of purpose.

Whenever possible, the patrol leader should have a Marine, such as the navigator, build a terrain model using dirt, sand, twigs, etc., explaining the concept of operations for movement to the objective area, actions at the objective area, and the return. Terrain models provide patrol members with a clear and simple layout of the area of operations and key terrain.

Supervise

Inspections and rehearsals, vital to proper preparation, are conducted even though the patrol leader and patrol members are well experienced in patrolling. Inspections determine the patrol's state of physical and mental readiness.

The patrol leader inspects before rehearsals to ensure completeness and correctness of uniform and equipment. The following areas are checked:

- Camouflage.
- Identification tags, Geneva Convention cards.
- Prescribed equipment, weapons, and ammunition are available and serviceable.
- Tape and other items are used to "silence" equipment (prevent noise produced during movement).
- Items that could provide information to the enemy (e.g., letters and papers) remain behind.
- Unnecessary equipment and excess weight remain behind.

The patrol leader questions each patrol member to ensure the following is known:

- The mission, planned routes (primary and alternate), and the fire support plan of the patrol.
- The individual's role: what to do and when to do it.
- What others are to do and how their actions impact.
- Challenges and passwords, codes, reporting times, radio call signs, frequencies, and any other pertinent details.

There is usually a period of time between final rehearsal and departure. The patrol leader reinspects just before departure to ensure all equipment is still in working order and the unit is ready to embark on the mission.

Rehearsals ensure the operational proficiency of the patrol. Plans are checked and needed changes are made. The patrol leader verifies the suitability of equipment. It is through rehearsals that patrol members become thoroughly familiar with the actions to take during the patrol.

If the patrol will operate at night, both day and night rehearsals are conducted. Terrain similar to that over which the patrol will operate is used. All actions are rehearsed. If time is limited, the most critical phases are rehearsed. Action at the objective is the most critical phase of the patrol and is always rehearsed.

An effective method is to talk the patrol through each phase, describing the actions and having each member perform individual duties. When satisfied, the patrol leader walks the patrol through all phases of the patrol using only the signals and commands to be used during the actual conduct of the patrol. Rehearsals continue until the patrol is thoroughly familiar with the plan. The rehearsal is also used to test the soundness of the patrol order and patrol organization.

After the rehearsal, the patrol leader makes final adjustments to the plan and patrol organization based on what was learned during the rehearsal and from other sources, such as the S-2 and adjacent patrols. When this is completed, the patrol leader issues final instructions to subordinate leaders noting any changes made in the patrol organization or plan. While the subordinate leaders are briefing the remainder of the patrol members, the patrol leader reports to the commander stating that the patrol is ready to begin the mission. The patrol leader also coordinates the location and time that the patrol can test fire all weapons prior to departure.

CHAPTER 11. MOVEMENT TO AND RETURN FROM THE OBJECTIVE AREA

This chapter provides guidance to patrol leaders for movement to and return from the objective area. Action in the objective area depends on whether the patrol is assigned a reconnaissance mission, combat mission or security mission. Chapters 12 and 13 provide detailed guidance on these types of missions.

11001. PASSAGE OF LINES

A passage of lines is an operation in which a force moves forward or rearward through another force's combat positions with the intention of moving into or out of contact with the enemy. (JP 1-02)

During the initial preparation for the patrol, the patrol leader selects a patrol assembly area and reconnoiters the area of passage designated by the commander. In coordination with the unit commander responsible for the area of passage, the patrol leader identifies gaps or lanes in minefields and wire obstacles and locates local security elements through which the patrol will pass. The patrol leader also checks the route from the patrol assembly area to the passage point or contact point where the patrol will depart friendly lines. If possible, both the route to the passage point and the route through the frontlines should be concealed from the enemy's view.

The patrol leader also reconnoiters the area for return passage of lines and coordinates with the unit commander responsible for the area of passage for passage points and lanes as necessary. The patrol leader observes these points from the direction that the patrol will use upon return to friendly lines, if possible, to aid in recognition upon return. The patrol leader provides the forward unit with information about the size of the patrol, general route, and expected time of return. The manner of challenge and recognition of the returning patrol should be coordinated in detail.

Upon return to friendly forward local security squads and/or frontlines, the patrol leader leaves the patrol in a covered position and moves forward with a radio operator and at least one Marine for security to make contact with the friendly unit in the manner agreed. After contact is made and recognized, the patrol leader rejoins the patrol and takes them to the passage point, personally checking in each member.

11002. ORGANIZATION FOR MOVEMENT

The patrol's task organization establishes the elements and teams needed to accomplish the mission in the objective area and to and from the objective. The patrol leader determines the formation(s) in which the patrol moves to the objective area.

Formations

The proper use of patrol formations is critical to the patrol's success. The squad and fire team formations described in FMFM 6-5 (proposed MCWP 3-11.2), *Marine Rifle Squad*, also apply to infantry patrolling formations. Because the movement of the patrol must be concealed from the enemy, the patrol normally moves through terrain that provides concealment. Control of the patrol in this type of terrain is difficult; thus the column formation, which is easily controlled, is normally used. However, as various types of terrain are encountered, the patrol leader uses the same considerations in determining the appropriate formation used in other infantry operations.

The standard squad and fire team formations are adaptable to a patrol. The patrol may change formations en route to match the situation and terrain. The patrol leader may have to sacrifice some control for better dispersion or give up some speed for greater stealth and more security. Other considerations include—

- Visibility, weather, terrain, and vegetation will influence dispersion and control of individuals and units. These factors may also affect the enemy; if visibility is good for the patrol, it is also good for the enemy. Two pieces of luminous tape worn on the back of the collar will aid in control and movement on dark nights. The collar is turned down when near the enemy. The tape can also be worn on the back of the cap, but should be covered or removed when near the enemy.

- Preserving the integrity of fire units (fire teams and, if attached, machine-gun teams) is of primary importance. If team members are detached from a machine-gun team, the unit integrity is lost and effectiveness is reduced. The patrol leader must also position fire units so as not to mask their fires.
- All-around defense of the patrol must not be sacrificed under any circumstances. The conventional squad and platoon formations provide adequate firepower in any direction required. When attachments are made to the patrol, the attachments are positioned within the formation to enhance the firepower of the patrol. If a fire unit, such as a machine-gun team or squad, is attached, it is incorporated into the all-around defense of the patrol by modifying the conventional formation (see fig. 11-1). It is permissible to employ machine-gun teams individually during movement. However, assault rocket squads and teams should be employed primarily as rifle fire units for movement. Firing the SMAW or AT-4 from within a patrol formation can present a serious back-blast danger to patrol members and should only be fired on the patrol leader's direction.
- Time allotted for mission accomplishment is also a major consideration. In selecting the formations, the patrol leader must consider the speed of the movement required to meet the time constraints (if any) imposed on the patrol. If required to meet a time schedule, a formation that permits rapid movement should be used. Speed, however, must never be permitted to force the patrol leader to make rash tactical decisions.

Exercise of Control

The patrol leader is positioned for best control over the patrol. The assistant patrol leader moves at or near the rear of the patrol. Other subordinate leaders move with their elements. All patrol members assist by staying alert and passing on signals and orders. A signal to halt may be given by any patrol member, but the signal to resume movement is given only by the patrol leader.

Arm-and-hand signals are the primary means of communication within a patrol and should be used exclusively when near the enemy. All members must know the standard infantry signals (refer to FMFM 6-5), as well as any special signals required, and be alert to receive and pass them to other members.

The patrol leader should speak just loudly enough to be heard. At night, or when close to the enemy, the

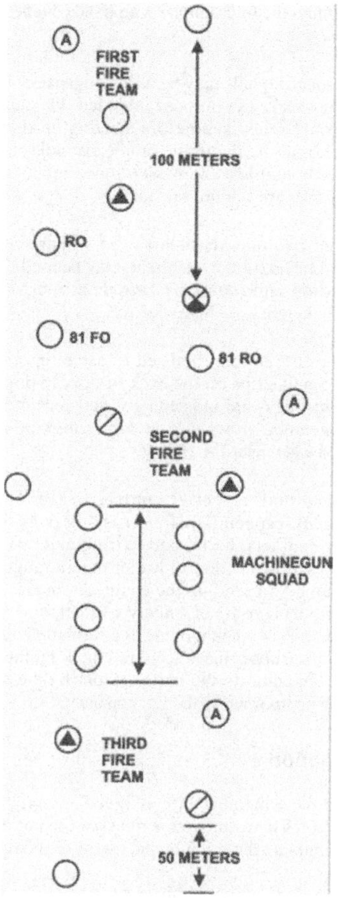

Figure 11-1. Squad Patrol with Attached Mortar Forward Observer Team and Machine Gun Squad.

patrol leader halts the patrol and has subordinate leaders come forward. They speak in a low voice and then pass the information to their subordinates by moving from member to member.

Radios provide a means of positive control within a large patrol; however, radios should be used only when arm-and-hand signals or face-to-face contact between the patrol leader and subordinate leaders is impractical. When close to the enemy, words are spoken into the microphone with hands cupped over it in a low voice.

Other sound signals may be used if the patrol leader is sure they serve the purpose intended. Planned sound signals are rehearsed before starting on the patrol. Sound signals used must be simple, natural sounds that are few in number and easily understood. Bird and animal calls are seldom satisfactory.

Infrared equipment, such as the sniper scope and infrared filters for the flashlight, may be used as means of sending and receiving signals and maintaining control at night.

Luminous tape may be used to assist in control at night. Small strips on the back of the cap or collar of patrol members aid in keeping visual contact with the front member. However, the luminous tape must be covered when near the enemy.

An important aspect of control is accounting of personnel, especially after crossing danger areas, enemy contacts, halts, and exiting and re-entering friendly lines. The patrol leader may arrange for the last member to send up the count automatically after crossing danger areas, enemy contact, and halts. In large patrols or when moving in a formation other than a column, subordinate leaders check members and report the count to the patrol leader by the quickest method appropriate to the circumstances.

Navigation

One or more members are assigned as navigators for the patrol. Their function is to assist the patrol leader in maintaining direction by the use of the compass.

The patrol leader assigns at least two members as pacers to keep track of the distance from point to point. The average of their count is used to approximate the distance traveled. Pacers are separated so they do not influence each other's count. Pacers must know how to convert their own paces to meters.

The route is divided into increments, with each increment starting at a recognizable point on the ground. The pacers begin their counts from zero at the beginning of each leg. This makes the pace count easier to keep and checks for accuracy.

The pace count is sent forward when the patrol leader turns to the member behind and in a low voice says, "send up the pace" or uses the arm-and-hand signal of tapping his boot. This signal is passed to both pacers, who in turn send up the pace count in meters; for example, "two-hundred" or "one-seven-five."

Patrol members must understand that the counts of both pacers are sent forward. The patrol leader must know the counts of both men in order to check them.

Security

The patrol is dispersed consistent with control, visibility, cover, and concealment. Scouts are employed to the front, flanks, and rear of the patrol to provide security. For the patrol members traveling in the main body of the patrol, areas of responsibility are assigned to the front, flanks, and rear. Scouts are the eyes and ears of the patrol leader. They move when and where directed by the patrol leader and maintain contact with the patrol leader at all times, except when momentarily obscured by vegetation or other terrain features.

Front

Small patrols (squad size) may employ from one scout up to a fire team as the point, depending on the enemy situation, terrain, and route being followed by the patrol. Normally, squad-size patrols will use two scouts as the point. The point is responsible for investigating the route of advance immediately to the front of the patrol.

The point moves as far ahead of the patrol as visibility and terrain permit. When visibility is good, the point may precede the main body by as much as 100 meters. The point travels right and left ahead of the patrol, searching the area over which the patrol will pass.

The point maintains direction by knowledge of the general route to be followed and visual contact with the patrol leader. The patrol leader or the navigator ensures that the point is proceeding correctly.

The point, which stays far enough ahead of the patrol to provide security, is not a trail breaker for the patrol. If the point loses contact with the patrol, the point waits for the main body to catch up or moves rearward if contact is not quickly regained.

One of the navigators may be positioned with the point. One or more members works as the point while the other is the navigator.

Flank

Flank security for a patrol of squad size or less may be provided by using one or two members on either flank. If two scouts are assigned to a flank, one is positioned to observe the patrol leader and the other works farther out from the patrol. The scout who must observe the patrol leader remains within a maximum distance of 100 meters. The scout farther out remains in sight of the inside scout but normally does not move more than 20 or 25 meters away and remains prepared to relieve flank security regularly. Moving through dense woods or jungle may render the use of flank security impractical because of reduced visibility. In such cases, it moves with the patrol itself, but maintains observation to its assigned flanks.

Rear

A small patrol normally has only one rifleman assigned as rear security. An interval between the member assigned as rear security and the last Marine of the patrol is maintained at the limit of visibility, up to 50 meters. This member maintains rear security for the patrol by constantly observing to the rear.

Halting

Speed of movement is slower at night than in day patrols and reduces the danger of a Marine becoming separated from the patrol. The patrol occasionally halts to observe and listen for enemy activity; this is called a security halt. Upon signal, when reaching a danger area and periodically throughout movement en route, every member freezes in place, remains quiet, observes, and listens. It may be necessary to call a security halt just after departing friendly areas and just before entering friendly areas.

The patrol may halt briefly to send a message, eat, rest, check direction, or make a reconnaissance. The area selected should provide adequate concealment and cover, as well as favor the defense. All-round security is established and the patrol leader ensures all members move out when the patrol resumes movement. (For extended halts, see paragraph 11005.)

Infiltration

The disposition of enemy forces may sometimes prevent a patrol from entering the enemy occupied area as a unit; however, pairs of scouts or fire teams may slip through without being discovered. (Refer to FMFM 6-5 for infiltration techniques and procedures).

11003. CONTROL MEASURES FOR MOVEMENT

Checkpoints

A checkpoint is a predetermined point on the surface of the Earth used as a means of controlling movement, a registration target for fire adjustment or reference for location (JP 1-02). Checkpoints are a means of control between the parent unit and the patrol. These locations are decided upon and coordinated before the patrol leaves, so that both the patrol members and parent unit will know the patrol's location when it reports in. The parent unit can follow the progress of the patrol without transmitting coordinates over the radio that the enemy could monitor.

Rally Points

A rally point is an easily identifiable point on the ground where units can reassemble and reorganize if they become dispersed (MCRP 5-12C). It should provide cover and concealment, be defensible for at least a short time, and be easily recognized and known to all patrol members. All rally points are considered and identified as tentative rally points until they are occupied, found to be suitable, and designated as rally points by the patrol leader.

Initial Rally Point

The initial rally point is within friendly lines where the patrol can rally if it becomes separated before departing the friendly area or before reaching the first en route rally point. It may be the assembly area where the patrol waits while the patrol leader contacts the last friendly position through which the patrol will pass. The location of the initial rally point must be coordinated with the forward unit commander.

En route Rally Points

En route rally points are between the initial rally point and the objective rally point; and from the objective rally point back to the point where the patrol reenters friendly lines. They are determined as the patrol passes through a likely area that is suitable for a rally point.

Objective Rally Point

The objective rally point is located nearest the objective where the patrol makes final preparations prior to approaching the objective. It also serves as a location where the patrol reassembles after completing actions on the objective. The objective rally point must be suitable to accommodate those activities accomplished prior to actions on the objective. This position must provide the patrol concealment from enemy observation and, if possible, cover from enemy fires. It may be located short of, to a flank or beyond the objective. It should be out of sight, sound, and small-arms range of the objective area. The patrol leader's reconnaissance of the objective is made from this position; it is the release point from which patrol elements and teams move to the objective to accomplish the mission.

Rally Point Selection

The patrol leader selects likely locations for tentative rally points during reconnaissance or map study. A tentative rally point must be confirmed and announced after examination proves suitability. A tentative initial rally point and a tentative objective rally point are always selected and identified in the patrol order. If necessary, the patrol leader selects additional rally points en route as suitable locations are reached. When the patrol reaches a danger area that cannot be bypassed, such as an open meadow or stream, the patrol leader selects a rally point on both the near and far side. If good locations are not available, the patrol leader designates the rally points in relation to the danger area. The patrol leader will say, for example, "50 meters this side of the trail" or "50 meters beyond the stream."

Rally Point Use

The initial rally point and en route rally points are designated to enable the patrol to reassemble if it is unavoidably separated or dispersed. Identifying features are pointed out. The patrol leader ensures that the information is passed to all patrol members. When crossing a danger area, a near side rally point and a far side rendezvous point are designated. These rally points should only be used when all other methods of retaining control of the patrol have failed. The success of the patrol may be jeopardized if it is dispersed and cannot rally expeditiously.

If the patrol has left the friendly area and becomes dispersed, patrol members return to the last designated rally point (the initial or an en route rally point) unless the patrol leader gives other instructions.

As previously noted, the patrol leader selects two rally points at the near and far sides of danger areas that cannot be bypassed. If the patrol becomes separated or dispersed at a danger area, and there has been no enemy contact, the patrol should reassemble at the rally point on the far side of it. If, however, the patrol is separated or dispersed at a danger area as a result of enemy contact, members who have already crossed the danger area assemble and reorganize at the rally point on the far side; members who have not crossed assemble and reorganize at the rally point on the near side. In the absence of the patrol leader and unless directed otherwise, the senior member at the rally point on the near side takes charge, attempts to move the rallied patrol members to the rally point on the far side, and rejoins the remainder of the patrol.

Rally Point Actions

The patrol leader plans the actions to be taken at rally points and instructs the patrol accordingly in the patrol order. Planned actions at the initial rally point and en route rally points must provide for the continuation of the patrol as long as there is a reasonable chance to accomplish the mission. Plans for actions at rally points should provide for—

- Recognition signals for assembly at rally points.
- Minimum number of members and maximum amount of waiting time required before the senior member at the rally point moves the rallied patrol members onward toward the objective or returns to friendly lines.
- Instructions for patrol members who find themselves alone at a rally point.

11004. PRECAUTIONS AT DANGER AREAS

A danger area is any place where the patrol is vulnerable to enemy observation or fire (open areas, roads, trails, and obstacles such as barbed wire, minefields, rivers and streams, and lakes). Any known or suspected enemy position the patrol must pass is also a danger area. The patrol leader plans for crossing each danger area and includes these plans in the order.

The patrol reconnoiters the near side of a danger area first, then the patrol leader sends scouts to reconnoiter the far side. Once the scouts report that the far side is clear of the enemy, the remainder of the patrol crosses the danger area. As each individual or group crosses the danger area, they are covered by those remaining and by those who have successfully crossed. Enemy obstacles are avoided since they are usually covered by fire.

In crossing a river, the near bank is reconnoitered first; then the patrol is positioned to cover the far bank. Scouts are sent across to the far bank. After the far bank has been reconnoitered and the scouts report that it is clear of the enemy, the patrol crosses as rapidly as possible. This may be done individually or in pairs. If crossing the river requires swimming, the patrol uses improvised rafts to float equipment, weapons, and ammunition across. (Refer to MCRP 3-02C, *Water Survival Handbook.)*

A road or trail is crossed at or near a bend or where the road is narrow. Observation is restricted and, if the enemy is present, the patrol is exposed as short a time as possible. The near side is reconnoitered first, then scouts are sent across to reconnoiter the far side. This includes reconnoitering the tentative rally point on the far side. Once the scouts report "all clear," the remainder of the patrol crosses rapidly and quietly.

If the patrol must pass close to an enemy position, it takes advantage of battlefield noises to cover the sounds of movement. If supporting fires are available, the patrol leader can call for them to divert the enemy's attention as the patrol passes.

11005. HIDE

When a patrol is required to halt for an extended period in an area not protected by friendly troops, the patrol moves into a location which, by the nature of the surrounding terrain, provides passive security from enemy detection. Such an assembly area is termed a *hide*. To establishment a hide—

- Cease all movement during daylight hours to avoid detection.
- Hide the patrol for an extended period while the patrol leader conducts a detailed reconnaissance of the objective area.
- Rest and reorganize after extended movement.
- Reorganize after a patrol has infiltrated the enemy area in small groups.

The patrol leader's plan must include tentative hide locations when the patrol's mission dictates an extended halt within enemy areas. These tentative locations must be confirmed by actual ground reconnaissance prior to occupation by the patrol. The plan for a hide includes both passive and active security measures.

Passive security measures are—

- Avoid built-up areas.
- Select an area remote from all human habitation.
- Avoid known or suspected enemy positions.
- Avoid ridge lines, topographic crests, valleys, lakes, and streams.
- Avoid roads and trails.
- Avoid open woods and clearings.
- Select areas offering dense vegetation, preferably bushes and trees that spread out close to the ground.

Active security measures—

- Establish security covering all likely avenues of approach into the site.
- Establish communications (wire, radio, signal, runner) with posted security to provide early warning of enemy approach.
- Select an alternate area for occupation if the original hide is compromised or found unsuitable.
- Plan for withdrawal in the event of discovery.
- Establish an alert plan with a certain percent of the personnel awake at all times.
- Organize the elements of the patrol so necessary activities can take place with a minimum amount of movement.

The size of the area physically occupied by a patrol in a hide and the number of security posts required are governed by the terrain, quantity and quality of cover and concealment, and size of the patrol.

If the situation permits, a hide can also be used as the final preparation position and/or objective rally point.

11006. IMMEDIATE ACTIONS UPON ENEMY CONTACT

A patrol may make contact with the enemy at any time. Contact may be through observation, a meeting engagement or ambush. Contact may be visual, in which the patrol sights the enemy but is not itself detected. When this is the case, the patrol leader can decide whether to make or avoid physical contact, basing his decision on the patrol's assigned mission and capability to successfully engage the enemy unit.

When a patrol's assigned mission prohibits physical contact (except that necessary to accomplish the mission), its actions are defensive in nature. Physical contact, if unavoidable, is broken as quickly as possible and the patrol, if still capable, continues its mission.

When a patrol's assigned mission permits or requires it to seek or exploit opportunities for contact (as in the case of a combat patrol), its actions are offensive in nature, immediate, and positive.

In patrolling, contacts (visual or physical) are often unexpected at very close ranges, and short in duration. Effective enemy fire often provides leaders little or no time to fully evaluate situations and issue orders. In these situations, immediate action provides a means for swiftly initiating positive offensive or defensive action, as appropriate.

Two types of physical contact with the enemy are meeting engagement and ambush. Meeting engagement is a combat action that occurs when a moving force, incompletely deployed for battle, engages an enemy at an unexpected time and place. It is an accidental meeting where neither the enemy nor the patrol expect contact and are not specifically prepared to deal with it. An ambush is a surprise attack from a concealed position.

Immediate Actions

Immediate actions are designed to provide swift and positive small unit reaction to visual or physical contact with the enemy. They are simple courses of action in which all Marines are well trained. Minimal signals or commands are required and they are developed as needed for the combat situation. The signals can, in many cases, be initiated by any member of the unit. It is not feasible to attempt to design an immediate action drill to cover every possible situation. It is better to know the immediate action drill for each of a limited number of situations that may occur during a patrol. Arm-and-hand signals associated with immediate actions—such as FREEZE, ENEMY IN SIGHT, and HASTY AMBUSH RIGHT or LEFT—are contained in FMFM 6-5 (proposed MCWP 3-11.2).

Immediate Halt

When the patrol detects the enemy but is not itself detected, the situation requires the immediate, in-place halt of the patrol. The first member visually detecting the enemy gives the silent signal for FREEZE. Every member halts in place, weapon at the ready, and remains absolutely motionless and quiet until further signals or orders are given.

Air Observation and/or Attack

These actions are designed to reduce the danger of detection from aircraft and casualties from air attack.

When an enemy or unidentified aircraft that may detect the patrol is heard or observed, the appropriate immediate action drill is FREEZE. The first member hearing or sighting an aircraft that may be a threat signals FREEZE. Every member halts in place until the patrol leader identifies the aircraft and gives further signals or orders. Members of the patrol must not look up at the aircraft as sunlight can reflect off their faces even when camouflaged.

When an aircraft detects a patrol and makes a low level attack, the immediate action drill air attack is used. The first member sighting an attacking aircraft shouts, "AIRCRAFT," followed by the direction of the incoming attack: FRONT, LEFT, REAR or RIGHT. The patrol moves quickly into line formation, well spread out, at right angles to the aircraft's direction of travel. As each member comes on line, the member hits the ground, using available cover, then positions the body perpendicular to the aircraft's direction of travel, to present the shallowest target possible (see fig. 11-2 on page 11-8). Between attacks (if the aircraft returns or if more than one aircraft attacks), patrol members seek better cover. Attacking aircraft are fired upon only on command of the patrol leader.

Meeting Engagement

Hasty Ambush. This immediate action is used to avoid contact and to prepare to initiate an unplanned ambush on the enemy. It may often be a subsequent action after the command freeze. When the signal HASTY AMBUSH is given (by the point member, patrol leader or another authorized patrol member), the entire patrol moves quickly to the right or left of the line of movement, as indicated by the signal, and takes up the best available concealed firing positions (see fig. 11-3 on page 11-9). The patrol leader initiates the ambush by opening fire and shouting, "FIRE"; thus ensuring initiation of the ambush if the weapon misfires. If the patrol is detected before this, the first member aware of detection initiates the ambush by firing and shouting. The patrol leader may decide not to initiate the ambush in order to avoid contact unless the patrol is detected. When used as an offensive

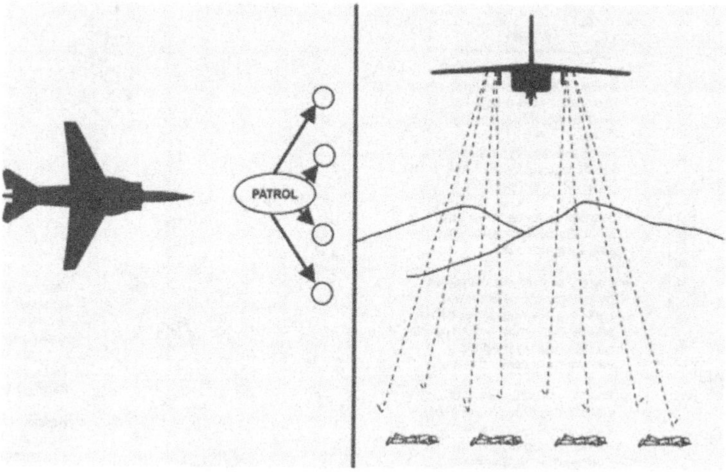

Figure 11-2. Immediate Action, Air Attack.

measure, the enemy is allowed to advance until he is in the most vulnerable position before the ambush is initiated. An alternate means for initiating the ambush is to designate an individual (for example, point or last member) to open fire when a certain portion of the enemy unit reaches or passes that member.

Immediate Assault. This immediate action drill is used defensively to make and quickly break undesired but unavoidable contact (including ambush) and offensively to decisively engage the enemy (including ambush). When used in a meeting engagement, members nearest the enemy open fire and shout, "CONTACT," followed by the direction of the incoming attack: FRONT, LEFT, REAR or RIGHT. The patrol moves swiftly into line formation and assaults (see fig. 11-4).

Defensive Measures

When used defensively, the assault is stopped if the enemy withdraws and contact is broken quickly. If the enemy stands fast, the assault is carried through the enemy positions and movement is continued until contact is broken.

Offensive Measures

When used offensively, the enemy is decisively engaged. Escapees are pursued and destroyed until orders to break contact are given by the patrol leader. If the patrol is fired upon from beyond 50 meters, the patrol must break contact as quickly as possible and continue the mission. If it engages the enemy any longer than necessary to break contact, it may put the mission in jeopardy.

Fire and Maneuver

Fire and maneuver is one means to break contact. One portion of the patrol returns the enemy fire while another portion moves by bounds away from the enemy. Each portion of the patrol covers the other by fire until contact is broken by all.

Clock System

The clock system is another means to break contact. Twelve o'clock is the direction of movement of the patrol. The patrol leader shouts a direction and a distance. For example: "TEN O'CLOCK-TWO HUNDRED," means the patrol should move in the direction of ten o'clock for 200 meters. Patrol members keep their same relative positions as they move so the original formation is not disrupted. Subordinate leaders must be alert to ensure that the members of their elements and teams receive the correct order and move as directed.

Counter Ambush

When a patrol is ambushed, the immediate action drill used depends on whether the ambush is a near ambush

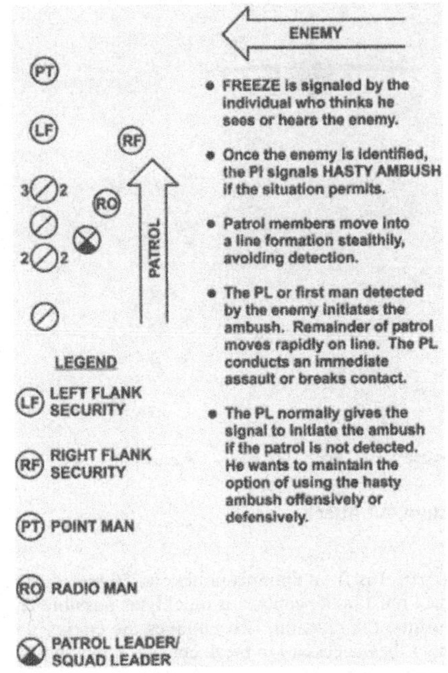

Figure 11-3. Immediate Action, Hasty Ambush.

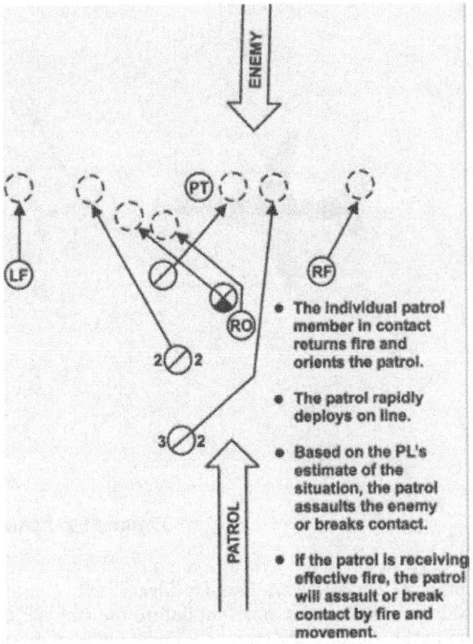

Figure 11-4. Immediate Action, Immediate Assault.

(the enemy is within 50 meters of the patrol) or a far ambush (the enemy is beyond 50 meters of the patrol). Fifty meters is considered the limit from which the ambushed patrol can effectively launch an assault against the enemy.

Near Ambush. In a near ambush, the killing zone is under very heavy, highly concentrated, close range fires. There is little time or space for members to maneu-ver or seek cover. The longer they remain in the killing zone, the greater the chance they will become casualties. Therefore, if members in the killing zone are attacked by a near ambush, they immediately assault without order or signal directly into the ambush position, occupy it, and continue the assault or break contact, as directed. This action moves them out of the killing zone, prevents other elements of the ambush from firing on them without firing on their own members, and provides positions from which other actions may be taken (see fig. 11-5 on page 11-10). Members not in the killing zone maneuver against the enemy as directed. The assault continues until all patrol members are outside of the killing zone.

Far Ambush. In a far ambush, the killing zone is also under very heavy, highly concentrated fires, but from a greater range. This greater range provides members in the killing zone maneuver space and some opportunity to seek cover at a lesser risk of becoming a casualty. If attacked by a far ambush, members in the killing zone, without order or signal, immediately return fire, take the best available positions, and continue firing until directed otherwise. Members not in the killing zone maneuver against the ambush force, as directed (see fig. 11-6 on page 11-10). The assault is continued against the enemy or until the order to break contact is given.

In each situation, the success of the counter ambush employed depends on the members being well trained in quickly recognizing the distance from which an ambush is initiated and well rehearsed in the proper reaction.

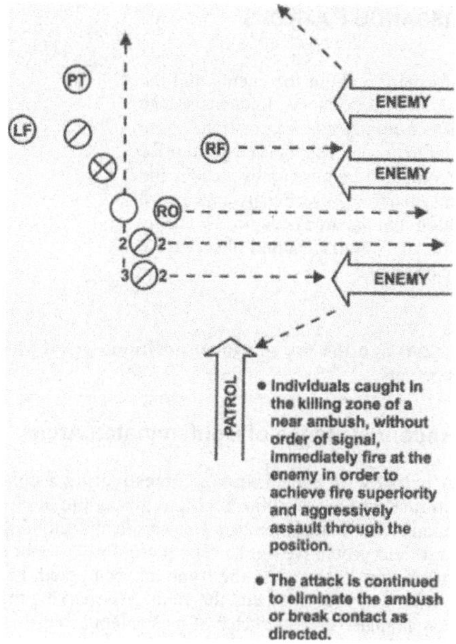

Figure 11-5. Near Ambush.

11007. PATROL LEADER'S ACTION IN A DEVELOPING SITUATION

While good patrolling depends on good planning, the patrol leader's plan must be flexible. Every combat situation develops differently than expected, and the patrol leader must be ready to quickly adapt to the situation as it develops. One of the most difficult tasks in battle is to recognize the correct moment for making a decision. Generally, it is more difficult to determine the moment for making a decision than it is to formulate the decision itself. When the situation demands, decisions must be made promptly without waiting for more information. In a developing situation, the patrol leader should use the following questions as a guide for battlefield decisionmaking:

- How has the situation changed?
- How does the change affect mission accomplishment and the immediate superior's mission?
- Must a decision be made now?
- What are the options?
- Which option best serves the mission and the unit as a whole?
- Which option offers the greatest chance of success?

11008. RETURN FROM OBJECTIVE AREA

After performing actions in the objective area, the patrol reassembles at the objective rally point. This phase of the patrol is perhaps the most difficult and dangerous. Patrol members are experiencing fatigue, emotional letdown, and wounds; they may be low on water and ammunition. Above all, the enemy was likely alerted if not in pursuit. At this point, the patrol leader must move the patrol rapidly but carefully and maintain patrol security at a high level. In returning to friendly lines, the patrol neither uses nor travels near the same route used to get to the objective area. The enemy may have that route covered. For information on reentry of friendly lines, refer to paragraph 11001.

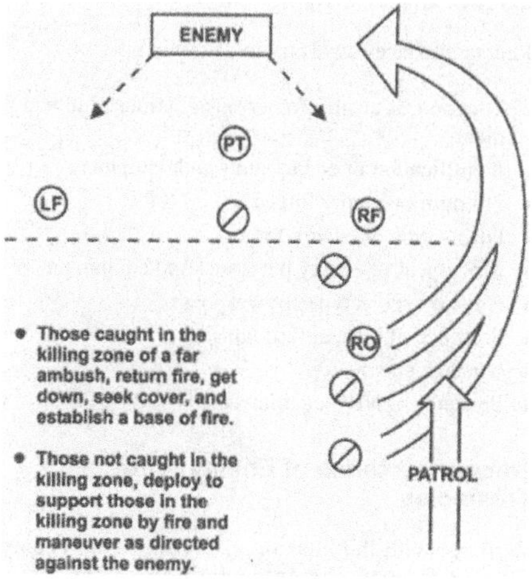

Figure 11-6. Far Ambush.

CHAPTER 12. RECONNAISSANCE PATROLS

The commander needs accurate, timely information about the enemy and the terrain the enemy controls to assist in making tactical decisions. Reconnaissance patrols are one of the most reliable means for obtaining this information. These patrols engage the enemy only when necessary to accomplish their mission or for protection. They generally avoid combat, accomplish the mission by stealth, and do not maintain contact with the unit that sent them out. A reconnaissance patrol is capable of carrying the search for information into the area occupied by enemy forces—usually beyond the range of vision of friendly ground observation posts—and examining objects and events at close range.

12001. GENERAL MISSIONS

General missions for reconnaissance patrols include gathering information about location and characteristics of friendly or hostile positions and installations, terrain (routes, stream crossings), and obstacles.

12002. SPECIFIC MISSIONS

Locate the Enemy

Reconnaissance patrols try to determine the—

- Location of enemy forces, installations, and equipment.
- Identification of enemy units and equipment.
- Strength of enemy forces.
- Disposition of enemy forces.
- Movement of enemy personnel and equipment.
- New or special types of weapons.
- Presence of mechanized units.
- Unusual enemy activity.
- Presence of NBC equipment.

Reconnaissance of Enemy Wire Obstacles

A patrol with the mission of investigating enemy-emplaced wire obstacles employs a formation providing for all-around security and takes precautions against being observed by the enemy. The patrol leader and one Marine inspect each gap and establish its location by means of compass bearings to prominent objects in the rear of enemy or friendly positions or through the use of global positioning satellite receivers.

Reconnaissance of Contaminated Areas

A patrol with the mission of investigating a contaminated area reconnoiters and marks the area's boundaries. Patrol members wear protective clothing and field protective masks. The patrol leader reports the extent of the area, the ttype of agent used, the terrain and vegetation and the method of marking the contaminated area. A sketch of the contaminated area, prepared by the patrol leader, should be included in the patrol report.

Reconnaissance of Enemy Minefields

Patrols assigned to reconnoiter enemy minefields are often composed of specially trained personnel. The type of mine is identified and the possibility of bypassing the area determined.

Reconnaissance of Terrain

The reconnaissance patrol is particularly suited for gathering information about the terrain within their area of operations. (Refer to MCRP 2-15.3B, *Reconnaissance Reports Guide*.) Commanders must know the location and condition of bridges, streams, and roads in order to make proper plans for the movement of troops. The following information should be obtained by a reconnaissance patrol assigned such missions:

- Bridges:
  - Maximum load capacity.
  - Material used in construction (wood, stone, concrete or steel).

- Material used for abutments (dirt, wood, masonry, concrete).
- Type and number of supporting members of the bridge.
- Condition and dimensions of bridge flooring.
- Presence or absence of mines and demolition charges on the bridge and at entrances/exits.
- Terrain crossed (ravine, stream, road).
- Primary use (rail, auto, foot).
- Location and type of an alternate crossing if bridge is unusable.

• Streams and fords:
  - Width and depth of stream.
  - Composition of the bottom (mud, sand, gravel, rock).
  - Speed of current in miles per hour.
  - Composition and gradient of banks.
  - Surrounding terrain and vegetation cover.

• Roads:
  - Width.
  - Composition of surface.
  - Condition.
  - Road blocks.
  - Gradient/degrees of slopes.
  - Curves (location and length).
  - Pitch and culvert locations.
  - Primary use and frequency of use.

12003. TYPES OF RECONNAISSANCE

Route

Route reconnaissance is a directed effort to obtain detailed information of a specified route and all terrain from which the enemy could influence movement along that route. (Refer to MCRP 5-12C.) It is also described as a form of reconnaissance focused along a specific line of communications—such as a road, railway, or waterway—to provide new or updated information on route conditions and activities along the route. (Refer to MCRP 5-12A, *Operational Terms and Graphics*.)

A route reconnaissance normally precedes the movement of forces. It provides detailed information about a specific route and the surrounding terrain that could be used to influence movement along that route. Considerations include trafficablility, danger areas, critical points, vehicle weight and size limitations, and locations for friendly obstacle emplacement.

The objective area is normally defined by a line of departure, a route, and a limit of advance.

Area

An area reconnaissance is a directed effort to obtain detailed information concerning the terrain or enemy activity within a prescribed area such as a town, ridge line woods or other features critical to operations. (Refer to MCRP 5-12C.) An area reconnaissance could also be made of a single point, such as a bridge or installation. (Refer to MCRP 5-12A.)

Emphasis is placed on reaching the area without being detected. Enemy situations encountered en route are developed only enough to allow the reconnoitering unit to report and bypass.

Zone

A zone reconnaissance is a directed effort to obtain detailed information concerning all routes, obstacles (to include chemical or radiological contamination), terrain, and enemy forces within a zone defined by boundaries. A zone reconnaissance is normally assigned when the enemy situation is vague or when information concerning cross-country trafficability is desired. (Refer to MCRP 5-12C.)

The commander specifies specific routes or areas of interest within the zone. The zone to be reconnoitered usually is defined by a line of departure, lateral boundaries, and a limit of advance.

12004. TASK ORGANIZATION

A reconnaissance patrol is organized around the current structure of the Marine rifle squad with one or more of the squad's fire teams assigned as the reconnaissance element to reconnoiter or maintain surveillance over the objective. At least one fire team acts as a security element whose functions are to—

- Secure the objective rally point (see para. 11003).
- Give early warning of enemy approach.
- Protect the reconnaissance unit.

Reinforcing the squad is based on METT-T. If two squads are required, then the mission should be assigned to a platoon.

A company may send a reconnaissance patrol to specifically reconnoiter a given area (a riverbed or a bridge), or to maintain surveillance on a particular streambed for the next four nights. Reconnaissance patrols perform three basic functions en route to and from the objective:

- Provide control.
- Provide security while moving and in the objective area.
- Conduct reconnaissance or surveillance.

Depending on the size of the reconnaissance patrol, an element must be tasked to reconnoiter the area or zone. In the case of a surveillance mission, two teams are recommended to share the same task (so one could lay up and rest, while the other lays low and hides to maintain strict silence and no movement while surveillance is being maintained on the objective). Whatever the case, each unit that has a surveillance mission or reconnaissance mission must be thoroughly briefed as to what essential elements of information are to be collected as well as their location in the objective area.

The patrol should be organized with one or two fire teams to actually conduct the reconnaissance mission and the remaining fire team to provide security. Each fire team should be prepared to assume either mission. A small area reconnaissance patrol needs only one fire team for the assigned mission. A patrol with a wider area reconnaissance mission should use two fire teams to physically conduct the mission and one fire team for cover and/or security.

The security element for a reconnaissance patrol should be organized to cover the likely avenue of approach into the objective area, to protect the units conducting the reconnaissance, and to cover the objective rally point.

The variations of the special organization for reconnaissance patrols illustrate the patrol leader's flexibility in organizing the patrol to meet mission requirements. The patrol leader is not limited to the variations reflected herein but can choose any special organization as long as that task organization accommodates the requirement for command and control, reconnaissance, and security.

12005. SIZE OF RECONNAISSANCE PATROLS

A reconnaissance patrol should be kept to the minimum number of personnel required to accomplish the mission. A mission requiring a patrol to remain away from its unit for a considerable period of time, or one requiring a patrol to send back information by messenger, increases the size of the patrol. Reconnaissance patrols seldom exceed a squad in strength. Unit integrity should be preserved whenever possible. Intelligence personnel, interpreters, and other specialists, such as radio operators or engineers, are assigned to a patrol if the particular mission requires.

12006. RECONNAISSANCE EQUIPMENT

Patrol members are armed and equipped as necessary for accomplishing the mission. The automatic rifle in each fire team provides a degree of sustained firepower in case of enemy contact. The patrol should have at least two: pairs of binoculars, pairs of wire cutters, maps, compasses, and watches. Night observation equipment may be used. Pencils and small notebooks are carried so notes and sketches can be made. A message book with message blanks and overlay paper is mandatory.

12007. RECONNAISSANCE PATROL ACTIONS AT THE OBJECTIVE AREA

Route

The patrol leader halts and conceals the patrol near the objective area in the final preparation position. The patrol leader then conducts leader's reconnaissance to confirm the plan for positioning the security teams and employing units assigned to the reconnaissance mission. The patrol leader returns to the patrol and positions the security to provide early warning of enemy approach and secure the objective rally point. The reconnaissance unit(s) then reconnoiters the objective area (route). The reconnaissance unit may move to several positions, along or adjacent to the specific route, in order to conduct a thorough reconnaissance. After completing the reconnaissance, each reconnaissance team moves to the objective rally point and reports to the patrol leader. The patrol then returns to friendly lines and the leader makes a full report.

Area

The patrol leader halts and conceals the patrol near the objective area in the final preparation position. The patrol leader then conducts leader's reconnaissance to pinpoint the objective and confirm the plan for positioning the security teams and employing units assigned the reconnaissance mission. The patrol leader returns to the patrol and positions security to provide early warning of enemy approach and secure the objective rally point. The reconnaissance unit(s) then reconnoiters the objective area. The reconnaissance unit may move to several positions, perhaps making a circle around the objective area, in order to conduct a thorough reconnaissance. When the reconnaissance is completed, the patrol leader assembles the patrol and tells members what has been observed and heard. Other patrol members contribute observations. The patrol then returns to friendly lines and the patrol leader makes a full report.

Zone

The patrol leader halts the patrol at the final preparation position, confirms the plan, and conducts leader's reconnaissance. The patrol leader positions the security team and sends out the reconnaissance team. When the entire patrol is used to reconnoiter the zone, it provides its own security. After completing the reconnaissance, each reconnaissance team moves to the objective rally point and reports to the patrol leader. The patrol then returns to friendly lines.

CHAPTER 13. COMBAT PATROLS

Combat patrols are assigned missions that usually include engaging the enemy. They are fighting patrols. Every combat patrol has a secondary mission: gaining information about the enemy and terrain. Combat patrols are employed in both offensive and defensive combat operations and they assist the parent unit in accomplishing its mission by inflicting damage on the enemy; establishing and/or maintaining contact with friendly and enemy forces; denying the enemy access to key terrain; and probing enemy positions to determine the nature and extent of enemy presence.

13001. TASK ORGANIZATION

A combat patrol is organized around the current structure of the Marine rifle platoon. A combat patrol leader should use the unit's normal organization (fire team, squad, and platoon) in assigning functions, patrol missions, and chain of command. Combat patrols must be able to perform the following four basic functions en route to and from the objective:

- Provide control.
- Provide security.
- Provide support by fire.
- Attack or assault the objective.

A rifle platoon could task-organize as a combat patrol as follows:

- Platoon headquarters (patrol headquarters).
- First squad (security).
- Second squad (support).
- Third squad (assault).

Every combat patrol must—

- Provide a control mechanism in the form of a headquarters.
- Designate a unit (a fire team or squad) to provide security while moving en route to the objective or while at the objective. At the objective area, this unit isolates the objective area, secures the objective rally point, and covers the withdrawal.
- Designate a unit to act as support. This unit provides the base of fire in the attack or covers withdrawals or advances.
- Designate a unit(s) to conduct the attack or assault. This unit(s) engages the enemy at the objective area by fire and maneuver or movement. It also operates immediately in the objective area (searching, demolition, prisoners of war, etc.). Paragraph 9001 outlines the general organization of combat patrols. As in the case with reconnaissance patrols, the task organization of a combat patrol depends on the specific mission assigned. If any special requirements are generated because of the specific mission, the patrol is task-organized to fit the needs of the mission.

13002. EQUIPMENT

Combat patrols are armed and equipped as necessary for accomplishing the mission. In addition to binoculars, wire cutters, compasses, and other equipment generally common to all patrols, it usually carries a high proportion of automatic weapons and grenades. Communications with higher headquarters is important as success of the mission may depend on being able to call for supporting fires. Also, internal radio communications with the units and teams may be useful. However, the patrol must not be so overburdened with equipment as to impede movement or mission accomplishment.

13003. RAID PATROLS

A raid is a surprise attack on an enemy force or installation with the attacking force withdrawing after accomplishing its mission. Raids destroy or capture enemy personnel or equipment, destroy installations, or free friendly personnel who have been captured by the enemy. Patrolling techniques are used in planning and when moveing to and from the objective. (Refer to MCWP 3-41.2, *Raids*.) Surprise, firepower, and violence of action are the keys to a successful raid. Patrols achieve surprise by attacking—

- When the enemy is least prepared (e.g., during periods of poor visibility such as darkness, rain, fog, or snow).
- From an unexpected direction. (This might be accomplished by approaching through a swamp or other seemingly impassable terrain.)
- With concentration of firepower at critical points within the objective.

Planning

A successful raid requires detailed planning. The leader of a combat patrol engaged in raiding must anticipate probable situations and decide upon definite courses of action to meet them. Rehearsals are imperative.

A raid patrol conducts such missions as destroying an enemy outpost or seizing prisoners from an observation post or lightly defended position.

While preparing for the mission, the patrol leader requests fire support required for the accomplishment of the mission. If practical, artillery and mortars should be employed to isolate the objective to prevent movement of enemy reinforcements into the area.

Execution

The leader's plan must be detailed and complete. All of the considerations outlined in chapter 11 must be covered. Patrol formations must provide for ease of control and all-around security while moving to and from the objective area and provide for rapid and coordinated deployment of the various units once the objective area is reached. The leader's plan usually includes the encirclement of the hostile position—either physically or by fire—in order to isolate it during the assault.

The final simultaneous assault against the objective develops when enemy defensive fires at the objective are suppressed by either friendly fire superiority or surprise. The assault is covered by the fire of the unit assigned the function of support by fire.

Grenades, SMAWs, and demolitions are most effective for clearing bunkers.

Security units are posted to isolate the objective. The patrol leader signals them when the withdrawal begins. As a minimum, security is on each flank and to the rear (at the objective rally point).

Actions at the Objective

The patrol leader halts the patrol near the objective at the final preparation position. Security is established and the leader's reconnaissance is made with appropriate subordinate leaders. When the leaders return to the patrol, they confirm previous plans or announce any changes. Movements are arranged so all units reach their positions simultaneously. This improves the patrol's capability for decisive action, if prematurely detected by the enemy.

The teams of the security element move to positions to secure the objective rally point, give early warning of enemy approach, block avenues of approach into—and prevent enemy escape from—the objective area. As the assault element moves into position, the security element informs the patrol leader of all enemy activity, firing only if detected or on the patrol leader's order. Once the assault element commences action, the security element prevents enemy entry into or escape from the objective area. The security element covers the withdrawal of the assault element (and support element, if employed) to the objective rally point, withdrawing only upon order or prearranged signal.

As the assault element approaches the objective, it deploys early enough to permit immediate assault if detected by the enemy. Each team uses stealth while moving into proper position. On command, or if one or more of the assault element is detected and fired upon by the enemy, the support element opens fire to neutralize the objective, then ceases or shifts fire according to prearranged plans and signals. As supporting fires cease or shift, the assault element assaults the objective. Demolition, search, and other teams are protected by the assault element while they work. On order, the assault element withdraws to the objective rally point.

If a support element is employed, its leader deploys teams to provide fire support for the assault element. Each member of the support element must know the scheme of maneuver to be used by the assault element, specific targets or areas to be neutralized by fire, and the signals that will be employed to commence, shift, and cease fires. The support element withdraws on order of the patrol leader. At the objective rally point, the patrol leader quickly reorganizes the patrol and begins the return movement to friendly lines.

13004. CONTACT PATROLS

General

Contact patrols establish and/or maintain contact to the front, flanks or rear by—

- Establishing contact with an enemy force when the definite location of the force is unknown.
- Maintaining contact with enemy forces through direct and/or indirect fires, or observation.
- Avoiding decisive engagement with the enemy.

Task Organization and Equipment

Task organization and equipment depend on the known enemy situation and anticipated enemy contact. A patrol sent out to establish contact with an enemy force is organized, armed, and equipped to overcome resistance of light screening forces in order to gain contact with the main enemy force. It is not organized and equipped to engage the main enemy forces in combat. Communication is paramount; radios must be reliable over the entire distance covered.

Actions at the Objective

The patrol leader selects a series of objectives. Once an objective is reached, the patrol leader initiates a planned set of actions in order to establish and maintain contact with the enemy. These plans and actions are guided by the missions to establish or maintain contact—not to engage in decisive combat. Contact with the enemy is maintained for the purposes of surveillance, applying pressure, and preventing seizure of the initiative. If the contact patrol becomes decisively engaged with the enemy, many of the tasks originally assigned to the patrol cannot be accomplished, since the enemy has seized the initiative and friendly forces are not forced to react.

13005. AMBUSH PATROLS

General

An ambush is a surprise attack from a concealed position upon a moving or temporarily halted target. It is one of the oldest and most effective types of military actions. Ambush patrols conduct ambushes of enemy patrols, resupply columns, and convoys. The intent of an ambush is to place the enemy in a dilemma where staying in the kill zone or attempting to move out of it prove equally lethal. The ambush may include an assault to close with and decisively engage the enemy, or the attack may be by fire only.

Purpose of Ambushes

Ambushes are executed for the general purpose of reducing the enemy's over-all combat effectiveness and for the specific purpose of destroying its units. The cumulative effect of many small ambushes on enemy units lowers enemy troop morale and harasses the enemy force as a whole.

Destruction is the primary purpose of an ambush because loss of men killed or captured, and loss of equipment and supplies destroyed or captured, reduces the overall combat effectiveness of the enemy.

Harassment, though less apparent than physical damage, is a secondary purpose of ambushes. Frequent ambushes force the enemy to divert men from other missions to guard convoys, troop movements, and carrying parties. When enemy patrols fail to accomplish their mission because they are ambushed, the enemy is deprived of valuable information. A series of successful ambushes causes the enemy to be less aggressive and more defensive minded. His men become apprehensive, overly cautious, reluctant to go on patrols, seek to avoid night operations, are more subject to confusion and panic if ambushed, and in general, decline in effectiveness.

Classification of Ambushes

A *deliberate ambush* is one in which prior information about the enemy permits detailed planning before the patrol departs for the ambush site. Information needed to plan a deliberate ambush includes the size, composition, and organization of the force to be ambushed; how the force operates; and the time it will pass certain points or areas. A deliberate ambush may be planned for such targets as—

- Any force if sufficient prior information is known.
- Enemy patrols that establish patterns by frequent use of the same routes or habitually depart and re-enter their own areas at the same point.
- Logistic columns.
- Troop movements.

An *ambush of opportunity* is conducted when available information does not permit detailed planning before the patrol departs. This is the type of ambush that an infantry unit normally conducts. An ambush of opportunity should not be confused with a hasty ambush. An ambush of opportunity is a planned ambush; a hasty ambush is an immediate action. In planning for an ambush of opportunity, the patrol must be prepared to execute any of several courses of action based on the types of targets that may be ambushed and must rehearse prior to departure. The course of action taken is determined when the opportunity for ambush arises.

The patrol leader may be directed to reconnoiter an area for a suitable ambush site, set up at the site selected, and execute an ambush against the first profitable target that appears.

The patrol may depart just after dark, move to a specific point, observe until a designated time, ambush the first profitable target after that time, and return before daylight.

A *hasty ambush* is an immediate action where the patrol makes visual contact with an enemy force and has time to establish an ambush without being detected. The actions for a hasty ambush must be well rehearsed and accomplished through the use of hand and arm signals given from the patrol leader.

Types of Ambushes

There are two types of ambushes: point and area. The *point ambush* is one where forces are deployed to attack along a single killing zone. The *area ambush* is one where forces are deployed as multiple related point ambushes.

A point ambush, whether independent or part of an area ambush, is positioned along the enemy's expected route of approach. Formation of the forces conducting the ambush is important because, to a great extent, the formation determines whether a point ambush is able to deliver the heavy volume of highly concentrated fire necessary to isolate, trap, and destroy the enemy.

The ambush formation to be used is determined by careful consideration of possible formations and the advantages and disadvantages of each in relation to—

- Terrain.
- Visibility.
- Forces.
- Weapons and equipment.
- Ease or difficulty of control.
- Target to be attacked.
- Combat situation.

For a detailed discussion of ambush formation, see appendix D.

Ambush Operation Terms

The *ambush site* is the location where an ambush is established.

The *killing zone* is that portion of an ambush site where fires are concentrated to trap, isolate, and destroy the target. On little-traveled roads, an obstacle placed in a defile, in the woods, on a bridge or on a steep upgrade can be used effectively to force vehicles to halt, and thus render the occupants vulnerable to attack. Antitank mines may be emplaced and the occupants of the wrecked vehicle killed or captured while still dazed by the explosion.

A *near ambush* is a point ambush where the attacking force is located within reasonable assaulting distance of the killing zone (50 meters is a guide). A near ambush is most often conducted in close terrain, such as a jungle or heavy woods.

A *far ambush* is a point ambush where the attack force is located beyond reasonable assaulting distance of the killing zone (beyond 50 meters is a guide). A far ambush may be more appropriate in open terrain offering good fields of fire or when the target will be attacked by fire only.

Factors for a Successful Ambush Patrol

There are many factors that give the ambush its best chance of success. The ideal situation would be to position the ambush on favorable terrain and have detailed planning completed beforehand.

The patrol leader of an ambush looks for the most favorable terrain in which the enemy is canalized between two obstacles with limited opportunity to attack or escape. Suitable areas include defiles, small clearings, bends in trails, and steep grades. Dense undergrowth adjacent to the ambush site permits observation from concealed positions. The ambush patrol should have maximum cover and concealment, not only for the firing positions, but for the routes of withdrawal. The enemy should be in an area offering as little protection from fire as possible. Favorable fields of fire include stretches of road, trail or open ground of at least 100 meters for machine guns and 15 meters for rifle fire and grenades. The ambush site can be improved by constructing obstacles—such as felled trees, wire, land mines, or booby traps—to impede the enemy.

Planning

A deliberate ambush or an ambush of opportunity requires thorough planning.

A deliberate ambush plan is based on extensive knowledge of the enemy and terrain, and is planned and rehearsed in great detail. A physical reconnaissance of the ambush site is made during the preparation phase and information gained is incorporated into the plan. All likely immediate actions of the enemy when ambushed are examined. Planned counteractions are developed and rehearsed.

In planning an ambush of opportunity, any available information on the enemy and terrain is used. A tentative plan for the ambush that incorporates all anticipated actions is developed and rehearsed. However, the bulk of planning is done concurrently during the patrol leader's reconnaissance of the prospective ambush site. In a rapidly developing situation, hasty ambush immediate action is employed.

The route and ambush site considerations apply to both deliberate ambushes and ambushes of opportunity. A primary route that allows the patrol to enter the ambush site from the rear is planned. Entering the prospective killing zone is avoided. If the killing zone must be entered to place mines or explosives, care is taken to remove any tracks and signs that might alert the enemy and compromise the ambush. If mines or explosives are to be placed on the far side of the ambush site, or if the appearance of the site from the enemy's viewpoint is to be checked, a wide detour is made around the killing zone. Care is taken to remove any tracks that might reveal the ambush. An alternate route from the ambush site to the objective rally point, as in other patrols, is planned.

Maps and aerial photographs are used to carefully analyze the terrain. When possible, an on-the-ground reconnaissance of the ambush site is made prior to occupation. Obvious ambush sites are avoided as the element of surprise is even more difficult to achieve in these areas. An ambush site must provide for—

- Favorable fields of fire.
- Occupation and preparation of concealed positions.
- Canalization of the target into the killing zone. (An ideal killing zone restricts the enemy on all sides, confining him to an area where he can be quickly and completely destroyed. Natural obstacles, such as cliffs, streams, embankments, or steep grades, are used whenever possible to force vehicles to slow down. Man-made obstacles, such as barbed wire, mines, and craters in the roads, are used to supplement natural obstacles.)
- Covered routes of withdrawal that enable the ambush force to break contact.
- Avoidance of enemy pursuit by fire.

Occupation of Ambush Site

The surrounding area is searched for enemy patrols prior to occupation of the ambush site. Ambush formations are used to physically deploy the patrol in a manner to inflict maximum destruction upon the enemy and to provide maximum security to the patrol. Ambush formations are contained in appendix D.

Positions

The patrol is moved into the ambush site from the objective rally point. Security is positioned first to prevent surprise while the ambush is being established. Automatic weapons are then positioned so each can fire along the entire killing zone. If this is not possible, overlapping sectors of fire are provided to cover the entire killing zone. The patrol leader then selects his position, located so he can tell when to initiate the ambush. Riflemen and grenadiers are positioned and sectors of fire are assigned to cover any dead space left by the automatic weapons. The patrol leader sets a time by which positions are to be prepared. Patrol members clear fields of fire and prepare positions in that order, with attention to camouflage for both.

Suitable Objective Rally Point

An easily located objective rally point is selected and made known to all patrol members. The objective rally point is located far enough from the ambush site so that it will not be overrun if the enemy assaults the ambush. Routes of withdrawal to the objective rally point are reconnoitered. Situation permitting, each man walks the route he is to use and picks out checkpoints. When the ambush is to be executed at night, each man must be able to follow his route in the dark. After the ambush has been executed, and the search of the killing zone completed, the patrol is withdrawn quickly but quietly, on signal, to the objective rally point where it reorganizes for the return march. If the ambush was not successful and the patrol is pursued, withdrawal may be by bounds. The last group may arm mines, previously placed along the withdrawal route, to further delay pursuit.

Local Security

Security must be maintained. Security elements do not usually participate in the initial attack, but protect the rear and flanks, and cover the withdrawal.

Patience

The Marines of the ambush force must control themselves so that the ambush is not compromised. Patience and self-discipline are exercised by remaining still and quiet while waiting for the target to appear, particularly if the patrol occupies the ambush site well ahead of the arrival of the enemy. Patience is necessary so as not to alert the enemy to the presence of the ambush.

Surprise

Surprise must be achieved, or the attack is not an ambush. If complete surprise cannot be achieved, it must be so nearly complete that the target is not aware of the ambush until too late for effective reaction. Surprise is achieved by careful planning, preparation, and execution so that targets are attacked when, where, and in a way for which they are least prepared.

Coordinated Fires

Properly timed and delivered fires contribute heavily to the achievement of surprise, as well as to destruction of the enemy. The lifting or shifting of fires must be equally precise; otherwise, the assault is delayed and the enemy has an opportunity to recover and react. All weapons, mines, and demolitions are positioned and all fires, including those of available artillery and mortars, are coordinated to achieve the following results:

- Isolation of the killing zone to prevent the enemy's escape or reinforcement.
- Surprise delivery of a large volume of highly concentrated fires into the killing zone.

Control

Close control of the patrol is maintained during movement to, occupation of, and withdrawal from the ambush site. This is best achieved through rehearsals and establishment and maintenance of good communications. When the enemy approaches, the temptation to open fire before the signal is given is resisted. The patrol leader must effectively control all elements of the ambush force. Control is most critical at the time the enemy approaches the killing zone. Control measures must provide for—

- Early warning of enemy approach.
- Fire control. Withhold fire until the enemy has moved into the killing zone, then open fire at the proper time.
- Initiation of appropriate action, if the ambush is prematurely detected. Individual patrol members must be prepared to react if detected by the enemy prior to the initiation of the ambush.
- Timely and orderly withdrawal of the ambush force from the ambush site and movement to the objective rally point.

It is important to remember that an ambush patrol should have four distinct signals: one to open fire (with an alternate signal to open fire to be used at the same time as the primary); a signal to cease fire or shift fire; a signal to assault or search the killing zone; and a signal to withdraw. The signal to open fire should meet two criteria: first, it should be the firing of a weapon that will kill the enemy; secondly, it should be a weapon reliable in any weather condition. A good primary signal is a Claymore mine, and an alternate signal would be a closed bolt weapon (M16A2). Open bolt weapons (M240G, M249) should not be relied upon to initiate an ambush.

Execution of an Ambush

The manner in which the patrol executes an ambush depends primarily on whether the ambush's purpose is harassment or destruction. To a lesser degree, the execution of the ambush is determined by whether the ambush is deliberate or an ambush of opportunity.

When the primary purpose is harassment, the patrol seals off the area with security teams to prevent enemy reinforcement and escape. Maximum damage is inflicted with demolitions and automatic weapons fire. The patrol delivers a very heavy volume of fire for a short time and withdraws quickly and quietly. The patrol avoids being seen by the enemy.

When the primary purpose of the patrol is destruction, the area is sealed off with security units. Maximum damage is inflicted with demolitions, antitank weapons, and automatic weapons fire from the support team or element. When these fires cease or shift, an assault is launched into the killing zone with heavy fire and violence to complete destruction. The assault unit provides security, while designated teams search and/or capture personnel and destroy vehicles and equipment. On the patrol leader's command, or by prearranged signal, all units withdraw to the objective rally point and move out quickly.

When the patrol's primary purpose is to obtain supplies or capture equipment, security units seal off the area. Demolitions and weapons are used to disable vehicles. The assault unit must use care to ensure its fire does not damage the desired supplies or equipment. Designated teams secure the desired items; other teams then destroy enemy vehicles and equipment.

The most successful ambush is one where the attacker is deployed and concealed in such a way that the enemy will unknowingly be surrounded by fire. The usual method is for the attackers to deploy themselves along a trail or route the enemy will travel. The enemy is permitted to pass by the center of the attacker's force so that the attack can be made from the front, flank, and/or rear. One or two men are posted well forward and to the rear along the route to prevent any enemy from escaping. All fires should be delivered simultaneously on a prearranged signal.

An effective method of luring the enemy is for an ambushing patrol to cut communication or electrical wire. The patrol then deploys and ambushes the line repair crew when it arrives. Since the line crew may be protected by riflemen, the attackers must be careful to engage the entire party.

Vehicles and foot personnel moving on well-established transportation routes can sometimes be captured by altering or moving directional signs so as to divert the enemy into an area where he can be more readily attacked. The attack can best be accomplished at an obstacle, such as a stream or gully, that forces the enemy to stop or slow down.

After the enemy has been ambushed and destroyed, the unit quickly withdraws over a prearranged route to the objective rally point. Speed is very important, since the noise of the ambush could alert other nearby enemy units.

13006. SECURITY PATROLS

General

Security patrols are assigned missions that may or may not require them to engage the enemy. They are used in proximity to defensive positions, on the flanks of advancing units or in rear areas. Purposes of security patrols are to detect infiltration by the enemy, destroy infiltrators, and protect against surprise and ambush.

In any situation where there is a threat of attack, such as a rear area threatened by guerrillas or a facility that is under threat of a terrorist attack, all Marines, not just the infantry, must know how to conduct a security patrol.

In just the offensive operations, infantry units provide security patrols to screen their flanks, areas, and routes. Whereas, in defensive operations, security patrols are used to prevent the enemy from infiltrating an area, detect and destroy infiltrators, and prevent surprise attacks. In rear areas, particularly when there is guerrilla or terrorist threat, the requirement to conduct security patrols increases for all Marine air-ground task force (MAGTF) units ashore, particularly aviation and combat service support units.

Task Organization and Equipment

Generally, a Marine rifle squad or similar sized organization is considered ideal for security patrols. Communications are important to higher headquarters so that they receive information from the patrol; and communications are important to the patrol to request fire support, etc. The radio the patrol carries must have the range necessary for higher headquarters to be able to receive transmissions from anywhere along the patrol route, and the patrol must have a secondary means of reporting (i.e., flare signals upon contact).

Patrol Procedures

All of the procedures presented in previous sections are to be used in security patrols.

Patrol Planning

Security patrol planning includes—

- Rehearsing prior to departing friendly lines.
- Maintaining communications.
- Support by organic weapons.
- Reinforced if necessary.
- Using varied routes and never establishing a routine pattern.
- Staying within proximity of friendly units.

Patrol Techniques

Within rear areas, an irregular pattern of patrol is established and changed daily. Outside of friendly lines it would be prudent to establish a definite preplanned route for the patrol, of which all adjacent units know the route. The parent unit commander establishes frequent checkpoints for control. If checkpoints are designated, the patrol leader treats them as individual objectives to be searched and cleared.

The patrol has a definite plan as to what to do if contact with enemy is made, how to break contact, how to defend itself, and how to call for supporting fires. It is imperative that patrol members know what to do if they become split or separated; i.e., location of rally points and how to be recovered.

13007. URBAN PATROLS

General

As national strategy continues to focus on regional vice global conflicts, the Marine Corps will continue to conduct urban operations in various operational environments. Cities and towns are often the center of economic and political power and are therefore extremely vulnerable to urban insurgent activities and violence. The lessons learned from recent operations in Somalia, as well as experiences gained by British forces in Northern Ireland, provide the foundations of urban patrolling.

Enemy Forces

Enemy forces in urbanized areas range from organized military forces to low intensity engagements with insurgents, such as terrorists or local gangs. For Marine Corps doctrine and tactics, techniques, and procedures (TTP) for dealing with the higher intensity threat in urbanized areas, refer to MCWP 3-35.3, *Military Operations on Urbanized Terrain.*

Common Threat Tactics

Cities provide cover and concealment for both friendly forces and enemy forces. However, enemy forces generally find active support only in certain areas of the town or city. The urban-based insurgent or terrorist usually lives in a friendly community or in one where the people are too frightened to withhold support or inform anyone about the situation. The insurgent or terrorist normally maintains close contact with leaders and others friendly to the cause. The enemy will often have efficient communication and intelligence systems, sometimes involving women and children to provide cover for its activities.

The urban insurgent normally cannot, like his rural counterpart, establish bases and recruit large military units. He is generally an individual or a member of a relatively small group. He relies on the cover afforded by the city's populace and terrorizing them to coerce loyalty or support.

Urbanized areas tend to give the insurgent and/or terrorist many opportunities to initiate action and gain advantage. The normal presence of large numbers of people in cities provides the foe an opportunity to mass crowds quickly and manipulate demonstrations easily. The presence of women and children during mass demonstrations may restrict the courses of action available to friendly forces. Major incidents stemming from overreaction or excessive use of force by friendly forces may provide the insurgent with propaganda material. Publicity is easily gained in an urbanized area because major incidents can't be completely concealed from the local population. Insurgent successes can be exploited to discredit the ability of host nation police, friendly forces, and the civil government, and gain recognition for the insurgents' cause.

The urban insurgent or terrorist can usually be expected to operate more boldly than his rural counterpart. This is reflected in the enemy's tactics. A single sniper or bomber may be the norm in the urbanized area, whereas the rural threat is generally the more conventional ambush. In urbanized areas, explosive devices can be easily emplaced and used effectively against large groups or select individuals. Enemy forces may be expected to employ the following tactics in urbanized areas:

- Using local communications, such as radio and newspapers, for propaganda purposes.
- Disrupting industry and public services through strikes and sabotage.
- Generating widespread disturbances designed to stretch the resources of the security force.
- Creating incidents or massing crowds in order to lure the patrol or reaction force into a trap.
- Provoking security forces in the hope that they may react improperly, therefore discrediting the security force by means of propaganda.
- Sniping at roadblocks, outposts, sentries, and patrols.
- Attacking friendly bases with rockets and mortars.
- Planting explosive devices, either against specific targets or indiscriminately, to cause confusion and destruction, and lower public morale and confidence.
- Using ambush patrols.
- Firing on friendly helicopters.

Principles of Urban Patrolling

Patrolling in an urban environment often presents conditions considerably different and often more complex than those encountered in rural and less inhabited areas. While the principles of patrolling are still relevant in an urban situation, the nature of urban patrolling has led to the development of six specific urban patrolling principles. They are—

1. **Depth.** The restrictive, canalizing nature of urbanized terrain usually limits a patrol's ability to disperse laterally. To prevent the patrol from bunching up, patrols normally maintain dispersion along the length of a patrol formation.

2. **Mutual Support.** The positioning of units in-depth within the patrol enables one unit to cover another unit's movement and facilitates immediate action during various situations. Aircraft, vehicles, and snipers also provide good mutual support.

3. **Deception and Pattern Avoidance.** Deception and pattern avoidance are normally a planning consideration of the headquarters directing the entire patrolling plan. By varying patrol routes, durations, and departure times, hostile actions commonly used against urban patrols—such as enemy ambushes, and roadblocks—are more difficult to plan and may be preempted.

4. **Intra-patrol Communication.** Elements of an urban patrol must have the means to communicate with each other. Ideally, each element will possess a radio enabling it to remain in continuous communication that facilitates rapid response and reporting to higher headquarters, reaction force coordination, and coordination of actions with other patrols or fire support agencies.

5. **Establishment of a Reaction Force.** The nature of urbanized terrain (its compartmentalization) makes urban patrols more vulnerable to a wider range of hostile actions. The requirement for immediate, coordinated reinforcement of a patrol is best satisfied by employment of an established reaction force. The reaction force requires superior mobility (relative to the enemy's) and fire support to be effective for this task.

6. **Three-Dimensional Threat**. Patrolling in an urban environment requires constant attention to its three-dimensional aspect; hostile actions can originate from rooftops, streets, subsurface levels or combinations of all levels at once.

Classification of Urban Patrols

Mission

The vast majority of urban patrols are overt in nature, with their presence readily apparent to the local populace. Most urban patrols are combat vice reconnaissance patrols. The vast majority of urban patrols are security patrols. However, units may be assigned secondary tasks of reconnoitering specific or general areas along the patrol routes.

Raids normally involve a swift penetration of an objective to secure information, confuse the enemy or destroy installations. Raids include a planned withdrawal upon completion of the assigned mission.

Movement

Means of movement are as follows:

- Dismounted: movement on foot.
- Vehicular: movement by motorized, mechanized or armored vehicle.
- Helicopterborne: movement by helicopter, however, helicopterborne patrols will usually involve dismounted or vehicular movement after patrol insertion.
- Combination: movement using a combination of methods.

Task-Organization

Units task-organize to meet the specific requirements of the mission and situation. The Marine rifle squad is ideally suited for urban patrolling and can easily integrate attached specialists required by specific missions. Specialists who may accompany urban patrols include—

- Interrogator-translator team (ITT) and counterintelligence team (CIT) Marines.
- Explosive ordnance disposal (EOD) personnel.
- Members of host nation or allied military forces.
- Interpreters.
- Local community leaders.
- Local law enforcement officers.
- Public affairs personnel escorting media representatives.

Dismounted Patrol Organization

Patrols should maintain unit integrity (fire team and/or squad) in organizing elements for an urban patrol. Similar to reconnaissance patrols, the vulnerability of urban patrols necessitates that all elements must provide for their own security in addition to the combat aspect of the mission. Combined assault and security teams are an effective method to organize for all-around security.

Urban Patrol Base Operations

Urban patrols may operate from an established patrol base that may be located within the unit's assigned area of operations or an area designated for a patrol. The patrol base should be located in a building used exclusively for this purpose. Patrol bases may also be located within a larger site that houses other agencies, such as a higher headquarters' command echelon.

If located within a larger site, the patrol base will be included in the overall facility security plan. If isolated, the patrol base must consider the following security factors:

- External security:
  - Barrier plan.
  - Sentry posts.
  - Local security patrols.
- Internal security:
  - Covered positions for all Marines.
  - Contingency plan for hostile actions against patrol base.

The Reaction Force

The high probability of an urban patrol becoming involved in a hostile or volatile incident requires the establishment of a dedicated reaction force for rapid reinforcement, support or extraction of the patrol. Ideally, the reaction force is—

- Large enough and task-organized in a manner that it can meet and quickly defeat the expected threat. Reaction forces are normally tiered with a lead element (normally one-third of the unit's size) and a main body force (the other remaining two-thirds of the force). Reaction force response times routinely are determined in advance by higher headquarters.
- Ready to respond immediately.
- Motorized or mechanized and supported by close air support and other fire support.
- Familiar with the area of operations.
- Briefed on the patrol's plans and monitors the status of patrols in progress.
- Task-organized to be multimission capable.
- Able to communicate with the higher headquarters, fire support assets, patrol base, and the patrol.
- Controlled by higher headquarters, once employed.

Patrol Preparation

Planning

Higher headquarters will—

- Designate the area for patrol.
- Provide intelligence briefs and updates.
- Ensure liaison with allied forces and the civilian populace.
- Provide special equipment and personnel required for the mission (scout snipers, public affairs officer, interpreters, etc.)
- Provide urban maps, photos, terrain models as required.
- Consider deception and pattern avoidance when issuing mission.
- Prescribe rules of engagement (ROE).

Intelligence Brief

An intelligence brief is conducted by the S-2 officer or representative prior to a patrol conducting its mission. The brief adresses the situation relevant to the specific patrol (e.g., routes, areas, updated enemy situations).

Coordination

Higher headquarters will effect liaison with adjacent and allied forces, as well as civil authorities and other agencies, having a possible effect on the patrol. The patrolling unit generally follows the same procedures as those used during patrol planning and execution in a jungle or forest environment.

The Urban Patrol Order or Warning Order

The Urban Patrol or Warning Orders use the same format and considerations as noted in this publication for patrolling. They rely heavily on a detailed terrain model, photographs, and subterranean construction to ensure complete understanding of the plan.

Rehearsals

The limited size of the patrol base usually precludes the need for full-scale rehearsals. Immediate action drills, such as crossing danger areas, are rehearsed in as much detail as possible, despite the limited available space.

Inspections

Initial and final inspections are conducted in the same manner as other patrols. Attached personnel must be fully integrated into the patrol and familiar with the plan and unit standing operating procedures (SOPs).

Conducting an Urban Patrol

Movement

Individual and unit movement considerations are generally the same as those for other patrols. However, urban environments require consideration of additional factors. Because of these factors, an urban patrol leader should—

- Ensure that each movement within a patrol takes place under the observation or cover of another individual or element of the patrol.
- Know where cover can be taken in the event of a hostile incident or action.
- Be prepared for contact with civilians, especially children, during the patrol and be aware that they may intentionally attempt to distract patrol members.
- Expect the presence of vehicles (both moving and stationary) along the patrol route.
- Expect members of the patrol to be approached by dogs and what action to take if threatened.

Patrol Formations

Squad-sized Patrols. The need for immediate fire power outweighs the dangers of becoming canalized. In contrast to other types of patrols, the headquarters element of an urban patrol will normally locate at the lead of the patrol column. This allows the patrol leader greater flexibility through control of two combined assault and security (A&S) teams. The leaders of these elements tactically stagger their members on each side of the road (see fig. 13-1 on page 13-12).

A&S teams follow in trace of the headquarters element and maintain unit integrity on separate sides of the street. One unit will remain slightly to the rear to create a staggered interval between Marines on either side of the street. This allows A&S teams to take lateral routes in support of headquarters element without having to cross a street to do so.

Platoon-sized Patrols. Squads will generally travel abreast of each other, moving along parallel routes. The interval between squad-sized units and/or teams is situation-dependent, but is usually between 100 and 150 meters (roughly two city blocks; this often prevents visual contact between the units). The intent is to create less of a target to an aggressor, yet still allow the patrol to quickly react to an incident. Individuals within units or teams will move in a staggered column as in a squad-sized patrol. See figure 13-2, which is located on page 13-13).

Night

Night patrols will generally be at least squad-sized and will generally use the same formation as that for day patrols. At night, it may be necessary to close distances between individuals or elements to maintain control. Consideration should be given to the use of night vision devices and thermal weapons sights.

Navigation, Control, and Security Measures

The patrol leader is ultimately responsible for the navigation. The headquarters element normally functions as the base unit during movement. The designated navigator is normally assigned from within the headquarters element. City maps are often inaccurate or outdated; however, when used with aerial photographs and other navigational aids, they can be effective for urban navigation.

Checkpoints and phase lines should be related to major streets (alleys, buildings, bridges) for easy identification. Arrival at checkpoints and crossing of phase

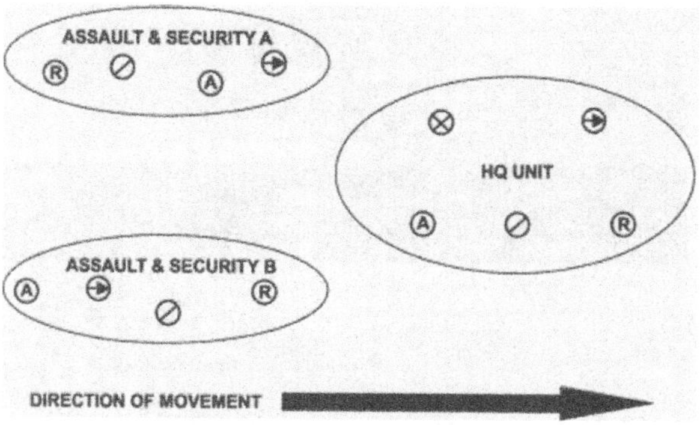

Figure 13-1. Squad-sized Dismounted Urban Patrol Formation.

lines should be relayed to higher headquarters using established brevity or codewords. A detailed patrol overlay is issued to both the reaction force and higher headquarters to keep them advised of routes and control measures used. A terrain model should be maintained at higher headquarters to aid in supervision and control of the patrol.

The use of camouflage should be limited to avoid frightening and confusing the local populace. The often overt nature of urban patrols may negate the need for camouflage. Patrols are sometimes deployed to show force presence and usually move on the urban street in plain view.

During daylight, patrols will routinely vary their rate of movement ranging from short halts to brief periods of double-timing. The British term for this urban patrolling technique is *hard-targeting*, meaning it makes the patrol harder for an enemy to target. Altering the rate of movement is intended to frustrate the enemy's ability to coordinate an attack or ambush against a targeted patrol.

Patrols should use short security halts, with Marines taking up mutually supporting firing positions. Marines must always work in pairs, ensuring mutual support. The last Marine in the element will provide rear security, but stays in his buddy's sight.

Individual Tasks

Individuals may be assigned collateral tasks performed throughout the patrol that may increase the patrol's situational awareness. Individual tasks may include—

- Vehicle spotter: looks for suspicious or known insurgent vehicles.
- Personnel spotter: observes and attempts to recognize previously identified enemy in crowds.
- Talker: attempts to gain information from casual conversations with the local populace. (Talkers are usually subordinate leaders or Marines with foreign language skills.)
- Searcher: conducts physical searches of vehicles and personnel while other patrol members provide cover and security.
- Marksman: engages point targets when the tactical situation does not permit massed or high volume fire. Other patrol members provide security to cover the marksman's engagement.

Departure of Friendly Lines

Urban patrols must vary their departure times to prevent being ambushed while exiting the patrol base.

Individual elements will usually depart exits at staggered times and at different movement rates, especially where sentries cannot provide cover. (This technique is used when the environment adjacent to the patrol base is dangerous.)

Once an element has exited friendly lines, a short halt is conducted in a predetermined, covered initial rally point 50 to 100 meters from the base. This ensures all elements are in position before the patrol continues.

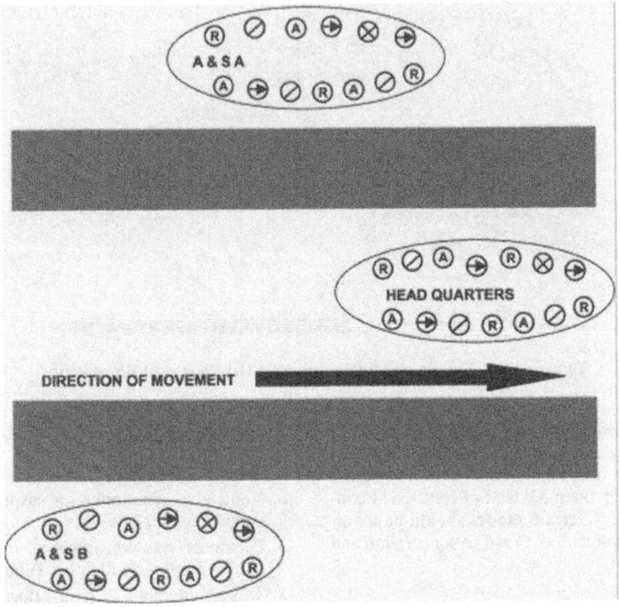

Figure 13-2. Platoon-sized Dismounted Urban Patrol Formations.

Exit points, routes from the base, departure techniques, and locations of IRPs should be varied constantly to avoid setting a pattern. This pattern avoidance may also include using vehicles to insert patrols away from the urban patrol base and employing empty vehicles as part of a deception plan.

Danger Areas

Urban patrols may encounter hundreds of danger areas during a single patrol. The three-dimensional threat requires keen situation awareness by every patrol member. Many danger areas can be dealt with simply by avoidance, while others require an adjustment of patrol formation, movement rate, etc. In the urban environment, places to be treated as danger areas are points that pose a major threat to the patrol, such as local political and religious headquarters, weapons containment areas, roads and routes that canalize movement and direct fire, and any area with a history of repeated contact.

Near and far side rally points are designated and briefed during the issuance of the patrol order. Squad-sized formations may use the A&S teams to provide flank security for the headquarters element and for each other. The headquarters element identifies the danger area and takes up a position on the near side of intersection. Individuals provide all-around security (see fig. 13-3 on page 13-14).

Two Marines (one from each A&S team) are designated to move through the headquarters element and establish respective firing positions on the near side of the danger area covering the patrol's near side flanks. They are followed by a second pair (again, one Marine from each A&S team) that moves across to the far side of the danger area and establishes respective firing positions covering the patrol's far side flanks (see fig. 13-4 on page 13-14).

Once near and far side flank security is established, the headquarters element moves across to the far side of danger area (see fig. 13-5 on page 13-15).

The remaining A&S team members then cross the danger area and join the headquarters element on the far side (see fig. 13-6 on page 13-15).

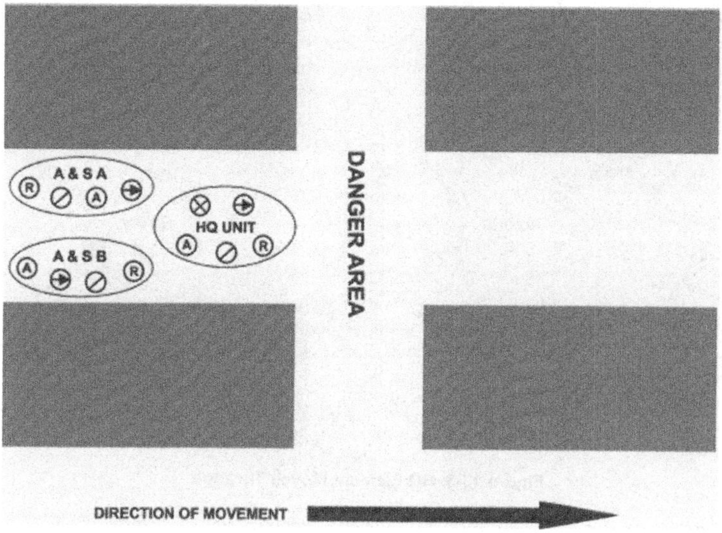

Figure 13-3. Approaching Danger Area.

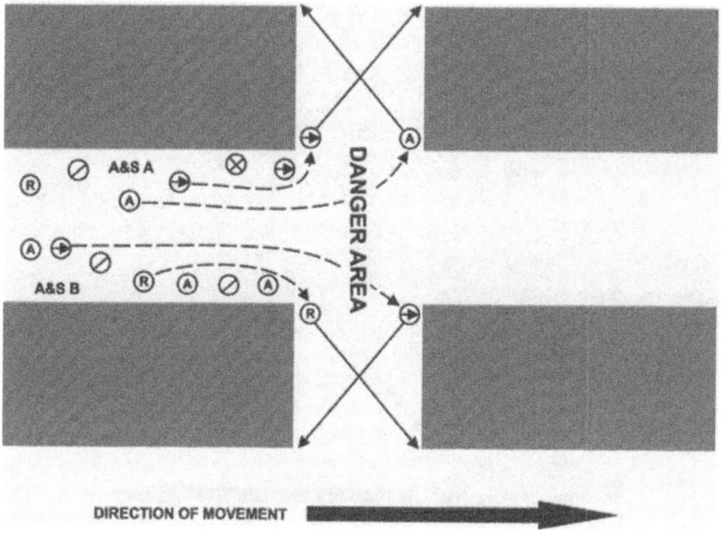

Figure 13-4. Securing Flanks.

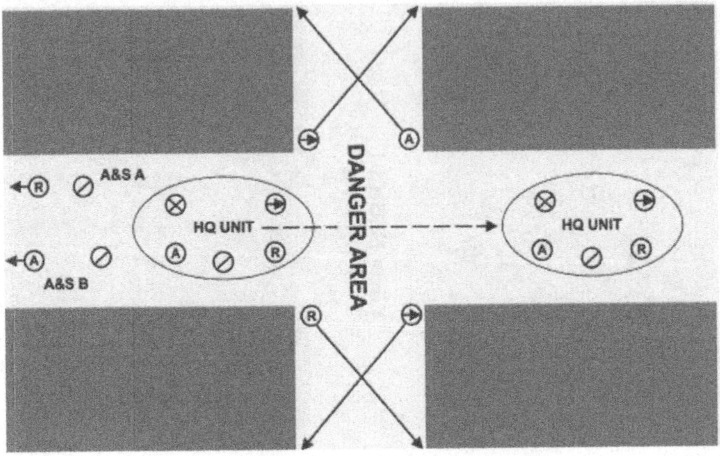

Figure 13-5. HQ Element Moves Through.

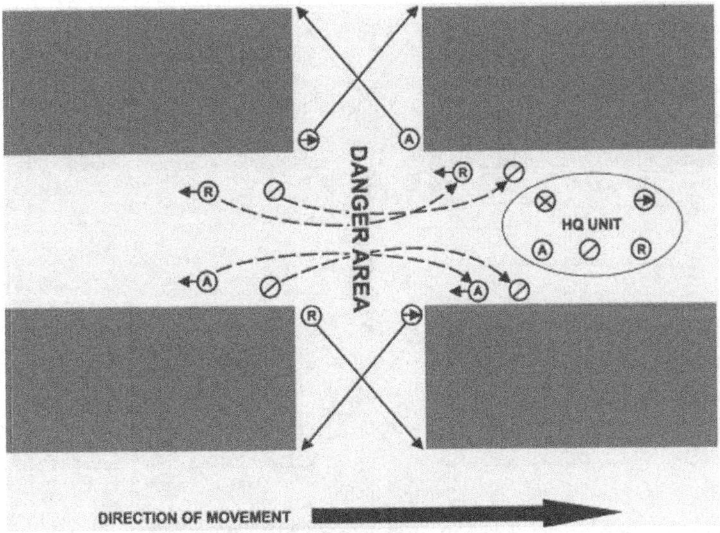

Figure 13-6. A&S Elements Move Through.

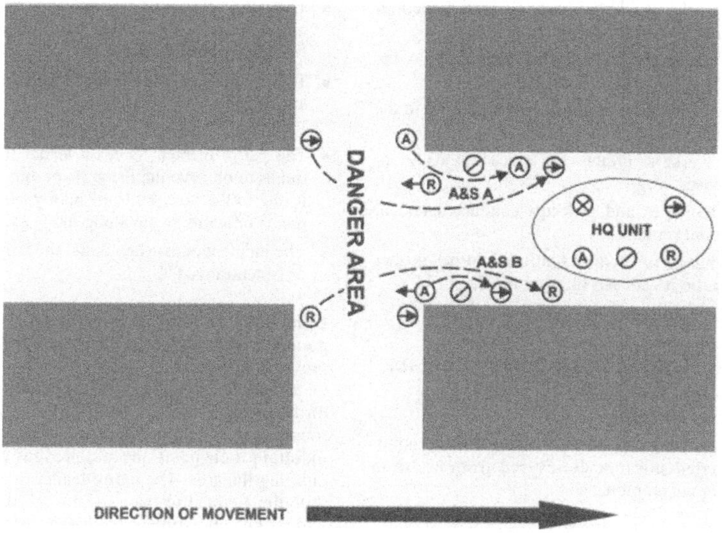

Figure 13-7. Patrol Resumes Movement Along Intended Route.

Once the trail A&S team members cross the danger area, they take up rear security and cover the movements of the near and far side flank security teams as they return to their positions in the patrol formation. The near side security team should collapse back first, followed by their far side counterparts. The patrol then resumes its advance away from the danger area (see fig. 13-7).

Interaction with Local Populace

Urban patrols must interact with the local populace. Patrols are at first a novelty to the civilians but can quickly become an unwelcome intrusion. The movement of the patrol must be fast enough to prevent the enemy from massing their fires upon it, but deliberate enough to ensure adequate security and mutual support. Patrol members must realize that they are usually the only Marines the local populace will encounter and that an inappropriate gesture, comment or act could lead to the deterioration of rapport between U.S. forces and the general population. Marines must remember that the vast majority of the individuals with whom they come in contact will be noncombatants attempting to survive in trying political, economic, and social situations.

Hostile incidents often seem to occur spontaneously, but there are usually indications that can alert Marines to imminent danger. The most obvious are the sudden alteration of normal routines, patterns, and attitudes of the local populace or other unusual activity. Some examples include—

- Observers on rooftops, in windows, etc., who are obviously tracking the patrol.
- The unusual absence of pedestrian traffic and people on porches.
- Stores, markets or street vendors closed suddenly or without explanation.
- Changes in civilian attitude toward patrol members.
- Unknown individuals or vehicles in the patrol area.
- Unfamiliar vehicles parked in the patrol area (possible car bomb).
- Roadblocks.
- Children throwing rocks at patrols to possibly draw the patrol's attention away from a more serious danger, such as a deliberate ambush.
- Vehicles riding unusually low due to overloading (possibly ferrying people, weapons, explosives).
- Agitators trying to provoke an incident with patrol members.

- Absence of the usual stray dogs (dogs are adept at sensing danger and avoiding it).
- Anti-American graffiti suddenly appearing in the patrol area.
- Pictures of enemy leaders and martyrs posted in the patrol area.
- Civilian workers failing to appear at U.S. or friendly bases.
- Normal deliveries and pick-ups conducted late or early without reason.
- Sudden change of civilian sentiment in newspaper articles, radio broadcasts or other media.
- Women and children leaving to live elsewhere.

Immediate Actions Upon Enemy Contact

Reaction to Sniping

Snipings are often executed from a single firing point, but coordinated snipings delievered from multiple points are not uncommon.

Patrol element or team leaders should constantly try to identify likely firing points and anticipate their own reactions to a possible shooting. Normally, snipers in an urban environment have a detailed withdrawal plan. Once a patrol comes into contact with a sniper, the patrol leader must immediately assess the situation and maneuver his patrol accordingly. The patrol's mission, location, size, ROE, and location of the threat often determine whether the patrol will attempt to neutralize the targeted sniper. If the patrol leader decides to kill or capture the sniper, he uses planned and rehearsed immediate actions to maneuver and counter the sniper's assault. The goal is to kill the sniper or cut off his escape and capture him. There are three immediate reactions to neutralizing a sniper: initial contact, immediate follow-up, and subsequent follow-up.

Initial Contact. The initial contact is made when the sniper fires the first shot. The patrol must react immediately and positively to get behind the firing position in order to kill or capture the gunman. The period of contact ends when the gunman is killed or captured, or the patrol element or team leader on the scene ends it. The following technique is the same for both squad- and platoon-sized patrols:

- The element or team in contact attempts to identify the firing position and maneuvers designated marksman into position to return well-aimed and controlled fire. Other members of the patrol take up positions to cover the marksman's engagement.
- The patrol element or team leader in contact sends initial contact report to the patrol leader, who notifies higher headquarters.
- The element or team leader in contact determines appropriate cut-off positions and relays them to flanking elements or teams.
- The patrol element or team leader in contact continues to observe the firing point, but does not enter it due to the possibility of booby traps. Flank elements or teams set up along likely escape routes.
- The incident ends when either the sniper ceases fire or is neutralized.

Immediate Follow-Up. Regardless of the fate of the gunman, isolation of the firing point is necessary to prevent reinforcement and preserve forensic evidence (scent, spent casings, etc.). If not under fire, members of the patrol element or team cordon off the area surrounding the firing point. Flanking elements or teams maintain their positions and prevent civilians from entering the area. The patrol leader moves to link-up with the element or team in contact (if not his own), and makes an estimate of the situation. The patrol leader sends a SPOTREP to higher headquarters.

Subsequent Follow-Up. The aim of the subsequent follow-up is to use follow-on forces to clear the building of remaining resistance or to obtain evidence that can be used to capture the gunman. The patrol leader establishes a position where he can brief arriving units (reaction force commander, S-2 representative, EOD personnel, etc.). Once the arriving units have been briefed, recommendations are made to higher headquarters via radio. No one is allowed into the cordon without the patrol leader's approval.

Reaction to Becoming Decisively Engaged

If a patrol becomes decisively engaged from numerous firing positions, the following immediate action should be taken:

- All patrol members move to available cover and return accurate fire on identified firing points.
- The patrol leader assesses the situation and makes a decision to either request the reaction force or break contact.
- If the reaction force is requested, the patrol will maintain its position until the reaction force arrives. The patrol should use fire and maneuver to gain better tactical positioning and support the arrival of the reaction force. When the reaction force arrives,

its commander may decide to either clear occupied buildings or cover the patrol during its extract.

Reaction to Bomb Threat or Discovery

The use of command-detonated explosive devices is a common ambush tactic employed by a terrorist or insurgent in an urban environment. The appropriate response to a reported threat or an actual discovery is generally involves four steps (known as the four Cs)—

1. The patrol leader **CONFIRMS** the presence of the suspicious item.

2. Without touching or moving anything suspicious, patrol elements or teams **CLEAR** the immediate danger area to a minimum of 100 meters. The area is cleared from the suspected device outward, inform civilians as to the reason for evacuation.

3. A&S teams establish a **CORDON** to secure the cleared area. Avenues of approach are cordoned off to keep people out and to protect EOD or engineer personnel clearing the device. The assistant patrol leader acts as the cordon commander and informs the patrol leader when the cordon is secure. An effective cordon technique is to tape off the area with engineer tape, creating both a physical and psychological boundary.

4. **CONTROL** of the area is maintained throughout the bomb clearing operation by the patrol leader. The patrol leader sends a report to a superior concerning details of the device (if known) and the area affected. The patrol leader coordinates with arriving personnel (EOD, engineers, etc.). The patrol leader maintains communication with the assistant patrol leader and keeps the Marines informed of the progress of the clearing operation.

Reaction to a Bomb Detonation

Bombs may be used by an insurgent as a means of initiating an ambush on mounted or dismounted patrols, in which case the actions for decisive engagement apply. Immediate action in response to an isolated explosion is similar to that used in reaction to a sniping and breaks down into the same three phases:

Initial Contact. The patrol leader attempts to identify the likely initiation point and sends an initial contact report to higher headquarters. If the bomb was command-detonated, the patrol leader sends his A&S teams deep to cut off the bombers' escape routes. Any casualties are moved a minimum of 100 meters from explosion and out of the line of sight to it.

Immediate Follow-Up. The A&S teams may need to maneuver to positions behind the likely initiation point to cut off escape. Once in position, personnel checks are conducted and any suspects are detained. The patrol leader coordinates requests for required support (MEDEVAC, reaction force, etc.)

Subsequent Follow-Up. Due to the possibility of secondary detonations, the four Cs (confirm, clear, cordon, and control) can be conducted as in reaction to a bomb discovery or bomb threat.

Civil Disturbances

Urban patrols must prepare to react to spontaneous aggression by the local populace. In many cases, civil disturbances are organized by the enemy to draw dismounted patrols into a targeted area, or to distract them from enemy activity occurring elsewhere. Civil disturbances are generally divided into two categories: minor aggressive actions, and full-scale rioting.

Minor aggressive actions are activities characterized by rock-throwing or use of devices such as Molotov cocktails and may either be directed at the patrol or take place between different ethnic factions of the population. Minor aggressive actions are normally spontaneous in nature and may have minimal or limited objectives for the insurgents.

Full-scale rioting events are usually in response to another major event or incident that may enflame the populace. Full-scale riots are well-planned and orchestrated, with clear objectives or targets in mind. At times, patrols will need to attempt to maintain control of a civil disturbance situation; however, dismounted and mobile small unit patrols should generally avoid potential flashpoints. Procedures to handle civil disturbances are as follows:

- The patrol leader reports the incident to headquarters and attempts to diffuse the crowd by talking to crowd leaders.
- If the patrol leader determines the size of the disturbance is too large for the force to handle, the patrol should move away from the disturbance to a safer, more remote covered area and occupy positions to observe and report the situation to higher headquarters. To prevent the patrol from being pursued by the crowd, the patrol should move quickly and change direction, often at road junctions, to gain distance from it.

Patrol members should maintain dispersion to create a more difficult target. They should face the crowd at all times to see and avoid any projectiles thrown. Individual self-discipline must be maintained throughout the disturbance. Marines charging into the crowd or throwing objects back at the crowd will only worsen the situation. If pursued or trapped, the patrol leader may consider using riot control agents (combat support, pepper gas, etc.) to disperse or slow a crowd's movement. The use of riot control agents must be authorized under the established ROE. If the patrol leader believes the situation is deteriorating beyond the patrol's ability to control it, the patrol leader should request the reaction force, which may be better equipped to handle a large riot or mob.

Break Contact

As with patrols in rural areas, the patrol leader may be forced to break contact as a result of decisive engagement with the enemy. On the basis of his estimate of the situation, the patrol leader will normally break contact in one of the following ways:

- As a patrol, with elements providing cover for movement as defined by clock direction and distance.
- As individual units/teams taking separate routes out of the area, then linking up at a designated rally point a safe distance away from the engagement.

As in any contact with enemy forces, smoke may be employed to screen movement. Fire support agencies can be utilized to suppress targets; riot control agents can be employed to disrupt enemy movement.

Reentry of Friendly Lines

The reentry of a dismounted patrol into an urban patrol base is no different from that of a patrol conducted in a rural area. The same planning considerations and control methods apply.

Missions Related to Urban Patrolling

House Calls

House calls missions are usually part of a coordinated effort to collect information within the area of operations. They involve obtaining up-to-date information on particular houses and occupants. When possible, local police should accompany patrols to do the actual talking to the occupants. If this is not possible, a technique that may be used by an urban patrol is—

- A&S elements move to provide cover around the target house.
- The headquarters element provides security just outside the house.
- The patrol leader and one Marine for security enter the house, if invited. If not invited, they talk to the occupants on the doorstep. Patience and tact are required in requesting information. An interpreter should be present when language differences exist.

Vehicle Checkpoints

Urban terrorists or insurgents commonly use vehicles to transport personnel, weapons, explosives, and equipment. Civilian vehicles are often used for these purposes, creating the requirement to check as many vehicles as possible. While permanent, fortified checkpoints may be conducted along approaches into an urbanized area, dismounted patrols can be employed to establish hasty vehicle checkpoints to stop vehicles and to keep the enemy off-guard. The two common types of vehicle checkpoints patrols establish are hasty and deliberate.

Hasty checkpoints are deployed anywhere based upon the decision of the patrol leader. Patrols must not set patterns through the frequent use of the same sites.

Deliberate checkpoints are tasked by higher headquarters to achieve a specific purpose. Time and locations are carefully considered to avoid setting patterns.

The general layout for a squad-sized, two-way dismounted checkpoint is depicted in figure 13-8 on page 13-20. The technique is as follows:

- The A&S teams are positioned stealthily in blocking positions on both sides of the road.
- Both the patrol leader and assistant patrol leader act as "talkers" for each direction of traffic (with local police or interrogator or translator Marines acting as interpreters) while a two-man team from the headquarters element physically searches the vehicles.
- Obstacles or parked vehicles may be employed to create a staggered roadblock in center of the checkpoint to slow approaching vehicles.
- The checkpoint location should be sited so that approaching vehicles cannot see it until they have passed a security team, and they have no escape route then available.
- Signs announcing the checkpoint should be displayed a safe distance from the search area for safety to both drivers and Marines.
- Normally higher headquarters will issue criteria that determines which vehicles are searched, but random checks of cars should normally be made as

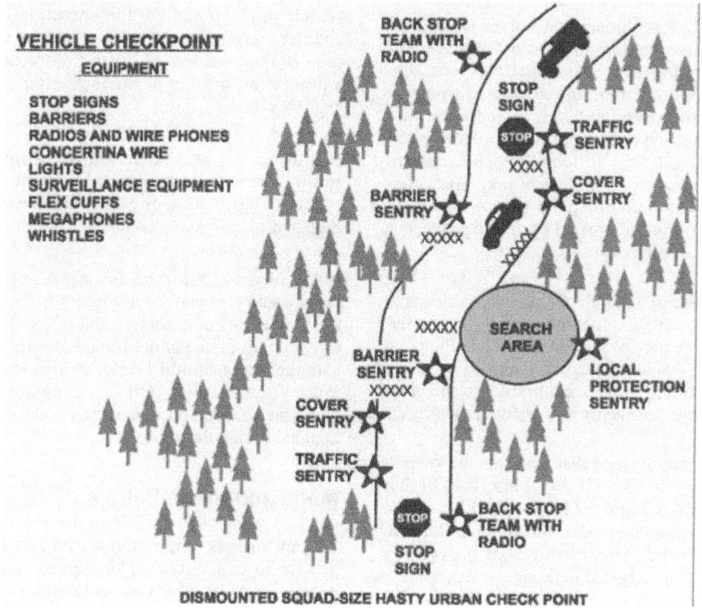

Figure 13-8. Dismounted Squad-sized Hasty Urban Vehicle Checkpoint.

well. While the vehicle is being searched the driver should accompany the searcher around the vehicle.

- Vehicle occupants should be made to exit the vehicle and then searched. Whenever possible, women should be searched by female police or female Marines.
- All patrol members should conduct themselves with courtesy and politeness. If nothing is found, an apology for the inconvenience is recommended.
- A hasty vehicle checkpoint should not be conducted any longer than 30 minutes for security reasons.
- The ROE should dictate whether or not action should be taken against vehicles that fail to stop at the checkpoint. Failure of a vehicle to stop does not automatically give authorization to fire.

Observation Posts

Urban observation posts are established to provide extended security, not only for patrol bases but also for patrols operating within the observation post's sector of observation. Observation posts can be established in conjunction with sniper operations and for providing overwatch for patrols operating within their sector of observation. Observation posts are nor-mally positioned on dominating terrain or in buildings outside the patrol base itself.

Insertion to the observation posts and conduct of observation may be either overt or clandestine in nature. Overt observation posts usually will be hardened positions to increase security. A patrol provides cover while the observation post is being inserted. A clandestine observation post relies on stealth of insertion and occupation for protection. It is normally positioned in abandoned buildings to cover sectors of observation that overt observation posts cannot. Because of their nature, clandestine observation posts are difficult to successfully establish and should not be manned for an extended period of time. Orders establishing observation posts (and patrols) must address the method of extraction as well as actions upon compromise/attack.

Cordon and Search

The cordon and search mission involves isolating a predesignated area by cordoning it off and systematically searching for enemy personnel, weapons, supplies, explosives or communications equipment. While large-scale cordon and search operations are planned and rehearsed in advance and normally entail extensive coordination with local law enforcement agencies, a squad-sized urban patrol may often conduct a cordon and search of a point target—searching one house or building identified by intelligence as a possible weapons cache.

The basic principle of a search of a populated area is to conduct it with limited inconvenience to the population. The populace may be inconvenienced to the point where they will discourage urban guerillas or insurgents from remaining in the area, but not to the point that they will assist the enemy as a result of the search.

Upon receiving intelligence that warrants the searching of a building or a specific tasking from higher headquarters, A&S elements of the patrol move to establish an inner cordon around the target building to seal it off, with the primary intent of preventing movement out of the targeted building.

On order, the designated reaction force deploys to establish an outer cordon, oriented outward some distance from the inner cordon and covering routes leading into the area in order to prevent outside interference/reinforcement. The reaction force maintains a reserve to reinforce either cordon or react to unfolding events (civil disturbance in response to the operation).

Once the cordons have been established, the patrol leader, with the assistance of local police or interpreters, informs the local populace that a building is about to be searched, that a house curfew is in effect (if permitted by higher headquarters), and that all occupants should remain indoors. Occupants of the target house are instructed to gather at a central location to stay out of the way of the search party.

The headquarters element, having linked up with any required assistance (explosive ordnance disposal (EOD), ITT, etc.) now acts as the search party and accompanies local police. A female searcher should be included in the party, if necessary.

Occupants are searched and screened first for possible enemy personnel. Apprehended persons are evacuated as soon as possible.

The head of the household should accompany the search party throughout the operation to be able to counter incriminating evidence and possible accusations of theft and looting against Marines. If possible, a prominent member of the local community should act as a witness.

Buildings are best searched from top to bottom. Ideally, the search is conducted with the assistance of combat engineers using mine detectors to locate hidden arms and ammunition.

If the targeted building is empty or the occupant refuses entry, it may be necessary to forcefully enter the premises to conduct a search of the dwelling. If an unoccupied house containing property is searched, arrangements should be made with the local community to secure it until its occupants return. Unnecessary force and damage to property should be avoided during the search.

Motorized Urban Patrols

The advantages of a motorized urban patrols is their ability to capitalize on the speed, mobility, and protection offered by various vehicles. They may be motorized, mechanized or armored vehicles or a combination. Generally, motorized urban patrols possess greater combat power than dismounted patrols and can cover larger areas faster than dismounted patrols.

The disadvantages of motorized urban patrols is that they are restricted to roads and are vulnerable to ambush by the enemy. They are also restricted in their ability to interact with the local populace.

Motorized patrols are generally organized in the same manner as dismounted patrols (see fig. 13-9). Unit integrity is maintained when assigning personnel to specific vehicles.

The urban patrolling principles apply to motorized patrols in much the same manner as dismounted patrols. Mutual support and depth are achieved by maintaining constant observation between vehicles and coordinating support with any dismounted patrols in the area. All-around security is achieved through the use of constant observation as well as the vehicle's mobility and firepower. Positive communications between units or teams are maintained through vehicle radios. Patrol routes and speeds are varied to promote deception or pattern avoidance.

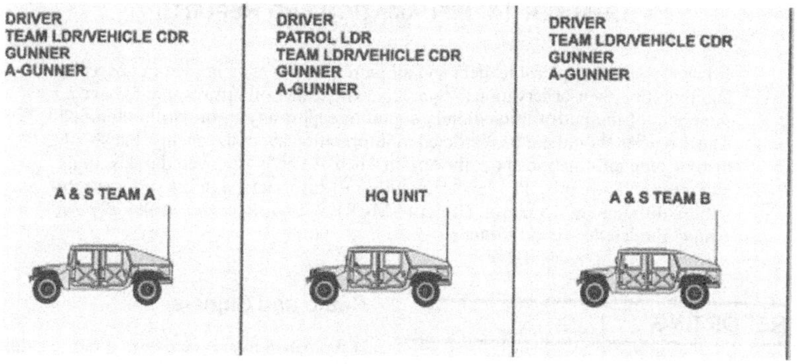

Figure 13-9. Organization of a Squad-sized Motorized Patrol.

The canalizing nature of streets and alleys force vehicular patrols to use a traveling overwatch movement technique to reduce vulnerability to ambushes. All vehicles travel at a moderate rate of speed with the lead vehicle stopping only to investigate potential danger areas. If vehicles must stop in danger areas, designated crew members' will dismount to provide security. The gunner will remain at the ready and in the turret while the driver remains in the driver's seat with the vehicle running.

Vehicles should move at a high rate of speed only when responding to an incident. At all other times, vehicle speed should be between 15 to 20 mph to allow for quick reaction and good observation. Distances between vehicles should be approximately 50 meters (one half to one city block) or such that visual contact and mutual support are ensured. Particular care is taken at major road junctions and other danger areas to ensure individual vehicles do not become isolated.

Vehicles with doors removed generally enhance observation and overall security, yet expose Marines to thrown objects, theft and concealment.

CHAPTER 14. INFORMATION AND REPORTS

It is necessary that patrol leaders and all patrol members be trained in observing and reporting their observations accurately. The leader of a patrol should have all members of the patrol immediately signal or report any information obtained. These reports should not be restricted to information about the enemy, but should also include information about the terrain, such as newly discovered roads, trails, swamps, and streams. The leader includes all information in the report to the officer dispatching the patrol. (Refer to MCRP 2-15.3, *Reconnaissance Reports Guide,* for detailed report formats.)

14001. REPORTING

The officer dispatching the patrol instructs the patrol leader on whether and when messages are to be sent back during the patrol and what communication means to use. Messages may be oral or written. They must be accurate, clear, and complete. Every message should answer the question what, where, and when. For a detailed discussion on reporting, see paragraph 6002.

Verbal Messages

A patrol leader sending a verbal message should make it simple, brief, and avoid using numbers and names. The messenger should accurately repeat the message back to the patrol leader before leaving.

Written Messages

In preparing written messages, the patrol leader must distinguish between fact and opinion. Information about the enemy should include: strength; armament and equipment; actions; location and direction of movement; unit destination, if known; time enemy was observed; and the patrol's location when the observation was made. Use of an overlay or sketch may often simplify the message.

Messengers

A messenger team is given exact instructions as to where to deliver the message and the route to take. Any information obtained along the route should be reported at the time the message is delivered. Messengers must be given all practical assistance. If in danger of capture, the messenger immediately destroys the message.

Radio and Signals

If the patrol is provided with a radio, a definite radio schedule for checking in must be arranged before departure of the patrol. The patrol leader takes every precaution to ensure that codes and copies of messages are not captured by the enemy. If a close reconnaissance of enemy lines is required, the radio should be left in a concealed location at a safe distance from the enemy. Once a report is sent by radio, the patrol should immediately leave the area to avoid the possibility of detection by enemy locating devices. Pyrotechnics (flares, colored smoke, grenades) and air-ground panels may also be used by patrols for reporting information by a prearranged signal.

SALUTE Report

Information must be reported as quickly, accurately, and as completely as possible. An established method to remember how and what to report about the enemy is to use the acronym SALUTE:

Size
Activity
Location
Unit
Time
Equipment

An example of such a report is: "Seven enemy soldiers, unit unknown, traveling SW, crossed road junction on BLACK RIDGE at 211300 August carrying one machine gun and one rocket launcher."

14002. CAPTURED ITEMS

Every patrol should make a practice of searching enemy casualties, prisoners, and installations first for booby traps, then for equipment, papers, maps, messages, orders, diaries, and codes. Search techniques should be practiced often in order to reduce the time exposed to potentially dangerous situations. Items found are collected by the patrol leader and turned in with the patrol report. The items found are marked as to time and place of capture. When possible, captured items should be linked to a specific prisoner who possessed the items or to the place where the items were found. When this is done, the enemy prisoner of war tag and item tag are marked accordingly. The patrol leader must impress upon the members of the patrol the importance of turning in all documents and equipment. Furthermore, the patrol leader must ensure all information gathered by the patrol is quickly disseminated.

14003. PRISONERS

A patrol normally does not capture prisoners unless required by the mission. If prisoners are taken, the "5S and T" rule applies. This memory aid stands for: search, segregate, silence, speed, safeguard, and tag.

Search. Prisoners are body-searched thoroughly for weapons and documents as soon as they have been captured. This search must include the helmet, body armor, and gas mask. These items are left with the prisoner for protection until the patrol is completed. Weapons, equipment, and documents are tagged and immediately sent to the patrol leader.

Segregate. Prisoners are segregated into isolated groups: officers, noncommissioned officers (NCOs), privates, deserters, and civilians. By segregating prisoners, it makes it more difficult for leaders to organize escapes and issue orders to subordinates.

Silence. Silence is essential. Do not allow prisoners to talk to each other.

Speed. Speed is required in getting prisoners to the commander who dispatched the patrol. Timely information secured from prisoners is essential.

Safeguard. Prisoners are safeguarded as they are moved. They are restrained, but not abused. If the patrol will soon reach friendly positions, prisoners are not given cigarettes, food or water until they have been questioned by interrogators. If the patrol will not return to a friendly position for a lengthy time, the prisoners are given food, water and medical aid.

Tag. Equipment and personal effects recovered from prisoners are tagged so that they may later be matched back to specific individuals.

14004. PATROL REPORT

Every patrol leader makes a report when the patrol returns. Unless otherwise directed, the report is made to the person ordering the patrol. If the situation permits, the report is written and supported by overlays and/or sketches. The patrol leader's report should be a complete account of everything of military importance observed or encountered by the patrol while on the assigned mission. It should include the following information:

- Size and composition of patrol.
- Tasks and purpose (mission).
- Time of departure.
- Time of return.
- Routes, out and back (show by sketch, azimuth, trace on map).
- Terrain (general description to include any man-made or natural obstacles and critical terrain features which, if occupied by either enemy or friendly forces, would allow them to control the surrounding area).
- Enemy (size, activity, location, unit, time, equipment).
- Any map corrections (show on map).
- Miscellaneous information not covered elsewhere in report.
- Results of enemy encounters.
- Condition of patrol, including disposition of any dead or wounded.
- Conclusion and recommendations.

14005. PATROL CRITIQUE

After the patrol has rested and eaten, the patrol leader should hold a critique. Constructive criticism is made. It is an excellent time to prepare for future patrols by going over lessons learned as a result of the patrol.

Appendix A. Patrol Warning Order

The warning order is issued as soon as practical with all available information included to assist patrol members in preparation.

1. Situation. Friendly and enemy situation information necessary for initial preparation.

2. Mission. Statement of what the patrol is to accomplish, and the purpose for accomplishing it. When, how, and where will be discussed in the patrol order.

3. Execution

a. Task Organization. General patrol organization and assignment of responsibilities if known; otherwise, promulgate in the patrol order.

b. Tasks. Alert subordinate leaders to patrol tasks (e.g., stream crossing, helicopter rappelling, demolitions) requiring preparation prior to departure.

c. Coordinating Instructions

(1) Time schedule and location for individual preparation, rest, briefings, in-spection, rehearsal, and departure.

(2) Time, place, uniform, and equipment for receiving the patrol order.

(3) Tasks for subordinate leaders to direct and supervise the initial preparation which may include drawing ammunition, rations, and special equipment; conducting immediate action drills or other necessary individual or unit training; meeting and briefing attachment personnel; reconnoitering the area for passage of lines; and coordinating with the necessary unit leaders.

(4) Preliminary guidance to specialists and key individuals regarding their roles and organization within the patrol.

4. Administration and Logistics

a. Individual uniform, equipment, weapons, and prescribed load of rations, water, and ammunition.

b. Crew-served weapons (if required) and guidance regarding distribution of weapons and ammunition during movement.

c. Special equipment requirements (wire cutters, demolitions, radios, flashlights, infrared equipment, mines, binoculars) and their distribution during movement.

d. Restricted or prohibited items.

5. Command and Signal

a. Designation of assistant patrol leader and his role in preparation.

b. Designation of navigators and radio operators (if required).

c. Brief outline of patrol leader's schedule for preparation, and where he can be reached.

Appendix B. Patrol Order

A patrol order follows a warning order. Any of the following subjects that have been addressed in the warning order may be omitted with the exception of the mission statement. A patrol order is more detailed than a 5-paragraph squad order in that a great deal of attention is given to individual duties.

The patrol order also provides orientation information, both enemy and friendly, that impacts the patrol and enemy forces. This information can include, but is not limited to, weather, terrain, visibility, NBC considerations, local population situation, terrain model and/or map orientation, and behavior.

1. Situation

   a. Enemy Forces

      (1) Composition, disposition, and strength are based on size, activity, location, unit, time, and equipment (SALUTE).

      (2) Capabilities and limitations to defend, reinforce, attack, withdraw, and delay (DRAW-D).

      (3) Enemy's most probable course of action.

   b. Friendly Forces

      (1) Mission of next higher unit (task and commander's intent).

      (2) Adjacent unit missions (task and intent). Also identify left, front, right, and rear.

   c. Attachments and Detachments. (date and time effective).

2. Mission. The mission is the task to be accomplished, and its purpose (who, what, where, when, and why). For patrols, specify if the mission or time has priority.

3. Execution

   a. Commander's Intent and Concept of Operations

      (1) Commander's intent.

      (2) The concept of operations tells the where, how, and who and lays out the patrol leader's general scheme of maneuver and fire support plan. It outlines the following:

(a) Task organization of the patrol.

(b) Movement to the objective area, to include navigation method.

(c) Actions in the objective area.

(d) The return movement, to include navigation method.

(e) Use of supporting forces (including illumination, if required).

b. Subordinate Element Missions. Subordinate element missions (task and purpose) are assigned to elements, teams, and individuals, as required.

c. Coordinating Instructions. This paragraph contains instructions common to two or more elements, coordinating details, and control measures applicable to the patrol as a whole. At a minimum, it includes—

(1) Time of assembly in the assembly area.

(2) Time of inspections and rehearsals (if not already conducted).

(3) Time of departure and estimated time of return.

(4) Location of departure and reentry of friendly lines and the actions associated with departure and reentry.

(5) Details on the primary and alternate routes to and from the objective area.

(6) Details on formations and order of movement.

(7) Rally points and actions at rally points.

(8) Final preparation position and actions at this position.

(9) Objective rally point and actions at this point.

(10) Actions at danger areas.

(11) Actions in the event of enemy contact.

(12) Details on actions in the objective area not covered elsewhere.

(13) Estimated time of patrol debriefing upon return.

4. Administration and Logistics

a. Changes/additions to uniform, equipment, and prescribed loads from that given in the warning order.

b. Instructions for handling wounded and prisoners.

5. Command and Signal

a. Command Relationships. Identify key leaders and chain of command.

b. Signal. Challenge and password, arm and hand signals, special signals, and radio frequencies and call signs.

APPENDIX C. PATROL EVALUATION CHECKLIST

This appendix provides a comprehensive checklist of critical patrolling steps, techniques, and procedures to aid unit leaders to critique patrol performance during training. Unit leaders may use the list as they observe the performance of a patrol to provide the patrol leader with a detailed analysis of the performance.

Warning Order	Yes	No
Did the PL ensure all patrol members were present before issuing the warning order?		
Did the PL issue a brief statement on the enemy situation?		
Did the PL issue a brief statement on the friendly situation?		
Did the PL state the mission in a clear, positive manner?		
Did the PL list all members of the patrol including attachments?		
Was the chain of command for the patrol covered fully?		
Were all members of the patrol asigned positions and duties in particular squads and teams?		
Were all the necessary individual duties assigned?		
Did the PL follow established principles in organizing the patrol into squads and teams?		
Was each patrol member assigned a particular weapon to carry on patrol?		
Did the PL list all special equipment needed to accomplish the mission?		
Was required special equipment assigned to the proper element to carry?		
Did the PL select uniform and equiment common to all based on METT-T?		
Coordination with Adjacent Units		
Did the PL coordinate with other patrols operating to the right and left?		
Was the route out and back coordinated?		
Was the time of departure and return cocordinated?		
Were call signs and frequencies coordinated?		
Was a signal for the FPF coordinated so as not to approach friendly lines during this time?		
Coordination With Front Line Units		
Did the coordinator pass the size of the patrol?		
Was the time of departure and return coordinated?		
Did the coordinator give a general area of oeprations for the patrol?		
Did the coordinator ask for information on known or suspected enemy positions and/or obstacles?		
Did the coordinator ask about information on the latest enemy activity?		
Did the coordinator ask for detailed information on friendly fire support available and the unit's barrier plan?		
Was the location of the IRP established and coordinated?		
Did the coordinator ask the forward unit to monitor their patrol frequency?		
Was the current challenge and password confirmed?		
Did the coordinator request that all information coordinated be passed on to any relieving unti?		
Patrol Order		
General		
Did the PL check to ensure all patrol members were present before issuing the patrol order?		

	Yes	No
Did the PL issue the patrol order in a forceful, confident manner?		
Did the PL make maximum use of available visual aids in issuing the patrol order (i.e., terrain model, sand table, map board, chalkboard)?		
Did the PL issue the patrol order in correct sequence?		
Did the PL issue the entire patrol order without allowing interruptions by patrol members?		
Did the PL adequately answer all questions asked by patrol members		
Did paragraph 1a (Enemy Situation) include—		
A weather forecast for the period of operation?		
A description of the terrain over which the patrol was to operate?		
Identification or description of enemy units known to bein the area of operations?		
Known locations of enemy units?		
Recent activity of enemy units?		
Stength of enemy units in the operating area?		
Did paragraph 1b (Friendly Situation) include—		
Mission of the next higher unit?		
Locations, missions, and planned actions of units on right and left?		
Fire support available to support the patrol (as per the coordination)?		
Missions and routes of other patrols operating in the immediate area?		
Did paragraph 1c (Commander's Intent) include—		
All views and ides of what the commander wants to be accomplished?		
The final result desired by the commander?		
Did paragraph 1c (Attachments and Detachments) include—		
All attachments to patrol and effective time of attachment?		
All detachments from patrol and effective time of detachment?		
Did paragraph 2 (Mission) include, at a minimum—		
Who was to conduct the patrol?		
What the patrol was tasked to do (i.e., conduct a point recon patrol)?		
Where the action was to take place?		
Why the action was to take place (i.e., the purpose)?		
Did paragraph 3 a (Commander's Intent and Concept of Operations) include—		
A complete concept of operations?		
A detailed description of the mission of elements?		
A detailed description of the mission of teams?		
A detailed description of the duties of specific individuals (i.e., navigator, compass man, pace m an, corpsman, APL)?		
Time of departure and time of return?		

	Yes	No
Type of formations and order of movement to be used?		
Description of the route and alternate routes(s) to include azimuths and distances between checkpoints as a minimum?		
Techniques to be used in the departure from friendly areas?		
Techniques to be used for the reentry into friendly areas?		
Location of the IRP and a tentative ORP, to include grid coordinates and recognizable terrain features?		
Method for designating and passing rally points?		
Actions to be taken at rally points if their use became necessary?		
Actions to be taken in the event of enemy contact?		
Actions to be taken at danger areas?		
Complete, detailed description of actions to be taken at the objective?		
Times and locations for rehearsals and the order of priority for rehearsals?		
Times and lcoations for inspections and the methods of conduct (to include uniform and equipment to be worn and/or carried)?		
Where, when, and by whom the debriefing is to be conducted?		
Did paragraph 4 (Administration and Logistics) include—		
Rations to be carried or reference to the warning order if there is no change?		
Arms and ammunition to be carried or reference to the warning order if there is no change?		
Uniform and equipment to be worn and carried or reference to the warning order if there is no change?		
Method for handling wounded or dead?		
Method for handling prisoners?		
Did paragraph 5a (Signal) include—		
Complete description of all signals, code words, frequencies, and call signs to be used within the patrol?		
Were signals adequate?		
Radio call signs to be used?		
Primary and alternate frequencies to be used?		
Call signs and frequencies to be used with other units in the area of oeprations (if applicable)?		
Required reports to be transmitted to higher headquarters, along with occasions for reporting?		
Code words and brevity codes to be used between the patrol and higher headquarters?		
Current challenge and password to be used in friendly held areas?		
Current challenge and password to be used forward of FEBA?		
Did paragraph 5b (Command) include—		
Chain of command or reference to the warning order if there is not change?		
Location of the PL during all stages of operation?		

	Yes	No
Location of the APL during all stages of operation?		
Inspection Rehearsal		
General		
Did the P L adequately use the allotted inspection time?		
Was the inspection conducted in an orderly manner?		
During the inspection, did the PL check the following:		
Completeness and correctness of uniform?		
Completeness of all equipment necessary to accomplish the mission?		
Operational condition of equipment?		
Did the PL question members to ensure they knew—		
The mission of the patrol?		
The concept of operations?		
Their individual duties and responsibilities?		
Chain of command and succession?		
Duties and responsibilities of key personnel?		
Did the PL utilize the rehearsal area to conform as much as possible to area of operations?		
Were the following major actions rehearsed:		
Actions at the objective?		
Actions at danger areas?		
Actions on enemy contacts?		
Departure and reentry of friendly lines?		
Did the PL critique each action after it had been rehearsed?		
Were interpatrol communications and control measures checked at rehearsal?		
Did the PL maintain control of the patrol during the rehearsal?		
Movement		
Did the PL use proper formations for movement?		
Did the PL adequately control the point team?		
Did the PL check the compass men?		
Was the PL aware of his position at all times?		
Was light and noise discipline enforced?		
Was the rate of movement appropriate for the mission?		
Did the PL make full use of pace?		
Did the PL make full use of count?		
Was proper security maintained during movement?		
Did the PL select and pass proper rallying points?		
Did the PL ensure that all members knew location of rallying points?		

	Yes	No
Did the PL properly use arm and hand signals?		
Did the PL recognize and halt the patrol a safe distance from a danger area?		
If the PL conducted a map check, did he conduct it properly (light discipline, security, etc.)		
Did designated individuals reconnoiter the far side of a danger area properly?		
Were support teams emplaced properly?		
Did the PL use a proper formation in crossing the danger area?		
Were reports made on enemy contact and at checkpoints?		
Actions on Enemy Contact		
Did the PL halt patrol a safe distance from a tentative ORP?		
Did the PL issue an adequate frag order before departing to look for a tentative ORP?		
Did the PL take appropriate personnel to look for the ORP?		
Did the PL select a suitable ORP?		
Did the PL secure the site adequately?		
Did the patrol occupy the ORP as stated in the patrol order or subsequent frag order?		
Did the patrol occupy the ORP in an orderly manner?		
Did the PL maintain control of the patrol during occupation of the ORP?		
Did the PL issue a satisfactory contingency plan to the APL before departing on a leader's recon?		
Did the PL take appropriate personnel on a leader's recon?		
Did the PL maintain or ensure that the patrol leader had communications with the patrol?		
Did the patrol avoid being detected by the enemy during the leader's recon?		
If contact was made, did the PL take appropriate action?		
Was security maintained during the leader's recon?		
Did the PL ensure the objective was kept under surveillance?		
Were appropriate orders given to surveillance teams?		
If a leader recon proved a tentative ORP unsuitable, did the PL move the ORP?		
Was a satisfactory leader's recon conducted?		
Was the ORP move conducted in a satisfactory and orderly manner?		
On return to the ORP, did the PL issue appropriate frag order and allow enough time for dissemination?		
Actions at Objective		
Did the PL issue a frag order for action at the objective?		
Was security emplaced prior to the teams departing from the ORP?		
Did the PL employ the terrain at the objective to the best advantage?		
Was stealth maintained while moving into the objective?		

	Yes	No
Did the Pl exercise positive control of squads, teams, and individuals at the objective?		
Were actions at the objective in accordance with the details outlined in the patrol order?		
Did the PL make use of supporting arms at the objective?		
Was the action at the objective successful?		
Was withdrawal from the objective accomplished quickly and orderly?		
Did units withdraw according to the details reflected in the patrol order?		
Was reorganization at the ORP completed in an expeditious and orderly manner?		
Did the patrol withdraw from the ORP quickly and quietly?		

APPENDIX D. AMBUSH FORMATIONS

Formations discussed here are identified by names which correspond to the general pattern formed on the ground by deployment of the assault element.

Line Formation

The assault element is deployed generally parallel to the route of movement of the enemy. This positions the assault element parallel to the long axis of the killing zone and subjects the target to heavy flanking fire. The size of the force which can be trapped in the killing zone is limited by the area the assault element can effectively cover. The enemy is trapped in the killing zone by natural obstacles, mines, demolitions, and direct fires. (See fig. D-l.)

A disadvantage of the line formation is the chance that lateral dispersion of the target may be too great for

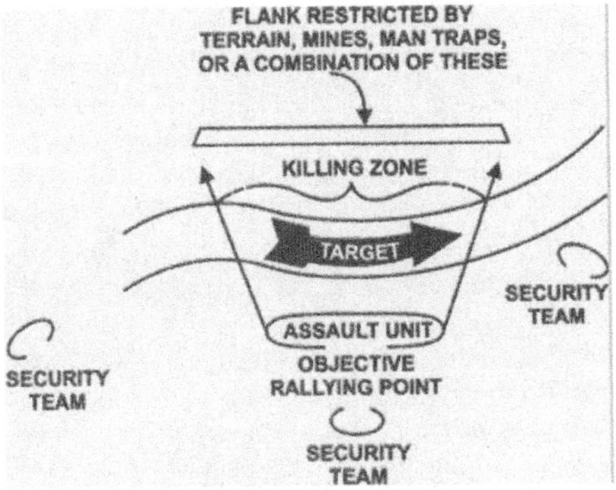

Figure D-1. Line Formation (Harassing or Destruction Ambush).

effective coverage. The line formation is appropriate in close terrain that restricts enemy maneuver, and in open terrain where one flank is protected by natural obstacles or can be protected by mines and demolitions. Similar obstacles can be placed between the attack force and the killing zone to provide protection from possible enemy counter attack. When an ambush is deployed in this manner, access lanes are left so that the enemy can be assaulted. (See fig. D-2.)

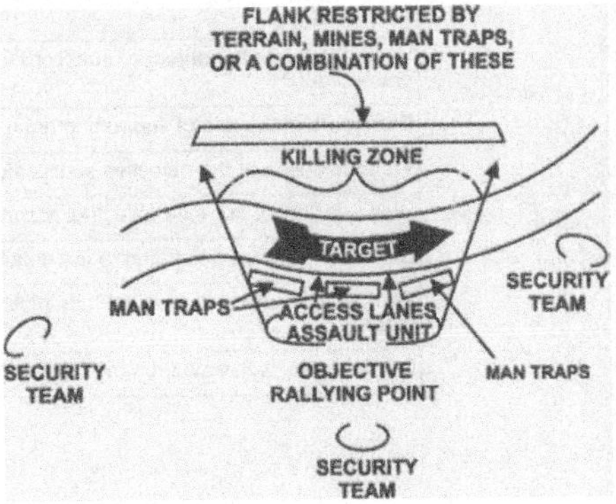

Figure D-2. Line Formation (Access Lanes for Assault of Target).

The main advantage of the line formation is its relative ease of control under all conditions of visibility.

L Formation

The "L" shaped formation is a variation of the line formation. This formation is very flexible because it can be established on a straight stretch of a trail or road (see fig. D-3) or at a sharp bend in a trail or a road (See fig. D-4 on page D-2).The long side of the assault element is parallel to the killing zone and delivers flanking fire. The short side of the attack force is at the

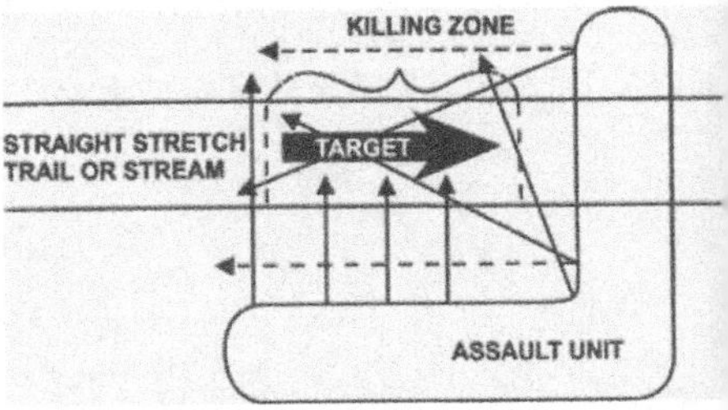

Figure D-3. "L" Formation (Destruction Ambush).

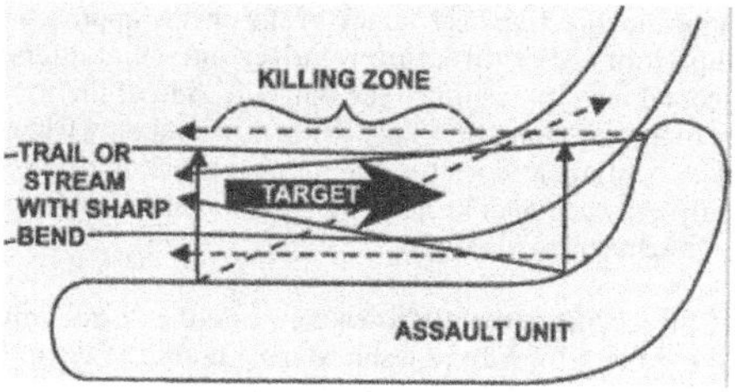

Figure D-4. "L" Formation (Bend of Trail or Stream).

end of, and at right angles to, the killing zone and delivers enfilading fire that interlocks with fire from the long side. When appropriate, fire from the short side can be shifted to parallel the long side if the enemy attempts to assault or escape in the opposite direction. In addition, the short side prevents escape and reinforcement. (See fig. D-5.)

"Z" Formation

The "Z" shaped formation is another variation of the line formation.

The assault element is deployed as in the "L" formation, but with an additional side so that the formation resembles a "Z". (See fig. D-6).

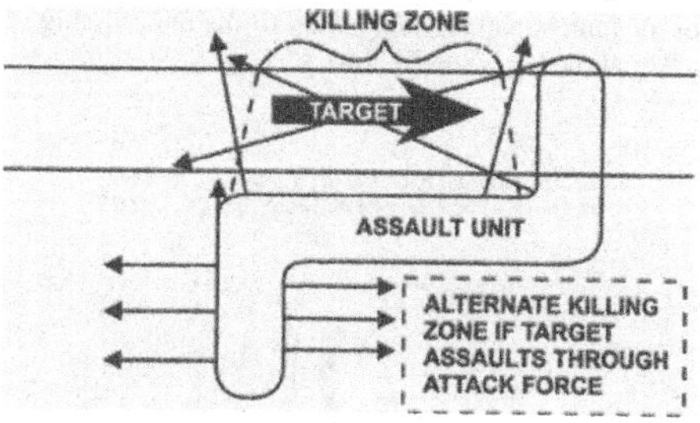

Figure D-6. "Z" Formation.

The additional wing may serve any of the following purposes:

- To engage an enemy force attempting to relieve or reinforce the enemy unit engaged in the kill zone.
- To seal the end of the killing zone.
- To restrict a flank.
- To prevent an envelopment.

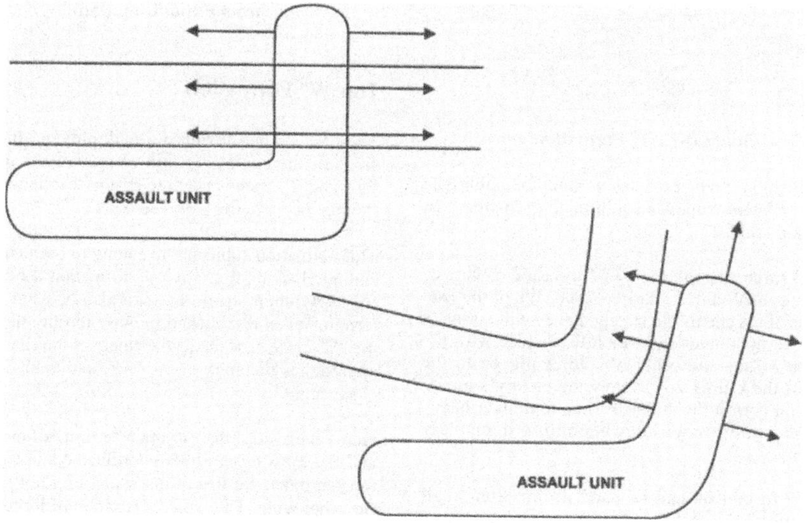

Figure D-5. "L" Formation.

"T" Formation

In the "T" shaped formation, the assault element is deployed across (perpendicular to) the enemy's route of movement so that its position forms the crossing of a "T" at the top. (See fig. D-7.)

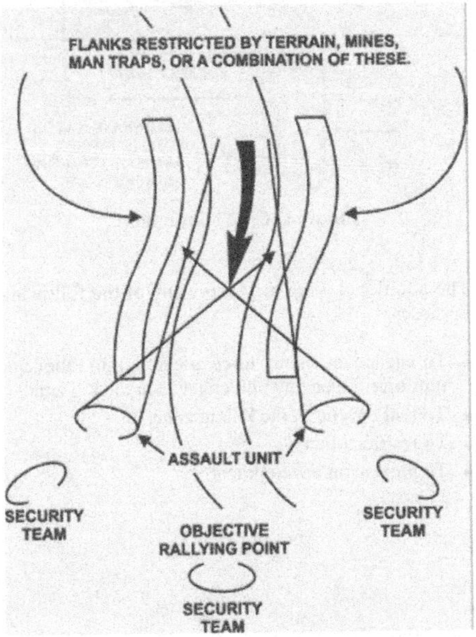

Figure D-7. "T" Formation.

This formation can be used day or night to establish an ambush to interdict movement through open areas that are hard to seal off.

A small force can use the "T" formation to harass, slow, and disorganize a larger force. When the lead elements of the enemy are engaged, they will normally attempt to maneuver right or left to close with the ambush. Mines and other obstacles placed to the flanks of the killing zones slow the enemy's movements and permit the ambush force to deliver heavy fire and withdraw without becoming decisively engaged.

The "T" formation can be used to interdict small groups attempting night movement across open areas. For example, the assault element is deployed along an avenue of approach with every second man facing the opposite direction. The attack of the enemy approaching from either direction requires only that every second man may shift to the opposite side of the formation. Each man fires only to his front and only when the enemy is at a very close range. Attack is by fire only and each man keeps the enemy under fire as long as it remains to his front.

If the enemy attempts to escape in either direction along the killing zone, each Marine takes the enemy under fire as the enemy comes into the Marine's sector of fire. The "T" formation is very effective at halting infiltration. But it has one chief disadvantage; there is a possibility that the ambush will engage a superior force at night while spread out. (See fig. D-8.)

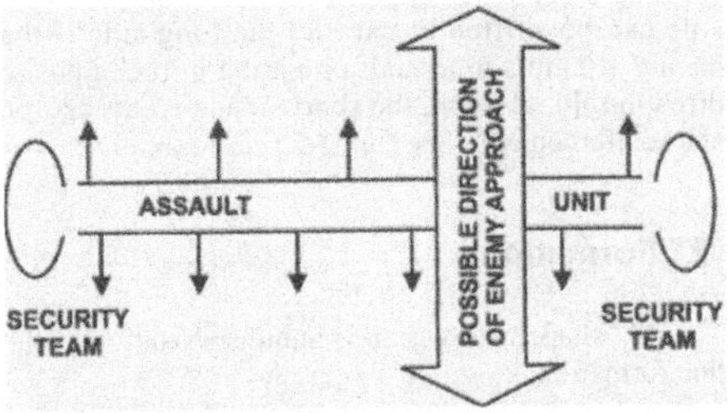

Figure D-8. "T" Formation (Target Approaching from Either Direction).

The "V" Formation

The "V" shaped formation is deployed along both sides of the enemy's route of movement so that it forms a "V"; care is taken to ensure that neither group (within the "V") fires into the other.

This formation subjects the enemy to both enfilading and interlocking fire. The "V" formation is best suited for fairly open terrain but can also be used in close terrain. When established in close terrain, the legs of the "V" close in as the lead element of the enemy force approaches the apex of the "V", and opens fire at a close range.

Here, even more than in open terrain, all movement and fire must be carefully coordinated and controlled to ensure that the fire of one wing does not endanger the other wing. The wider separation of forces makes this formation difficult to control, and there are few sites that favor its use. Its main advantage is that it is difficult for the enemy to detect the ambush until it is well into the killing zone. (See figs. D-9 and D-10.)

Triangle Formation

The triangle is a variation of the "V" formation and can be varied in three ways:

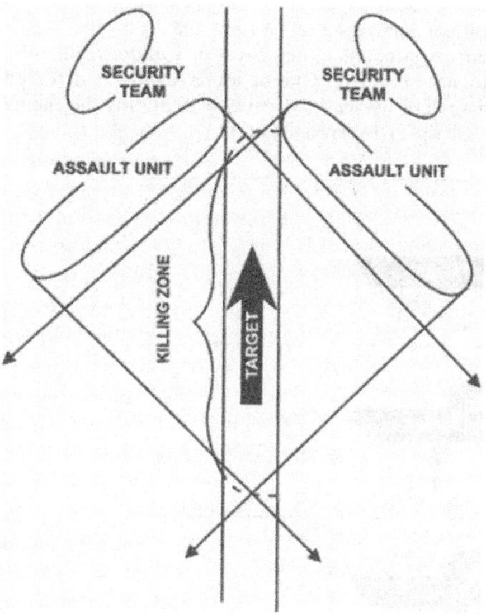

Figure D-9. "V" Formation (Open Terrain).

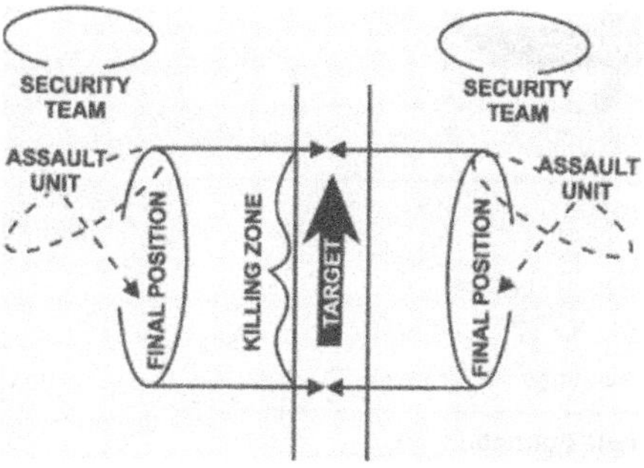

Figure D-10. "V" Formation (Close Terrain).

Closed Triangle Formation

(See fig. D-11.) The assault element is deployed in three teams, positioned so that they form a triangle (or closed "V"). An automatic weapon is placed at each point of the triangle and positioned so that it can be shifted quickly to interlock with either of the others. Men are positioned so that their sectors of fire overlap. Mortars may be positioned inside the triangle.

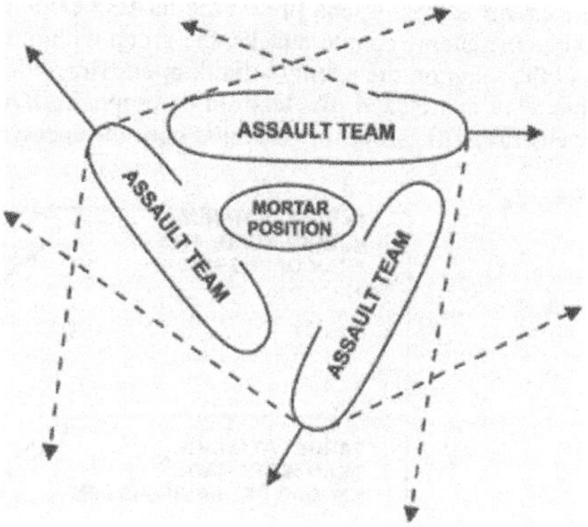

Figure D-11. Closed Triangle Formation (Night Ambush).

When deployed in this manner, the triangle ambush is used to interdict night movement through open areas. When enemy approach is likely to be from any direction, this formation provides all-around security, and security forces are deployed only when they can be positioned so that if detected by an approaching enemy, they will not compromise the ambush. Attack is by fire only, and the enemy is allowed to approach within close range before fire is initiated.

The advantages of the closed triangle formation are ease of control, all-around security, and the enemy can be brought under the fire of at least two automatic weapons, regardless of the direction they approach.

Disadvantages are that it requires a force of platoon size or larger to reduce the danger of being overrun by an unexpectedly large force and that one or more sides of the triangle may come under enfilade fire. The lack of dispersion, particularly at the points, increases the danger from enemy mortar fire.

Open Triangle Harassing Formation

This variation of the triangle ambush enables a small force to harass, slow, and inflict heavy casualties upon a large force without itself being decisively engaged. The assault element is deployed in three teams, positioned so that each team becomes a corner of a triangle containing the killing zone. (See fig. D-12.)

When the enemy enters the killing zone, the team to the enemy's front opens fire on the leading element. When the enemy counterattacks, the group withdraws and the team on the enemy's flank opens fire. When this team is attacked, the team to the opposite flank opens fire. This process is repeated until the enemy is pulled apart. Each team reoccupies its position, if possible, and continues to inflict the maximum damage possible without becoming decisively engaged.

Open Triangle Destruction Formation

The assault element is again deployed in three teams, positioned so that each team is a point of the triangle, 200 to 300 meters apart. The killing zone is the area within the triangle. The enemy is allowed to enter the killing zone; the nearest team attacks by fire. As the enemy attempts to maneuver or withdraw, the other teams open fire. One or more teams, as directed, assault or maneuver to envelop or destroy the enemy. (See fig. D-13 on page D-6.)

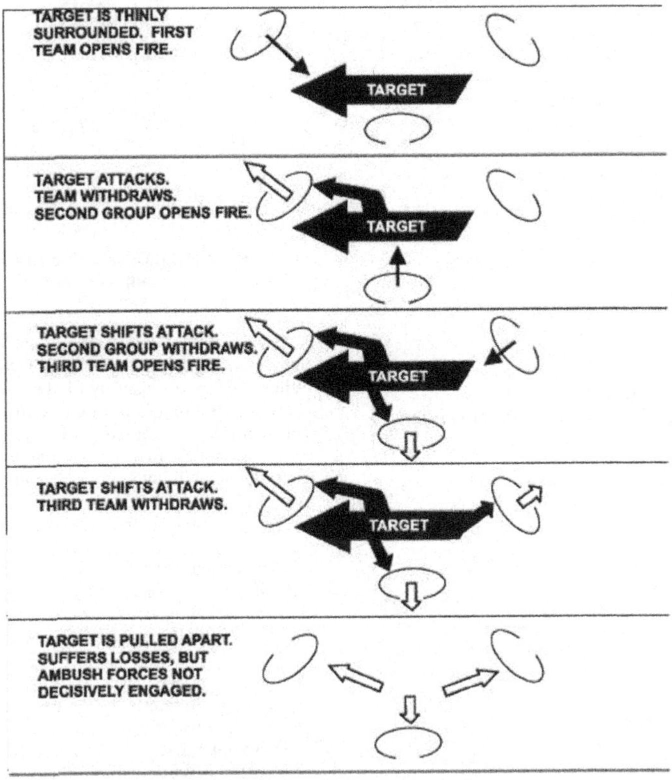

Figure D-12. Open Triangle Formation.

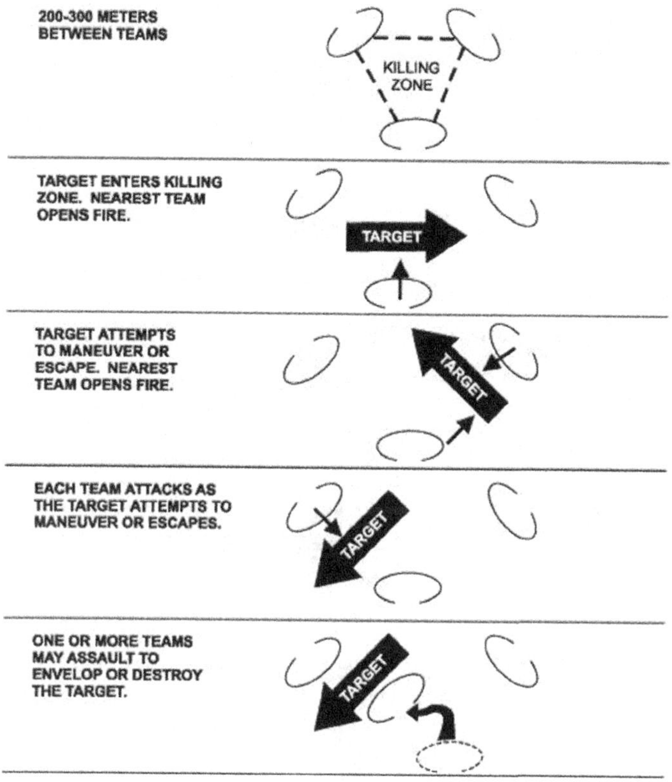

Figure D-13. Open Triangle Formation (Destruction Ambush).

This formation is suitable for platoon-size or larger ambush forces. A smaller force would be in too great a danger of being overrun. Another disadvantage is that control, in assaulting or maneuvering, is very difficult. Very close coordination and control are necessary to ensure that assaulting or maneuvering teams are not fired on by another team. The ambush site must be a fairly level open area that provides concealment around its border for the ambush force.

The Box Formation

The "box" formation is similar in purpose to the open triangle ambush. The assault element is deployed in four teams, positioned so that each team becomes a corner of a square or rectangle containing the killing zone. It can be used for a harassing or destruction ambush in the same manner as the variations of the open triangle formation. (See figs D-14 on page D-7 and D-15 on page D-8.)

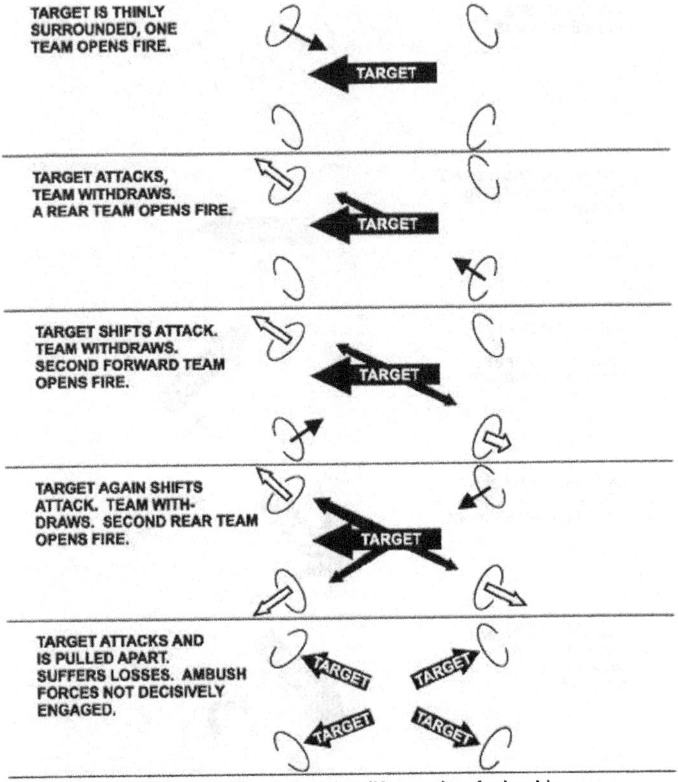

Figure D-14. Box Formation (Harrassing Ambush).

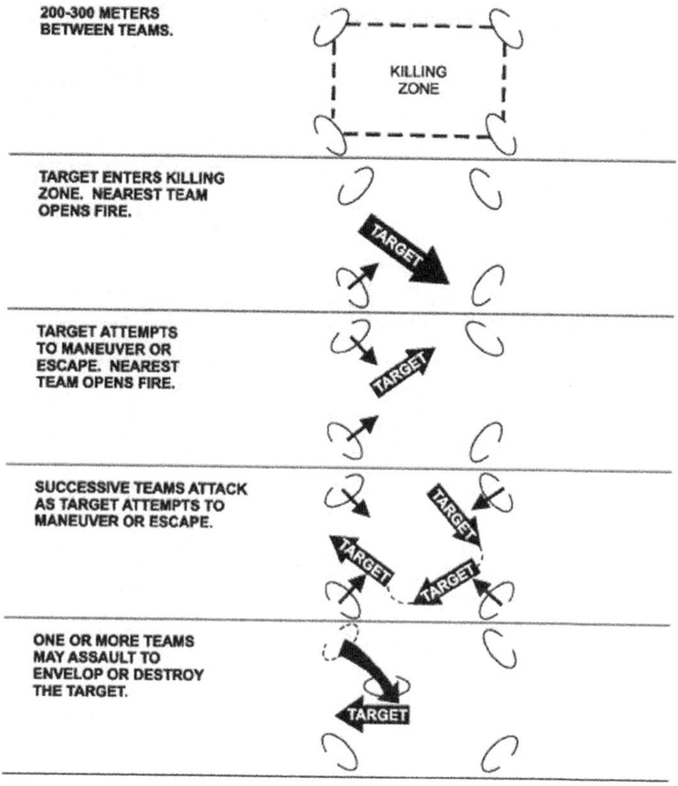

Figure D-15. Box Formation (Destruction Ambush).

APPENDIX E. ACRONYMS

A&S....................assault and security

CIT....................counterintelligence team

EOD....................explosive ordnance disposal

ITT....................interrogator-translator team

MAGTF....................Marine air-ground task force
MEDEVAC....................medical evacuation
METT-T.........mission, enemy, terrain and weather, troops and support available

NBC....................nuclear, biological, chemical
NCO....................noncommissioned officer

OTL....................overserver-target line

RCA....................riot control agents
ROE....................rules of engagement

SMAW....................shoulder-launched multipurpose assault weapon
SMEAC....................situation, mission, execution, administration and logistics, and command and signal

SOP....................standing operating procedures
SPOTREP....................spot report

TTP....................tactics, techniques, and procedures

Appendix F. References

Joint Publication (JP)

1-02 DOD Dictionary of Military and Associated Terms

Fleet Marine Force Manual (FMFM)

6-5 Marine Rifle Squad (proposed MCWP 3-11.2)

Marine Corps Warfighting Publications (MCWP)

3-15.3 Scout Sniping (under development)

3-35.3 Military Operations on Urbanized Terrain (MOUT)

3-41.2 Raids (under development)

Marine Corps Reference Publications (MCRP)

2-15.3A Reconnaissance Patrol Leader's Planning Handbook (under development)

2-15.3B Reconnaissance Reports Guide

3-02C Water Survival Handbook (under development, currently exists as Fleet Marine Force Manual 0-13)

5-12A Operational Terms and Graphics

5-12C Marine Corps Supplement to the DOD Dictionary of Military and Associated Terms

Made in United States
Orlando, FL
04 November 2026

90206693R00262